MATERIALS SCIENCE AND TECHNOLOGIES

HEAVY METALS

SOURCES, TOXICITY AND REMEDIATION TECHNIQUES

MATERIALS SCIENCE AND TECHNOLOGIES

Additional books in this series can be found on Nova's website
under the Series tab.

Additional e-books in this series can be found on Nova's website
under the eBooks tab.

MATERIALS SCIENCE AND TECHNOLOGIES

HEAVY METALS

SOURCES, TOXICITY AND REMEDIATION TECHNIQUES

DEEPAK PATHANIA
EDITOR

nova publishers
New York

Library of Congress Cataloging-in-Publication Data

Names: Pathania, Deepak, editor.
Title: Heavy metals : sources, toxicity and remediation techniques / editor, Deepak Pathania (professor & head, School of Chemistry, Shoolini University, Solan, Himachal Pradesh, India).
Other titles: Heavy metals (2016)
Description: Hauppauge, New York : Nova Science Publishers, Inc., 2016. | Series: Materials science and technologies | Includes index.
Identifiers: LCCN 2016004802 (print) | LCCN 2016015108 (ebook) | ISBN 9781634847407 (hardcover) | ISBN 9781634847667 (Ebook) | ISBN 9781634847667 ()
Subjects: LCSH: Heavy metals--Environmental aspects.
Classification: LCC TD196.M4 H45 2016 (print) | LCC TD196.M4 (ebook) | DDC 628.5/2--dc23
LC record available at https://lccn.loc.gov/2016004802

Published by Nova Science Publishers, Inc. † New York

CONTENTS

PREFACE

The drinking water supply throughout the world is threatened by environmental pollution. The pollution was caused by various hazardous chemicals such as heavy metals and organic toxicants. The term heavy metal refers to any metallic chemical element that has high density and toxic at low concentrations. Heavy metals are the natural components of the earth's crust. These heavy metals are of specific concern due to their toxicity, bio-accumulation tendency and persistency in nature. Contamination of heavy metals in the environment is a major global concern, because of toxicity and threat to the human life and ecosystem. The levels of metals in all environments, including air, water and soil are increasing to toxic levels, with contributions from wide variety of industrial and domestic sources.

These metal ions become hazardous to food chain causing mental retardation, hemoglobin deficiency and interference with normal cellular metabolism. Their strong exposure may cause the immediate gastric pain, nausea, vomiting, severe diarrhea. Consequently, in many countries, more strict legislation has been introduced to control the heavy metals contamination. The monitoring of heavy metal is important issue so that reliable methods would be initiated for the removal. Remedial techniques include the adsorption, coagulation, flocculation, ion exchange, membrane separation, electrochemical oxidation etc. have been used for the treatment of heavy metal ions. Photocatalytic degradation, advanced oxidation, aerobic degradation, filtration, ozonation, coagulation, flocculation, distillation, extraction, precipitation etc. has also developed for the removal of various pollutants from aqueous system. Biosorption process utilizes various natural materials of biological origin, including bacteria, fungi, yeast, algae, etc.

Because of their toxic nature, metals are not as amenable to bioremediation as organics. Unlike organics, metals are persistent in the environment and cannot be degraded through biological, chemical or physical means to an innocuous by-product. The chemical nature and bioavailability of a metal can be changed through oxidation or reduction; however, the elemental nature remains the same because the metals are neither thermally decomposable nor microbially degradable. Consequently metals are difficult to remove from the environment.

This edition is the outcome of wonderful contribution from 25 authors of different countries with desired specialization and great contribution to sciences through research and review work. This book thoroughly covers the latest findings in heavy metal, their sources, contamination and remediation technology. This book is very useful different discipline of Sciences, Nanotechnology, Engineering, Biotechnology, Biomedical etc. This edition will surely helpful to the researcher and scientist working in field of Environmental Sciences,

Industrial chemistry, Analytical chemistry, Applied sciences, Material sciences, Biological sciences etc. Based on the thematic topics, the edited book contains following 10 chapters:

Chapter 1 – This chapter deals with the technology available for the detection of mercury and latest remediation techniques. The future perspectives are also presented which help for the future researches in this field.

Chapter 2 – This chapter deals with general background to polymers and their composite materials. The application of composites for removal of specific toxic heavy metals from waste water is discussed.

Chapter 3 – This chapter explain the mining, mineral beneficiation and extractive metallurgy as source of heavy metals. The possible mitigation measures for metals recovery are also discussed.

Chapter 4 – In this chapter, different source of heavy metals and their adverse impacts on environmental are described. Various remediation technique such as phytoremediation, adsorption, ion-exchange, coagulation, precipitation, flocculation, ultra-filtration and electrochemical methods are discussed in details.

Chapter 5 – This chapter focuses on the available biosorption methods for the removal of heavy metals. It also describes the different adsorption isotherm models and adsorption mechanism.

Chapter 6 – This chapter concentrates on toxic effects of different heavy metals on living systems in account of their bonding interactions and strategies for the development of heavy-metal-detection devices on chemical recognition and specific affinity of receptor.

Chapter 7 – In this chapter, a detailed discussion on heavy metals pollution, sources and available remedial techniques have been highlighted.

Chapters 8 – This chapter insight the rice husk low-value agricultural by-products as a sorbent for the removal of heavy metals. The chapter also includes the different chemical activation process and utility of Dubinin-Radushkevich (DR) isotherm.

Chapters 9 – This chapter describes various natural and man-made sources of uranium, biological toxicity studies and outlines current methods for remediation. The future direction research has also been outlined.

Chapters 10 – This chapter discusses the characteristics and adsorptive properties of sludge ash adsorbent for heavy metals removal from water system.

ACKNOWLEDGMENTS

First and foremost I would like to thank Almighty who gives me the power to believe in my enthusiasm and pursue my dreams. I would like to express my gratitude to all those who provided support, talked things over, offered comments, allowed me to quote their remarks and assisted in editing, proofreading and design.

First and foremost, I would like to thank my wife Sarita Kumari for standing beside me throughout my career and editing this book. She has been my inspiration and motivation for continuing to improve my knowledge and move my career forward. I dedicate this book to her. I also thank my wonderful children Divanshu Pathania for always making me smile and for understanding on those weekend mornings when I was editing this book instead of playing games. I hope that one day they can read this book and understand why I spent so much time in front of my computer. My families, including my father, Sh. Om Parkash Pathania mother, Smt. Shakuntla Devi and mother-in-law, Smt. Brahmi Devi always supported me throughout my career and authoring this book and I really appreciate it. My father, who rests in peace, forged my celebrity and share admire on every goal I attain.

I would like to thank each of the authors of this book for devoting their time and effort towards publishing in time. I think that it will be a great asset to the public. It was a long and difficult journey for them. I'm grateful to work with them and hope that this book will be useful for everyone.

I am particularly indebted to Professor P.K. Khosla, Vice Chancellor, Shoolini University, Solan, Himachal Pradesh for all persuade my work and encourage me to do better work at all times.

I would like to thank Dr. Rajesh Sharma, Dr. Gaurav Sharma, Dr. Amit Kumar, Mr. Arush and Ms. Manita for helping me in editing the book. I am deeply appreciative and most grateful to Nova Publishers, New York for their decision to publish this book.

Thanks to all my friends for sharing my happiness when I started this project and following with encouragement when it seemed too difficult to be completed.

Last and not least, I beg forgiveness of all those who have been with me over the course of the years and whose names I have failed to mention.

Deepak Pathania
Editor

In: Heavy Metals
Editor: Deepak Pathania
ISBN: 978-1-63484-740-7

Chapter 1

MERCURY: SOURCES, TOXICITY AND REMEDIATION

Mohammad A. K. Khan and Zia Mahmood Siddiqi*
Department of General Studies, Jubail University College,
Jubail Industrial City, Kingdom of Saudi Arabia

ABSTRACT

This chapter deals with human and natural sources of mercury in the environment, its toxicity in biological systems and remediation techniques in detail. The chemical mechanism of the toxicity is examined in terms of inorganic and organic mercurial. Latest remediation techniques are explained with the evidences based on field experience and use of most modern instruments or Methods. Lastly, a section is dedicated for the technology available for the detection of mercury in different samples. Finally, future perspectives are presented that may help for further research initiatives.

Keywords: mercury, source, toxicity, estimation, remediation

ABBREVIATIONS

AAS	atomic absorption spectroscopy
AuNPs	gold nano particles
BBB	blood-brain barrier
CNS	central nervous systems
CV AFS	cold vapor atomic fluorescence spectroscopy
CV AAS	cold vapor atomic atomic absorption spectroscopy
CPE	cloud point extraction

* Corresponding author: Zia Mahmood Siddiqi. Department of General Studies, Jubail University College, Jubail Industrial City 31961, Kingdom of Saudi Arabia. Tel: +966 (1-3) 342-9000 ext. 1409, 1412, fax: +966 3 3429000-1064, e-mail: siddiqizm2005@hotmail.com.

Cys	cysteine
DNA	deoxyribonucleic acid
EDTA	ethylene di-amine tetra acetic acid
FI-VGAAS	flow injection Vapor generation atomic absorption spectrometry
GC	gas chromatography
GEM	gaseous elemental mercury
GOIM	gaseous oxidized inorganic mercury
G.I.	gastrointestinal
GPx	glutathione peroxidase
Hb	hemoglobin
Hg_T	total mercury
$Hg_{(P)}$	particulate mercury
HPLC	high performance liquid chromatography
RNA	ribonucleic acid
RP-HPLC	reverse phase high performance liquid chromatography
ICP	inductively coupled plasma
ICP-MS	inductively coupled plasma mass spectrometry
ICP-AES	inductively coupled plasma atomic emission spectroscopy
IRB	iron-reducing bacteria
LC	liquid chromatography
Mg	megagram
MeHg	methylmercury
MeHgSe	methylmercury selenium
MS	mass spectrometry
PDMS	poly- dimethyl siloxane
SPME	solid phase metal extraction
SRB	sulfate-reducing bacteria
UV-Vis	ultravoilet-visible
USEPA	united state environmental protection agency
w/w	weight by weight
ZVI	zero valent iron

1. INTRODUCTION

Mercury (Hg on the periodic table and hereafter) is a naturally occurring chemical element that is found in air, water, and soil. Mercury is a liquid silvery highly toxic heavy metal that poses a major public health threat. The symbol Hg that mercury is known by comes from its Greek name, *hydrargyrum*, which means "liquid silver" — to reflect its shiny surface. The element is also known as quicksilver for its mobility. Mercury has been known to humanity for ages. In fact, evidence of its use has been found in China, India and Egypt, and traces of mercury were found in 3,500-year-old Egyptian tombs.

Mercury is released into the air from power plant smokestacks and other sources. It can fall to the ground either as dry particles or with rain/snow and enter water bodies in a process known as deposition (wet and dry). Mercury makes more surface waters unsafe for fishing than any other toxic contaminant, and people are most often exposed to mercury by eating contaminated fish (see Figure 1). Mercury gets into the oceans from multiple sources, such as air and water pollution as well as naturally occurring minerals. Much of the mercury in the North Pacific, for instance, comes from air pollution, especially power plants in Asia, according to a study published in August 2013 in the journal Nature Geoscience. In another study, published in May 2009 in the journal Global Biogeochemical Cycle, researchers estimated that the levels of mercury in the North Pacific could double by 2050 if human-induced emissions continue at the current rate.

Even where the global sources are major contributors in mercury pollution it is important to recognize that the large global pool of mercury is not naturally occurring. The global pool is fed by the emissions that result from the combustion of coal the around the world [1].

1.1. Mercury as a Global Contaminant

Mercury (Hg) is a global contaminant due to its long-range atmospheric transport, bioaccumulation in aquatic ecosystems, and neurotoxicity to humans [2, 3]. Mercury occurs naturally and can be released to the surface environment by weathering of geological materials, volcanic eruptions, and natural forest fires [4].

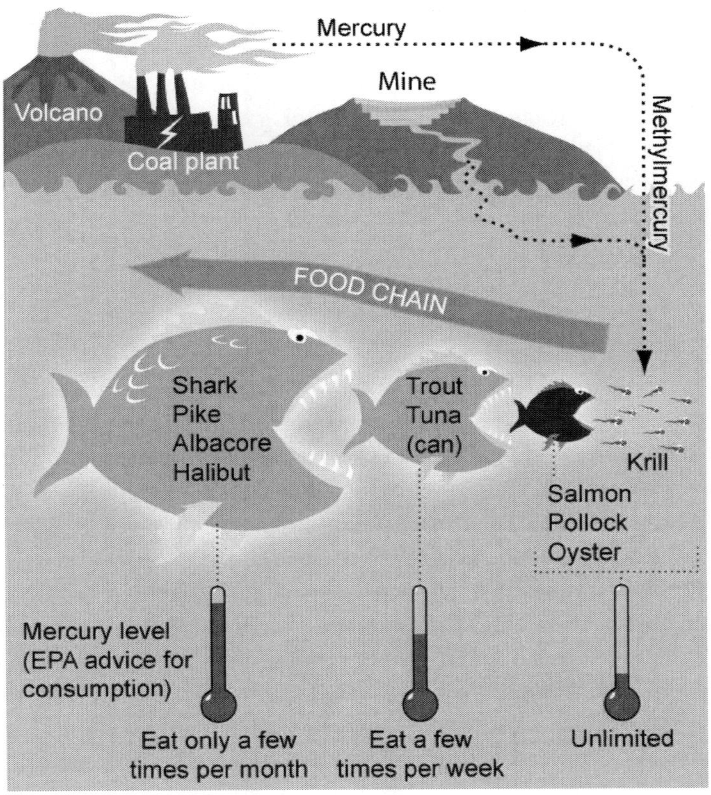

Figure 1. Mercury: source to food chain.

However, human activities have greatly increased Hg emission to the environment since the industrialization. Major anthropogenic Hg sources include coal combustion, gold and non-ferrous metals production, and biomass burning [5].

Mercury exists in the environment in two major oxidations states: Hg^0 and Hg (II). The volatility and low reactivity of Hg^0 is primarily responsible for the long-range transport via the atmosphere, accounting for its widespread distribution even in remote regions that are far away from localized Hg sources. Once in the aquatic or terrestrial systems, inorganic Hg (II) may undergo methylation to form methylmercury (MeHg). Methylmercury in the aquatic environment is thought to be produced primarily by microbial methylators such as sulphate-reducing bacteria (SRB) [6] and also by iron-reducing bacteria (IRB) [7]. SRB which thrive in the aerobic-anaerobic transition environment utilize methylcobalamin to convert inorganic Hg into MeHg. Intracellular methylcobalamin is proposed to catalyze the methyl group transfer [8, 9]. SRB such as *Desulfovibrio desulfuricans* and other closely related organisms were shown to utilize an acetylcoenzymeA pathway, with methyl group transfer from methyltetrahydrofolate [8, 10]. Chemical methylation of inorganic Hg (II) is also possible in the aquatic environment. Chemical reagents such as methyl iodide, dimethylsulfide, and dissolved organic matter such as fulvic and humic acids are thought to be able to methylate inorganic Hg [11-13]. Where present, organometallic complexes such as methylcobalamin, methyllead or methyltin in the aquatic environment, are known to methylate Hg via transmethylation reactions. Transmethylation reactions can proceed by transferring carbocationic Me^+, carbanionic Me^- or radical $Me^\bullet$, depending on the chemical properties of the metal component of the methylating agent [11].

While both inorganic Hg(II) and MeHg can bioaccumulate in the aquatic ecosystems, only MeHg biomagnifies, meaning increasing MeHg concentrations are found at higher trophic levels in the aquatic ecosystems due to the process of retaining MeHg more than excretion in the subsequent level of the food chain [14]. As a result, MeHg concentrations in top predators such as sharks can be as high as 4 mg/kg, which exceeds MeHg concentrations in seawater [15] (typically <0.1 ng/L) by a factor of 10^6 [16].

1.2. Physicochemical Properties of Mercury

Mercury looks pretty, in its shiny, fast-moving liquid form, but can be extremely poisonous to humans. The reactivity of Hg depends on its chemical and physical properties. Following Table 1 lists few chemical and physical constants of mercury.

Mercury (Hg) is a chemical element with an atomic number of 80. Mercury is a silver plated heavy metal, liquid and odorless at normal conditions. It easily alloys with many other metals like gold or silver producing amalgams, is insoluble in water and soluble in nitric acid.

The main source of Hg is cinnabar or mercury sulfide (HgS), a stable compound and insoluble usually recovered as a byproduct of ore processing. Mercury in this form is found in the earth's crust average concentrations of 0.5 ppm [17].

Mercury is one of the most toxic elements to human health and ecosystem. At temperatures above 40°C mercury produces toxic and corrosive fumes. It is harmful by inhalation, ingestion and contact, is a very irritating to skin, eyes and respiratory tract, even to nervous system, its gaseous form is absorbed by lung tissues [17, 18].

A wide variety of mercury species exist in the environment and its various chemical forms can differ in bioavailability, transport, persistence, and toxicity. Still, every mercury species is toxic with methyl mercury being the most toxic species. This element can exist in the environment as elemental (Hg^0), oxidized inorganic (Hg^{2+} -mercuric, Hg_2^{2+}-mercurous) or oxidized organic (methyl/ethyl mercury) forms. Mercuric and mercurous forms are more stable under oxidizing conditions.

In moderately reducing conditions, the organic or inorganic mercury can be reduced to its elemental form and be converted to forms released by biotic or abiotic processes: these are the most toxic forms of mercury, as well as being soluble and volatile. Hg (II) forms strong soluble complexes with a variety of organic and inorganic ligands. Hg sorption in soil, sediment and humic materials is an important mechanism for the removal of mercury from solutions [19].

Any form of mercury in the environment may evolve into a more toxic species (methyl mercury) under biogeochemical transformation processes. Due to these processes and the high mobility of mercury species, a good understanding of how mercury species transform and accurate monitoring are essential for assessing the risk of mercury in the environment. The impact of mercury depends strongly on its chemical species; understanding mercury trans- formations and the impact of its various chemical forms are vital to preventing harmful effects on humans and the environment [19, 20].

2. Sources and Uses of Mercury

Two major sources of mercury are (1) natural occurring process and (2) man made processes. The major naturally occurring mercury emission processes include: (1) Degassing from mercury mineral deposits (2) degassing from mercury contaminated aquatic and terrestrial systems (3) volcanic emissions, and (4) forest fires.

Table 1. Chemical and physical constant for mercury

Atomic radius, non-bonded (Å)	2.23
Covalent radius (Å)	1.32
Electron affinity (kJ mol^{-1})	Not stable
Electronegativity (kJ mol^{-1})	1.9
Density (g/cm^3)	13.5336
Melting point	-37.8°F (- 38.83°C)
Boiling point	674.11°F (356.73°C)
Ionization energies (kJ mol^{-1})	1st:1007.066, 2nd:1809.756, 3rd:3299.8
Common oxidation states	2, 1
Isotopes (natural abundance)	^{196}Hg (0.15%), ^{198}Hg(9.97%), ^{199}Hg (16.87%), ^{200}Hg (23.1%), ^{201}Hg (13.18%), ^{202}Hg (29.86%), ^{204}Hg (6.87%)
Half-life (^{196}Hg), mode of decay	>2.5 x 10^{18} y, α decay
Specific heat capacity (J kg^{-1} K^{-1})	140
Vapor pressure (400K) Pa	140
Young's modulus, Sheer modulus (GPa)	Unknown
Bulk modulus (GPa)	25

While the anthropogenic emission sources of mercury mostly result from: (1) solid waste incineration (municipal and medical wastes), (2) coal and oil combustion, (3) pyro-metallurgical processes (iron, lead, and zinc), and (4) production of mercury and gold.

Among these sources, coal combustion and solid waste incineration account for more than half of the total global emissions [21]. In 1995 about half of the total estimated worldwide emissions were from Asia (mostly from China, India, North and South Korea) [22].

The sources of mercury contamination in aquatic systems are placed in six categories:

1) atmospheric deposition,
2) erosion sources,
3) urban discharges,
4) agricultural sources,
5) mining discharges, and
6) combustion/industrial discharges.

2.1. Mining Sources

Mining activities including gold, silver, mercury, and lead have emitted historically significant amounts of mercury into the environment. In such scale of mining amalgamation method was generally used which resulted in a significant amount of mercury vapor release into the atmosphere. Moreover, discharge of the processing water in the environment further increases mercury contamination. Small scale gold mining activities at regional level are known to be causing serious mercury contamination problems. The mercury mine drainage contains high concentrations of inorganic mercury, MeHg, and sulfate. The high concentrations of sulfate (around 1000 mg/l) in mercury mine drainage favor the growth of sulfate reducing bacteria (SRB) - a mediator of mercury methylation. Thus, it is believed that mercury mine drainage has a high methylation rate [23]. In addition to gold and mercury mining activities, silver, lead mining and smelting processes also emit a large quantity of mercury into the environment. In the vicinity of the smelter and the center of the mining area, mercury concentrations were found 40 times higher than the background concentration [24]. Leaving other insignificant source of Hg contamination it is suggested that historical gold and silver mining activities dating back to the 19th century are responsible for the mercury contamination problem [25]. Based on the studies done, it is believed that inactive mines, some of which are over 100 years old, still pose an environmental threat to adjacent aquatic systems.

2.2. Atmospheric Deposition

Mercury exists in the atmosphere in gaseous, particulate, and aqueous (i.e., attached to water droplets) forms, but atmospheric mercury is mostly gaseous [26]. About 80% of total mercury (Hg_T) in the atmosphere is in Hg^0 gaseous form [27]. Due to its high volatility, Hg^0 could remain in the atmosphere for more than one year [27, 28], which makes long-range atmospheric transport of Hg^0 a major environmental concern. Although gaseous mercury is the predominant form in the atmosphere, particulate phase mercury can have a significant impact on atmospheric mercury deposition [29]. Figure 2 shows the comparative atmospheric particulate Hg concentration in the Artic, Toronto and Beijing cities [30]. While considerable amount of particulate Hg is present in the Artic the increasing amount particulate Hg in urban atmosphere is a cause of concern with respect to human health. To estimate the particulate Hg in atmosphere different techniques of sampling are adopted like comparative sampling [31]. Oxidation processes (e.g., photochemical reactions) transform the Hg^0 into Hg^{2+} in the atmosphere. It is believed that these processes take place mostly at the interface of aerosols and atmospheric water droplets [27]. Compared to Hg^0, Hg^{2+} in the atmosphere tends to precipitate faster to surface terrestrial and aquatic systems [32]. Studies performed throughout the world strongly support the theory that atmospheric deposition is an important (sometimes even the predominant) source of mercury contamination in aquatic systems [33-41].

Atmospheric mercury deposition is a complex phenomenon. Different factors, which are site-specific such as temperature, humidity, wind pressure and direction, and different seasons [42] may influence the transport and transformation of this metal in the atmosphere. High moisture content may increase the oxidation of Hg^0 substantially [27]. As a result, atmospheric mercury quickly settles into terrestrial or aquatic systems. The concentrations of mercury in the atmosphere and flux of mercury deposition also vary with the seasons. The total amount of mercury deposited in summer is higher than the value reported in winter [38, 43]. Since plants absorb atmospheric mercury through their foliage and then mercury is passed into terrestrial systems and watersheds by litter fall, forested watersheds can have atmospheric mercury deposition flux twice as high as un-forested watersheds [44]. Industrial activities have increased atmospheric mercury levels up to 24 times since the start of the industrial period [45].

2.3. Urban Sources

Human activities may break the natural mercury cycle, and significantly increase the release of mercury into the environment. Road construction, domestic sewage discharge, medical incineration, vehicular emission and other activities are believed to be responsible for the current mercury contamination in urban areas. Tillage, logging, deforestation, and other activities cause loss of vegetation and greater soil erosion. Mercury contaminated soil is carried into the nearby aquatic systems more easily by surface runoff.

Discharge of domestic sewage into aquatic systems without proper treatment also increases aqueous mercury concentrations.

2.4. Agricultural Sources

Pesticides and fungicides with mercury content were widely used in agriculture for a long period of time resulting in high concentrations of mercury in intensively cultivated soil. An estimate showed that 2100 metric tons of mercury was used in agriculture in the 1960s throughout the world, with 1600 metric tons used in Japan alone [46]. These pesticides and fungicides have been banned in many countries due to their toxic properties. As an example, these chemicals were banned in the United States in the 1970s.

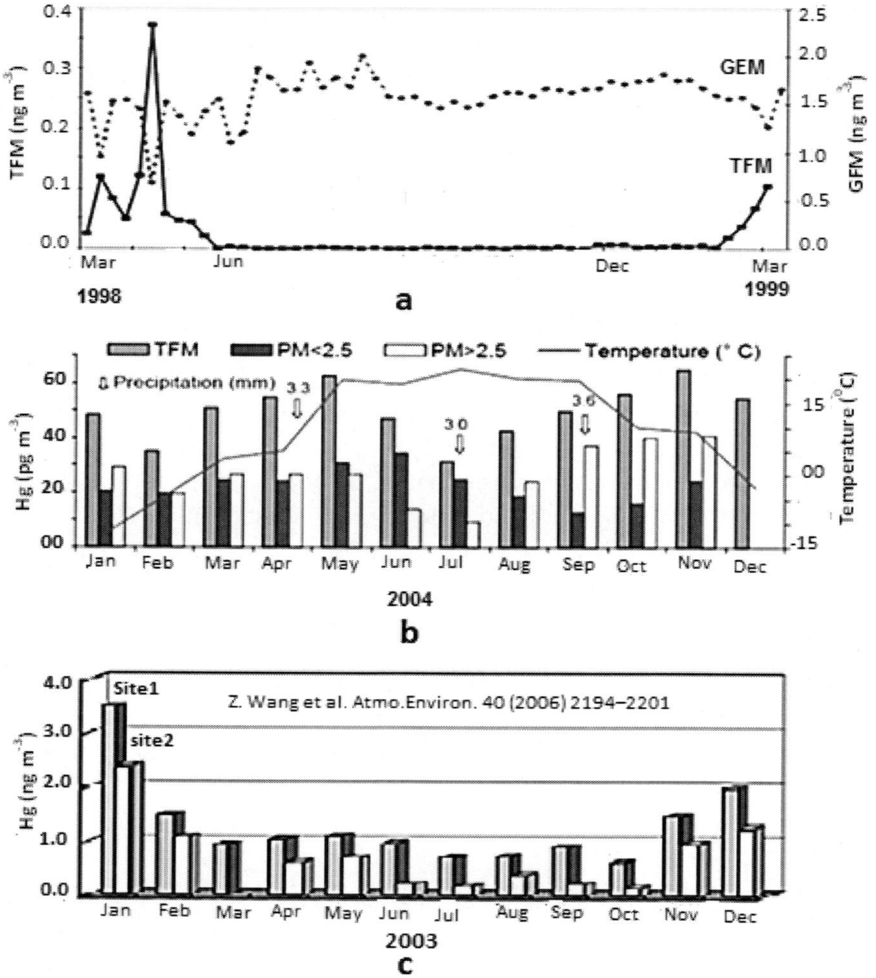

Figure 2. Total atmospheric particulate mercury in (a) the Arctic (b) Toronto (c) Beijing.

In spite of restrictive use of pesticides and fungicides containing mercury it is still widely distributed in many countries. Because of the strong persistence of mercury in the environment, mercury released long ago may still lead to serious contamination problems. An agricultural region cultivated long ago may still show elevated concentrations of mercury in soil, sediment, water, and fish samples.

2.5. Erosion Sources

One of the important source of mercury in aquatic environment is the contaminated soil and its erosion by surface run offs from rain or snow. The cause of mercury contamination in surface soil may be due to: (1) deposition of atmospheric mercury, (2) agricultural activities, (3) usage of sludge-amended soil, and (4) natural degradation of ferrallitic soils due to podzolization (the process by which the soil is becoming acidic).

Human activities, such as tillage and logging, make soil subject to further erosion. Large amounts of particulate mercury in aquatic systems are released from riverbed and bank erosion. In the St. Lawrence River in Canada, about 1.4k mol mercury is released from the erosion of riverbeds and banks each year. This erosion is the major source of particulate mercury contamination in this Canadian river [47]. Flooding also brings significant amounts of mercury into aquatic systems. Furthermore gaseous mercury can be emitted from contaminated soil and then enter into the atmosphere. One of the possibilities is that surface soil mercury may leak into underground waters. Municipal sewage sludge is widely recycled to amend soil. The sludge-amended soil in Tennessee contains 7.3 mg/g of mercury [48]. Depending on the geochemistry and soil composition, different reactions may take place after mercury enters the land areas. Generally, mercury tends to associate with organic chemicals (such as humic substances) and inorganic iron oxy-hydroxides in the soil [49, 50].

2.6. Combustion and Industrial Sources

Combustion and industrial sources are very important sources of mercury emission into the environment. Mercury from combustion sources usually first enters the atmosphere, and then part of it settles in nearby water bodies or land. The rest remains in the atmosphere and becomes part of the global or regional atmospheric mercury deposit. On the global scale, Asia (especially China and India) accounts for about 50% of total anthropogenic mercury emissions [22] because coal as an energy source is the major source of mercury emissions. In China, coal contains an average of 0.22 mg/kg of mercury, of which approximately 70% is emitted into the environment during combustion [51]. In United States, it is estimated that about 97% of total anthropogenic mercury emissions come from combustion and industrial sources. Important combustion fuels include coal, waste, and oil while major combustion devices include utility combustors, waste incinerators, industrial boilers, and residential combustors. Almost 50% (79 metric tons annually) of total anthropogenic mercury emissions (158 metric tons annually) in the United States come from coal combustion. Mercury-containing products, such as batteries, fluorescent bulbs, and paints constitute municipal waste. Since the efficiency of removing mercury by municipal waste incineration is quite low, from 30 to 45% [52], this becomes an important source of urban pollution. It is estimated that, in the United States, the release of 43 metric tons of mercury from municipal waste incineration is about 30% of the total mercury emissions per year. Following table (Table 2) lists important industrial Hg pollution sectors and its level of emission [53].

Industrial sectors have been important sources of mercury emission. Historically chlor-alkali plants were responsible of 18% of the total Hg emission from industrial sector. But due to greater restrictions and better treatment - technology of removing mercury in wastewater, contamination by chlor-alkali plants, has been significantly decreased. Metal production and recovery sectors have emitted large amounts of mercury into the environment. All over the world, other important industrial sources of mercury emissions into the environment include the manufacture of cement, electrical apparatus, pulp and paper, thermometers, and fluorescent lamps [51, 54, 55].

Table 2. Mercury emission by industrial sectors [53]

Industrial sector	Amount of Hg emission	Year reported	Major emitter
Iron steel manufacturing	43 Mg/year	1990-2000	Asia (14.4 Mg yr^{-1}), Europe and North America (12.5 Mg/year each)
Primary and secondary non-ferrous metal Smelters	310 Mg/year	2008	Developing countries including China
Caustic Soda production (Chlor-alkali plants)	162.9 Mg/year	2008	Global production including China and India
Cement production	232 Mg/year	2005	Global
Waste derived from industrial processes	183 Mg/year	2000	Global
Landfill process	0.29-0.83 Mg/landfill	-	Finland
Wastewater treatment process	0.6-3 mg/Kg of sewage sludge	1995/1998	7 EU countries

3. METALLOMICS OF MERCURY

3.1. Introduction to Metallomics

Biological systems take up and utilize many metal ions for various purposes and via many routes. While some of the metals are essential for biochemical functions and thus for the health of biota, others cause diseases or exert toxic effects. For example, some metalloenzymes contain a specific number of metal ion(s) (such as Fe, Mo, Cu, Se) at the active sites in specific proteins [56] where they act as biocatalysts to regulate the biochemical reactions and physiological functions including gene (DNA, RNA) synthesis, metabolism, antioxidation. Genes and proteins cannot be synthesized without the assistance of metalloenzymes containing zinc and other metals like Cu [57].

Whereas the absence of these essential metals results in adverse effects, exposure to elevated concentrations of non-essential metals may cause carcinogenic effects (e.g., As, Cr, Pt), nephrotoxicity (e.g., Cd, U), or neurotoxicity (e.g., Al, Hg) [58].

Therefore, to better understand how a cell functions, one needs to characterize the cell not only by its genome in the nucleus and by protein content (proteome) but also by the nature, distribution and speciation of the metals and metalloids in various cell compartments. Williams [59] first termed these biometals and biometal species as metallomes by analogy with proteome. To distinguish from other fields of studies and promote this area of research, Haraguchi [57] later coined the term “metallomics” and proposed it as a separate branch of -omics study complementary to proteomics and genomics.

3.2. Exposure Pathways of Mercury

The bioavailability and toxicity of Hg depend on its concentration and chemical form (species), the co-existence of other chemicals (e.g., organic matter, selenium (Se)), the pathway and length of exposure, and the sensitivity and age of the organism. The primary source for Hg in marine mammals, predatory birds and humans is via the dietary consumption of fish, which takes up Hg and MeHg either through food or directly from water. Whereas MeHg is almost completely absorbed from the gastrointestinal (G.I.) tract and incorporated into erythrocytes (red blood cells or RBC) in the blood, inorganic Hg (II) is absorbed at the microvilli interface which results in a very low uptake rate [60]. The absorption of inorganic Hg salts from the G.I. tract is one order less than that of MeHg [61]. Laboratory experiments showed that the transfer of MeHg from diatom to copepod is four times more efficient than that of inorganic Hg [62]. This is due to the fact that inorganic Hg, being more hydrophilic in nature, tends to be excreted more readily than being absorbed. In contrast, the more lipophilic MeHg tends to be more readily associated with the soluble fraction of the diatom cell, which allows it to be accumulated in the fatty tissue of copepod [62]. As a result, the average proportion of MeHg over total Hg increases rapidly from phytoplankton to fish (15% to 95%) [63].

3.3. Tissue Distribution of Mercury

For a given species, the total Hg content is known to vary among the different organs (e.g., brain, liver, kidneys, and spleen). Wagemann et al. [64] found that more than 90% total Hg in muscle tissues in marine mammals is present as MeHg, while in liver inorganic Hg(II) dominates. Mercury content has also been shown to vary across the species exposed to same source depending on their food intake. MeHg was found to be in seabirds like black-footed albatross, herring gull and arctic tern in 35%, 36% and 66% in liver, kidneys and muscle, respectively [65]. This is due to the complicated metabolic processes involved.

Once in marine mammals and humans, MeHg readily transfers into the blood, with the largest portion binding to haemoglobin (Hb) in red blood cell (RBC) [66, 67] and the rest in plasma [68]. Red blood cell acts as a carrier for the MeHg to various organs including the brain (Figure 1). When binding with L-cysteine, MeHg is thought to easily cross the blood-brain barrier (BBB) and affect the central nervous systems (CNS), giving rise to its neurotoxicity [69]. Elemental Hg can also cross the BBB and be oxidized to Hg(II) and damage CNS [70].

Methylmercury can be gradually demethylated to inorganic Hg in animal tissues [71, 72]. In fact, inorganic Hg in the liver and kidneys or other tissues including the brain is mainly due to biotransformation of MeHg [73, 74]. Fang et al. [75] reported that the C-Hg bond can be cleaved by dietary Se which accounted for the inorganic Hg species in rat liver and kidneys. Hirayama and Yasutake [76] showed that reactive oxygen species actively participate in the MeHg demethylation in the liver. Rowland et al. [77] demonstrated the involvement of gut flora in the MeHg demethylation in mice.

Methylmercury is excreted out mainly through the bile and through urine to some extent [78-80]. Methylmercury complexes excreted in bile are mostly associated with small molecular weight forms that can be reabsorbed from the intestine. This circulation is one of the major reasons responsible for the long half-life of MeHg in animals [79]. Inorganic Hg is mostly excreted (in general 50%) by feces in rats, which most probably is related to the biotransformation of MeHg in liver followed by the biliary excretion, or to metabolic transformation in the intestinal lumen [71].

Figure 3 is a simplified diagram showing the dynamics of mercurials in animals. Details of the metallobolic pathways remain poorly understood.

3.4. The Role of Thiols

Both inorganic and MeHg speciation in the intracellular environment is dominated by MeHg-SR complexes (-SR = amino acids contain sulphur where R is the rest part) [14], due to the abundance of S-containing proteins and their strong affinity to Hg^{2+} or $MeHg^{+}$. Despite the very high formation constants (10^{15}-10^{17} range [81]) of MeHg complexes with thiols, the MeHg-SR bonding is extremely labile, and can undergo rapid ligand exchange reactions in aqueous solutions [70, 82-84] and likely in biological systems [70].

Mercurials are specific in binding with thiols but not selective; thus almost every protein in the biological systems is a potential target for Hg [85]. Typically MeHg is found to bind with di- and tri-peptides containing cysteine moieties [86]. It can also bind with glutathione or methionine. Glutathione binds with MeHg making it portable [69] into different organs. This helps MeHg to distribute through all potential organs of a biological system where it can bind with all sulphur-rich amino acid environments in the proteins. Several enzyme systems are inhibited both by inorganic and organic Hg. MeHg disrupts the structure and function of any protein to which it binds.

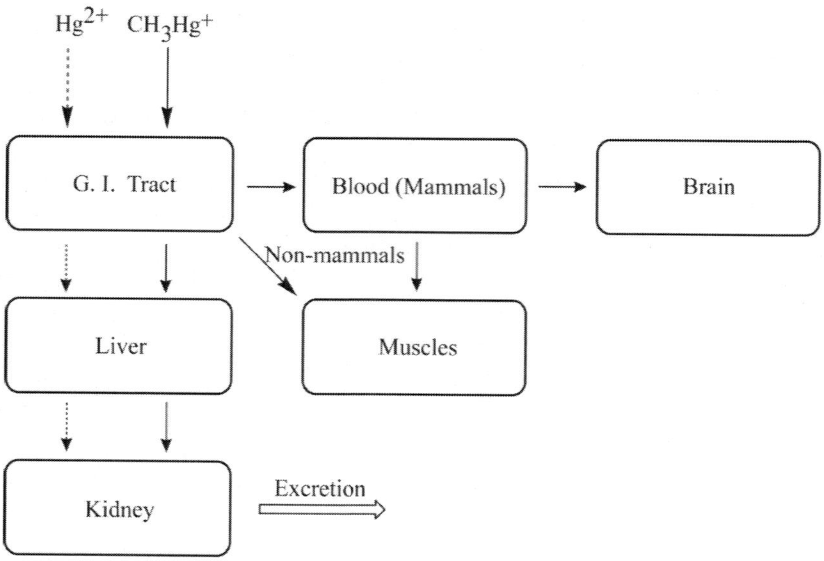

Figure 3. A schematic diagram showing the general transport of inorganic Hg (II) and MeHg in different organs of a mammal.

For example, MeHg blocks the assembly of tubulin and ultimately depolymerises microtubules, a component of the cytoskeleton, which is essential for cell structure and support, cell division, and cell migration [87-89]. It also inhibits protein synthesis in general (e.g., selenoprotein P). One most important feature of this extreme mobility is that after intravenous injection of MeHg compounds, it appears in the brain within 5 min [90, 91]. It was shown by Kerper et al. [69] that MeHg can form complex with cysteine or glutathione (act as a source for MeHg-cysteine) and cross the BBB through the L-type large neutral amino acid transporters (LAT 1 and 2). It was assumed that this occurs in part due to the resemblance of MeHg-cysteine complex with methionine that can easily cross the BBB to enter into the brain.

Inorganic Hg, irrespective of its chemical forms, tends to accumulate in the kidneys and little is found in the brain due to its inability to cross the BBB. It is assumed that this may take place through the endocytosis of filtered Hg-albumin complexes as albumin has a free-SH group on a terminal cysteinyl residue [92, 93] and inactivate them or form heavy-metal derivatives with crystalline proteins. This inhibition may partly account for the toxicity of mercurials [94-97]. Other primary sites of action of Hg may be the cell membrane [98, 99]. Mercury compounds may cause increased membrane permeability and changes in active membrane transport. Inorganic Hg is more effective than the organic one in disrupting membrane function [100-102]. Interaction with -SH groups might cause a disturbance stimulation of membrane receptors, which in turn may disturb immunoregulation [103].

3.5. Role of Selenium in the Metallomics of Mercury

Since Parizek and Ostadalova's [104] discovery of the protective effect of Se on kidney intoxication by Hg in laboratory rats, extensive studies have demonstrated that Se can modify the toxicity of both inorganic Hg and MeHg in the laboratory and in nature. The following possible pathways have been suggested for the Hg/MeHg-Se antagonism in biological systems:

3.5.1. Formation of MeHg-Se Compounds

Several MeHg-Se compounds such as bis(methyl mercuric)selenide (BMSe) [105, 106] and MeHg-selenoamino acid complexes are suggested to form *in vivo*, which could reduce the toxicity of MeHg. However, the presence of BMSe has never been analytically verified *in vivo*. Khan et al. [107] recently reported the synthesis and characterization of MeHg-selenoamino acid complexes of biologically relevant species.

3.5.2. Selenium-Aided Demethylation of MeHg

Se is suggested to be involved in the demethylation process. MeHg complexes with seleno amino acids are not stable and undergo rapid demethylation process forming Hg-Se as end product. A complete mechanism of this demethylation process is recently proposed [108].

3.5.3. Formation of Inorganic Hg-Se Compounds

Similar to the case with MeHg, binding between Se and inorganic Hg could also reduce the bioavailability and toxicity of inorganic Hg. The sparingly soluble HgSe(s) is likely to be the ultimate detoxification product. Inorganic mercury form pH reversible nanoparticles with glutathione peroxidase (GPx), which form insoluble Hg-Se as end products at low pH and can be discreted from the body through bile [109].

3.5.4. Redistribution of Inorganic Hg in the Presence of Se

It has been observed that Hg may redistribute itself in the presence of Se in biological systems. However, it is not clear how at molecular level Se induce redistribution of MeHg in different organs.

3.5.5. Selenium Inhibition of Methyl Radicals from MeHg

According to this hypothesis, oxygen-dependant metabolic reactions at the close proximity of lipids of the target tissues (e.g., the brain) break down MeHg and form methyl radical which results in the toxicity of MeHg [110]. This proposed mechanism can explain the protective effect of the anti-oxidants such as glutathione peroxidase (GPx) (Se in GPx decomposing peroxides which otherwise would initiate the MeHg breakdown) and vitamin E in the biological systems [110, 111]. However, no direct evidence has been reported to support this hypothesis [112].

3.5.6. Mercury-Induced Se Deficiency

Since MeHg has been shown to bind with Se very strongly, therefore it is possible that MeHg may make Se less bioavailable and cause Se deficiency in biological systems.

4. Recent Technologies for Estimation of Mercury

4.1. Handling of Samples

Most of the time the determination of mercury is done after extracting the samples from solid (Chemical digestion, ultrasonic, microwave, etc.) or pre-concentration from liquid and gases (by adsorption, ion-exchange, chelate formation, etc.) and utilization of the chromatographic and other techniques (GC, HPLC, LC, Ion-exchange, super-critical fluid chromatography, capillary zone electrophoresis) for separation of various forms of mercury (speciation) followed by detection of Hg by spectroscopic techniques sometimes coupled with more than one detection techniques (UV-Vis, AAS, CV-AAS, CV-AFS, ICP-MS, ICP-AES).

4.2. Extraction from Solid Samples

For mercury, two common extraction methods are used; acid lixiviation and alkaline digestion. The former was usually applied to sediment or soil samples using mineral acid (HCl or H_2SO_4) with KBr or NaCl [113, 114]. In some cases, Cu^{2+} was used additionally in order to release the mercury species from samples completely [115, 116]. For alkaline digestion, TMA (tetra-methylammoniumhydroxide) or KOH/CH_3OH was recommended to extract the mercury species especially from biological samples. Optimization of extraction technique must be done as reported for speciation of mercury in coal using HPLC-CV-AFS system [117]. TMA was chosen to be the optimized extractant to release the MeHg and EtHg simultaneously from coal under the condition of 5 min of extraction time, optimized power of 90W and optimized extractant of TMA, the recoveries for MeHg and EtHg from coal were 90.3% and 77.6%, respectively.

4.3. Estimation from Solid Samples

4.3.1. Determination of Hg in Coal

For determination of Hg species in coal sample combination of HPLC and CV-AFS techniques were utilized. Table 3 reports optimized parameters for Hg speciation in coal sample [117]. This method was applied to determine MeHg and EtHg in ten real coal samples.

4.3.2. Extraction and Determination of Hg in Fish Tissues

Mercury in Fish is mostly determined by Cold Vapor Atomic Fluorescence Spectroscopy. For each sample two separate aliquots of approximately 0.25 grams are transferred to 50 mL polypropylene test tubes. To each aliquot 5.0 ml of 3:1 (w/w) H_2SO_4:HNO_3 is added and then allowed to set at room temperature for 2 hours. Next the samples are heated to 800°C for 40 minutes. At this point all samples are liquefied. Then 15.0 ml of 6N HCl, 3.0 ml of 0.1N BrCl solution and 4.0 ml of de-ionized water are added. After mixing, the samples are heated to 600°C for 60 minutes. The sample solution becomes yellow in color and clear without any precipitate. Before analysis, each sample is diluted to 1:10 with 2% (w/v) HCl followed by 0.1 ml of hydroxylamine hydrochloride to remove free bromine. For calibration standards 4.0 ml of standard is added to polypropylene test tubes. All sample reagents were added to each standard cup except for the 4.0 ml of de-ionized water. The Hydra IIAF may be used in the direct fluorescence mode for the analysis. Instrument conditions have to be set. The calibration curve is to be drawn using 0.00, 1.00, 2.00, 6.00, 20.00, 50.00, and 100.0 ppb solutions. This calibration curve is employed to determine the concentration of mercury in digested sample solutions.

Table 3. Experimental condition of HPLC-CV-AFS for Hg speciation in coal sample [117]

Column	Shimadzu Shim-pack CLC-ODS, 6 x 150 mm, 5 μm
Mobile phase	0-11.7 min, 6% (v/v) CH_3 CN, 0.01% (v/v) 2-mercaptoethanol; 11.7-35 min, 25% (v/v) CH_3 CN, 0.01% (v/v) 2-mercaptoethanol
Flow rate of mobile phase	1.2 ml min^{-1}
Injection volume	20 μl
Oxidant solution	1% (m/v) K_2 S_2 O_8 in 10% (v/v), HCl, 2.4 ml min^{-1}
Reducing solution	0.5% (m/v) KBH4, 1.8 ml min^{-1}
Lamp	Hollow cathode mercury lamp, 253.7 nm
PMT voltage	280 V
Lamp primary current	50 mA
Carrier gas	Argon, 500 ml min^{-1}

To calculate the concentration in the original sample results were multiplied by 4.0 (the volume of standard added) and divided by the sample weight (approximately 0.25 gm).

The certified reference material is further diluted (1:2) with 2% HCl to bring it into calibration range [118].

4.3.3. Determination of Ultra-Trace Amount Methyl-, Phenyl- and Inorganic Mercury in Human Hair and Fish Tissue

For this analysis in environmental and biological samples, the cloud point extraction (CPE), and pre-concentration of ultra-trace amount of mercury species prior to reverse-phase high performance liquid chromatography (HPLC) with inductively coupled plasma mass spectrometry (ICP-MS) detection is done. Mercury species including methyl-, ethyl-, phenyl- and inorganic mercury are transformed into hydrophobic chelates by reaction with sodium diethyldithiocarbamate, and the hydrophobic chelates are extracted into a surfactant-rich phase of Triton X-114 upon heating in a water bath at 40°C. Ethylmercury is partially decomposed during the CPE process. Under the optimized conditions, detection limits of 13, 8 and 6 ngl^{-1} (as Hg) are achieved for $MeHg^+$, $PhHg^+$ and Hg^{2+}, respectively. Seven determinations of a standard solution containing the three mercury species each at 0.5 $ngml^{-1}$ level produced relative standard deviations of 5.3, 2.3 and 4.4% for $MeHg^+$, $PhHg^+$ and Hg^{2+}, respectively [119].

4.4. Determination in Liquid Samples

4.4.1. Mercury Speciation by Cold Vapor AAS

Mercury speciation analysis may be done by CVAAS method without chromatographic separations. Stannous chloride ($SnCl_2$) and sodium borohydride ($NaBH_4$) are the most common reducing reagents for the determination of mercury. A tungsten coil (150 W, 15 V) may be used as the atomizer for the proposed Flow Injection Vapor Generation Atomic Absorption Spectrometry (FI-VGAAS) system to determine inorganic mercury and total mercury concentrations. In this method inorganic mercury is measured when the W-coil is at room temperature and total mercury was measured when the W-coil temperature is set to 500°C. The organic mercury concentration was calculated by subtracting the inorganic mercury concentration from the total mercury concentration [120]. In order to validate the method, known concentrations of methyl mercury and inorganic mercury are spiked into deionized water and tap water samples.

4.4.2. Sensitive Mercury Speciation by Reversed-Phase Column High-Performance Liquid Chromatography with UV-Visible Detection for Sea Water

Solid-phase extraction using 6-mercaptopurine and dithizone are done before the Hg speciation. The method is based on the reaction of mercury with 6-mercaptopurine and solid-phase extraction (SPE) of the complex on an octadecylsilane (C18) cartridge. The complex is then treated with ammoniacaldithizone solution, and the complexes of inorganic and organic mercury with dithizone are eluted by methanol. The speciation analysis of methylmercury (MeHg), phenylmercury (PhHg), and inorganic Hg (II) is carried out by RP-HPLC. Some experimental variables influence the SPE and derivatization, such as pH, chelating and derivatizing agent concentration, and surfactant addition. The procedure was successfully applied to seawater and tuna fish samples. The method offers a good sensitivity and selectivity for the speciation of mercury compounds. Mercury speciation analysis is performed by using HPLC with UV-Vis detection. Taking into account that no coupling techniques are necessary, the methodology is simple and inexpensive [121].

4.4.3. Speciation Analysis of Mercury in Seawater by On-Line Pre-Concentration HPLC–ICP-MS

A method based on the coupling of HPLC with ICP-MS with an on-line pre-concentration micro-column has been developed for the analysis of inorganic and methyl mercury in the dissolved phase of natural waters. This method allows the rapid pre-concentration and matrix removal of interferences in complex matrices such as seawater with minimal sampling handling. Detection limits of 0.07 ngL^{-1} for inorganic mercury and 0.02 ngL^{-1} for methyl mercury have been achieved allowing the determination of inorganic mercury and methylmercury in filtered seawater from the Venice lagoon. Good accuracy and reproducibility is demonstrated by the repeat analysis of the certified reference material BCR-579 coastal seawater. The developed HPLC separation is shown to be also suitable for the determination of methyl mercury in extracts of the particulate phase [122]. Although this methodology does not reach the detection limits of GC–ICP-MS with SPME or purge and trap pre-concentration that have been used for the determination of mercury species in pristine environments; it is suitable for the determination of mercury species in large numbers of samples from polluted aquatic environments, as the time required for pre concentration and matrix removal is less than a minute per replicate.

4.5. Determination in Gaseous Samples

4.5.1. Simultaneous Removal of Hg^0, SO_2 and NO from Flue Gas Using $NaClO_2$ Solution

Elemental Hg may be determined along with SO_2 and NO in flue gases using bench-scale bubbling reactor. Removal of Hg is found to be higher in acid conditions and at higher NO concentration in the presence of sufficient $NaClO_2$ solution. However, SO_2 concentration does not affect the efficiency of Hg removal. Moreover under acidic condition, $NaClO_2$ solution was an excellent adsorbent [123].

4.5.2. Automated Monitoring of Gaseous Elemental Mercury (GEM) in Ambient Air

Automated monitoring of gaseous elemental mercury in ambient air could be done by using anHg speciation system, anHgvapor analyzer, a particulate Hgmodule, an oxidized Hg speciation module, and a pump module. In combination with all the above equipment separation and quantification of GEM, GOIM and Hg (p) < 2.5 in ambient air is done [42]. Ambient air is pulled in through particulate Hg module, an oxidized Hg speciation module by the pump module to remove particulate matter and GOIM from the air. The Hg speciation system is programmed to sample at 10 L min^{-1} with a 2-hour cycle (1 h for sample pre-concentration and 1 h for sample analysis) in which GEM is sampled at 1.0 L min^{-1} and determined every 5 min during the GOIM and Hg(p) sampling period. During the sample analysis period, the Hg (p) < 2.5 and GOIM sampling units are first flushed with zero air for 15 min. The Hg (p) filter in the particulate Hg module is then heated at 800°C for 20 min to convert the particle-bound Hg to Hg^0 and the released Hg^0 is quantified by Hg vapor analyzer. With the temperature of the quartz filter being maintained at 800°C, the KCl-coated denuder in the oxidized Hg speciation module is heated to 600°C for 15 min to convert gaseous Hg(II) to Hg^0 and the released Hg^0 is detected by the Hg vapor analyzer. Finally, the system is flushed with zero air for 10 min before the next cycle is started. The denuder (GOIM collector) is reconditioned and the quartz particulate filter (RPF, fine particle (diameter < 2.5 mm) collector) is replaced biweekly according to the manufacturers protocol. The analyzer is calibrated with a built-in elemental mercury permeation source and also by external manual Hg^0 injections to confirm the calibration. For total filterable particulate matter in ambient air, filtration based mini-samplers is used for collection [124]. A custom-built quartz mini-sampler containing the quartz fiber filter is positioned vertically and ambient air is pulled in using an air pump. The sampling flow-rate is measured using a mass flow. During the sample analysis, three heating steps controlled by a programmable timer are applied, viz., pre-cleaning of the analytical system, sample analysis (mercury detection and data acquisition) and cleanup after sample analysis. The mini-sampler, along with the collected sample, is heated at 900°C for 5 min using nickel-chromium alloy resistance heating ribbon. A stream of argon at a flow-rate of 100 $mLmin^{-1}$ is used to transport mercury species from the sampler to the pyrolyzer, where all mercury species are converted into their elemental forms, then to a gold trap where the elemental mercury is adsorbed. The elemental mercury is then thermally desorbed at 500°C from the gold trap and detected by a cold vapor atomic fluorescence spectrophotometer (CVAFS). The output signals from the detector are captured by an integrator. The sampling and analysis procedure have a detection limit of 14 pg, which is based on 3 times the standard deviation of the total blank (including sampling, sample handling, and analysis).

4.5.3. Measurement of Atmospheric Mercury Species with Manual Sampling and Analysis Methods

In a Case Study in Indiana: Ground-level concentrations of three atmospheric mercury species were measured using manual sampling and analysis to provide data for estimates of mercury dry deposition. Three monitoring stations were operated simultaneously during winter, spring, and summer 2004, adjacent to three mercury wet deposition monitoring stations in northern, central, and southern Indiana. The monitoring locations differed inland-use setting and annual mercury-emissions level from nearby sources. A timer-controlled air sampling system that contained a three-part sampling train was used to isolate reactive gaseous mercury, particulate-bound mercury, and elemental mercury. The sampling trains were exchanged every 6 days, and the mercury species were quantified in a laboratory. A quality-assurance study indicated the sampling trains could be held at least 120 h without a significant change in reactive gaseous or particulate-bound mercury concentrations. The manual sampling method was able to provide valid mercury concentrations in 90 to 95% of samples. Statistical differences in mercury concentrations were observed during the project. Concentrations of reactive gaseous and elemental mercury were higher in the daytime samples than in the nighttime samples. Concentrations of reactive gaseous mercury were higher in winter than in summer and were highest at the urban monitoring location. The results of this case study indicated manual sampling and analysis could be a reliable method for measurement of atmospheric mercury species and has the capability for supplying representative concentrations in an effective manner from a long-term deposition monitoring network [125].

4.6. A Non-Instrumental Method

A novel oligo-T-based gold nano probe for rapid and portable detection of Hg^{2+} with the help of a power-free poly (dimethylsiloxane) (PDMS) microfluidic device was developed. The PDMS device should be equipped with a microscope for visual detection since the channels are only of micrometer sizes [126]. An Au nanoprobe was prepared through modification of AuNPs with thiolated thymine oligonucleotides (a mixture of T6/T10). Because of the strong Coulombic repulsion between highly negatively charged DNA strands located at the AuNP surface, these nanoprobes were highly stable and remained dispersed (with a characteristic red color) even at high ionic strength. In the presence of Hg^{2+}, adjacent thymine probes at the surface of each AuNP formed theT–Hg^{2+}–T complex, which changed the charge distribution at the surface and destabilized AuNPs against aggregation at high ionic strength. As a result, the color of the gold nanoprobes turned purple as purple-colored aggregates formed, an effect originating from the inter particle coupled plasmon absorbance of aggregated AuNPs. This simple strategy forms the basis of visual approach for Hg^{2+} detection. The system can be used for rapid and visual detection of low micro-molar Hg^{2+} in real environmental samples.

5. Remediation Techniques

As the toxicological impacts of Hg are largely dependent on its speciation, understanding its transformations and the impacts of various chemical forms is vital to the prevention of harmful human and environmental health effects. The Hg species present in a given environment depends on the initial released form, the thermodynamic stability of this compound and the transformation rate of the released form to a more stable one. For effectively designing and evaluating appropriate remedial measures in Hg impacted areas the above mentioned issues must be well understood.

5.1. Soil Remediation

The most commonly used techniques for the remediation of mercury contaminated soils have been classified as either excavation techniques or containment techniques, and are grouped as follows [17]: (a) *Ex situ* treatments: physical separation, thermal treatments, hydrometallurgical treatments; (b) *In situ* recuperation: vapor extraction coupled with evaporation (soil), permeable reactive barriers; (c) *In situ* leaching and extraction: electrokinetic separation, interceptor systems, phytoremediation, passive remediation; (d) containment: pump and treat impermeable barriers, sealed surfaces and drainage, stabilization and solidification, sediment covering.

5.1.1. Excavation and Ex-Situ Treatment

Excavation and *ex-situ* (i.e., off-site or aboveground) treatment is essentially a well-understood and most frequently employed practice for Hg recovery. Sometimes it can be complicated if it extends below the water table or costly if the contamination is distributed over a large area. Mercury is subsequently liberated from excavated soil by one of the following processes.

5.1.1.1. Physical Separation

Most Hg compounds have a strong affinity for the fine fraction of soils. In wet screening, metallic Hg droplets behave as a separate liquid phase being concentrated into the fines. If analyze Hg grain size distribution is needed, dry sieving must be adopted. Rubble and coarse constituents of soils, such as gravel, are initially removed via sieving. Water is used to wash small to medium fractions (~50 mm to >0.1 mm diameter) and fines (i.e., silts, clays or organics) are separated from these fractions, generally using hydrocyclones, spiral classifiers, and fluidized beds [127]. The remaining Hg-enriched sludge or fine fraction is de-watered and subsequently isolated or treated further (for example, via thermal methods). This physical separation procedure does not require the use of other chemicals. This technique is most effective for soils dominated by coarse materials (i.e., sand and gravel) with some (<20%) fines.

5.1.1.2. Thermal Treatment

The volatility of Hg and its compounds increase with temperature therefore thermal heating of excavated soil is employed as effective means for Hg recovery from contaminated soils. Hempel and Thoeming [127] determined that all Hg compounds should volatilize at temperatures below 600°C and, in the absence of HgO, 250°C should be sufficient. Matsuyama et al. [128] determined that low-temperature thermal treatments could adequately recover Hg from soils, were less costly and less likely to negatively alter soil properties. At temperatures 200-210°C even stable compounds like HgS begin to volatilize. In the presence of a catalyst, specifically iron chloride, Hg (as HgS) removal increased from 50% to 99.5% in a one-hour period at 300°C. Soil is treated in batches as follows: first it is dried at a temperature of 100°C then transferred to a heating drum where temperatures of 600°C are maintained. Gas from these stages passes through an afterburner where Hg would be recovered from the gas phase using a gas washing system, charcoal filter, and iodine-impregnated scrubber or through condensation.

5.1.1.3. Hydrometallurgical Treatments

Chemical extraction of Hg from excavated soils can be induced through four primary mechanisms: desorption of adsorbed species; oxidation of metallic Hg; use of strong complexing agents; and through dissolution of precipitated Hg. Efficiency of any mechanism employed may decrease over time due to re-complexation and re-adsorption and removal of the most soluble compounds at early time. The two most promising hydrometallurgical techniques employed are electro-kinetic or electro-leaching and leaching methods. Leaching agents commonly applied include halide compounds, such as hypochlorite or hydrobromic acid, iodine in the form of potassium iodine, and a mixture of nitric acid and NaCl [127].

Sodium hypochlorite and sodium pyrophosphate are believed to be particularly effective at liberating metals from organics, but hydrogen peroxide releases more metals from oxides and sulfides [129].

5.1.2. In-Situ Recovery

In *in-situ* recovery, subsurface heterogeneity limits the effectiveness of processes besides clean-up times tend to be longer than *ex-situ* treatments. Thus methods for *in-situ* recovery of Hg are far less established than *ex-situ* techniques. Despite these factors, many *in-situ* technologies are very promising and mainly due to the fact that contaminated soil and groundwater remain in the subsurface may become more cost-effective and practical than excavation and treatment methods for many Hg-contaminated sites.

5.1.2.1. Soil Vapor Vacuum Extraction Coupled with Soil Heating

Soil Vapor Vacuum Extraction uses a vacuum to force air through the unsaturated zone of soil. Volatile and semi-volatile compounds easily partition into the vapor phase and are subsequently removed by the vacuum. The effectiveness of these systems depends on contaminant volatility and availability to air channels, tendency to adsorb on solid surfaces, and soil properties (e.g., grain size). It is well demonstrated that Hg fluxes into the vapor phase increase significantly with airflow and temperature [130, 131]. Thus, soil vapor vacuum extraction coupled with soil heating may be an effective technology for *in-situ* remediation of Hg.

5.1.2.2. Electro-Kinetic Separation

This process involves the generation of an electric field through application of a potential difference or current in a soil matrix. Metals like Hg, migrate towards electrodes placed in the soil where they accumulate and can subsequently be removed at a lower cost than excavating the entire impacted area.

5.1.2.3. Interceptor Systems

Interceptor Systems, such as trenches and drains, are extremely simple and effective at recovering Hg as metallic Hg. But these treatments are limited by topography, stratigraphy and Hg residual saturation.

5.1.2.4. Phytoremediation

Phytoremediation is remediation process in which plants are used. Few native species of plants or specially developed plants, possesses the resistance to the toxic metals. Such plants that accumulate substantial quantities of metals from soils are known as hyper-accumulators. The recovery of these metals occurs after collecting and incinerating the plants. Phytoremediation is a promising remediation and unproven technology. This technique holds much promise for the cost effect remediation of shallow soils over a fairly widespread area. Issues such as limited access to vegetation by wildlife and time required for cleanup must be addressed.

5.2. Surface and Ground Water Remediation

5.2.1. In-Situ Treatments

5.2.1.1. Permeable Reactive Barriers

In permeable reactive barriers, dissolved compounds react with constituents of barrier to precipitate contaminants. Permeable reactive barriers are engineered structures installed below the ground surface perpendicular to the flow of contaminant-laden groundwater. These barriers, so called walls, are geochemically engineered to transform contaminants to relatively benign and/or immobile forms and to operate passively for extended periods with little or no maintenance [132]. Two main criteria that must be addressed in effective design of wall include permeability and the level of reactivity with wall constituents. Wall materials have ranged from organic-rich composted sewage sludge or wood chips combined with neutralizing materials (e.g., limestone) for removal of metals from acidic mine drainage to zero-valent iron (ZVI) for various organic and inorganic contaminants. Other materials proposed for metals studies include hydroxyapatite, zeolites, hydrous ferric oxides and bone char phosphate [133].

5.2.1.2. *In-Situ* Leaching and Extraction

Used together with Pump-and-Treat systems, *In-situ* Leaching and Extraction involves the injection of chemicals to enhance Hg solubility in groundwater. Pump-and-treat is a frequently practiced, cost-effective remedial alternative employed either for removal of contaminants from the subsurface and/or hydraulic containment of a contaminant plume [132]. In many natural water systems, Hg is strongly adsorbed to soils and/or is present as a precipitate – in either situation; Hg may dissolve or desorb over time and thereby provide a long-term source to groundwater. To reduce clean-up time and increase the efficacy of the clean-up system, solubility-enhancing chemicals may be injected up-gradient from the zone of contamination. This method is limited because of (1) it is not well demonstrated and (2) often unacceptable to regulators.

5.2.1.3. Passive Remediation – Wetlands

The use of wetlands for Hg immobilization is a somewhat controversial issue as some wetlands contain microorganisms that can convert mercury into even more toxic species-methylmercury (MeHg). Despite this potential hazard, wetland may be used for remediation of Hg by controlling amount and form of sulfur. US Department of Energy [134] has had notable success in significantly limiting the amount of MeHg generated (<1 ng/L) in an artificial wetland. Mercury bound to decomposed organic matter descends to the bottom of anaerobic cells where it reacts with gypsum to form relatively insoluble sulfide.

5.3. Containment

Cleanup of many contaminated sites is often not feasible, generally due to financial or technical reasons. Several methods of containment are commonly employed to inhibit mobilization of contaminants and minimize ecological and human health risks of exposure.

5.3.1. Pump-and-Treat

The characteristics of the certain contaminant or system do not allow removing and hydraulic containment may be necessary to protect the surrounding environment. For the contaminant mass remaining in the subsurface, pump and treat systems can prevent site - contamination. Basically, pump-and-treat systems involve the installation of extraction wells below the water table within or slightly down-gradient from the zone of contamination. As extracted water must be treated at the surface, well placement and pumping rate should be chosen to ensure capture of contaminated groundwater and limit recovery of clean water. Contaminant-monitoring wells should be installed around the contaminant plume to assess containment and evaluate hydro-geochemical conditions.

5.3.2. Impermeable Barriers (Sealed Surfaces and Drainage)

Several geotechnically-engineered approaches are possible for prevention of the off-site migration of contaminants. Mud walls are low permeability barriers generally composed of excavated material mixed with bentonite or cement bentonite mixture. Walls can be constructed as a trench, which are typically 0.5 to 2 m wide and a maximum depth of 50 m [132]. There are other types of barriers such as grout walls and sheet pile walls, which are constructed by injection molding or by vibratory forces respectively.

On the other hand surface seals and drains are used to control infiltration and limit mobilization of contaminants into groundwater [132]. Low permeability materials, like compacted clays, natural soils mixed with stabilizers or bentonite, or geosynthetic membranes, are frequently used to inhibit infiltration.

5.3.3. Stabilization and Solidification

Stabilization and solidification techniques involve both *ex-situ* and *in-situ* mixing of impacted soils with additives that intend to reduce the mobility or leachability of contaminants [129]. Stabilization specifically attempts to bind contaminants to the solid, which often reduces soil permeabilities. On the other hand solidification improves the physical characteristics of materials, such as sediments or mud, so they can be excavated and transported more easily. Common stabilizing agents are organic polymers, cement, calcium carbonate spiked with binders, furnace slag silicates or fly ash, asphalt and bitumen.

5.3.4. In-Situ Capping

In situ capping or covering involves placing an insulating layer over the contaminated material. This is done for three basic purposes (1) physical isolation of the contaminated sediments from benthic organisms (2) prevention of sediment re-suspension and transport, and (3) reduction of the contaminant migration into the water column [135]. For mercury in sediments, the increased solubility and diffusability of methylmercury must be considered during evaluation of this alternative. Site-specific issues such as qualities of the watercourse, functions of the waterway and geo-environmental properties must be incorporated in design. Sufficient understanding of groundwater flow is particularly important, as it will influence the flux of contaminants into the water body.

5.4. Electro-Remediation of Mercury Polluted Soil: A Case Study

Electro-remediation has been successfully applied in a variety of soil restoration studies, this methodology having the advantage of exhibiting simultaneous chemical, hydraulic and electrical gradients. The main advantages of electro-remediation, as compared with other soil treatment procedures, are (1) electroosmotic flow is not dependent on either pore or particle size, (2) hydraulic gradient is enhanced by electro-migration, (3) treatment can be applied in situ, (4) it can be applied to low permeability soils, (5) there is minimal disruption of normal activities at the site, (6) the required investment is usually lower than that for other conventional treatments, and (7) it can be applied in conjunction with techniques such as pumping, vacuum extraction or bioremediation.

Indeed, for efficient mercury removal from a saturated soil with electro- remediation, application of either an electric field or direct current through two electrodes (anode and cathode) is required. These are usually inserted in wells containing a supporting electrolyte made from inert salts, leading to improved electric field conductive properties [136-138]. Specifically, for mercury polluted soil electro-remediation, the use of complexing agents like ethylene diamine tetra acetic acid (EDTA), KI, and NaCl under a constant potential gradient has been reported [139]. Based on the above precedents, the electro-remediation was developed aided by extracting agents for mercury removal from San Joaquin's Sierra Gorda soil samples (Figure 4) [140]. Electro-remediation of mercury polluted soil, facilitated by the use of complexing agents, proved to be an attractive alternative treatment for the removal of mercury from polluted soil in mining areas located at Sierra Gorda in Queretaro, Mexico (Figure 4A). Implementation of this remediation protocol is expected to improve the living conditions and general health of the population in the Mine "El Rincón" in San Joaquin (Figure 4B). Experimental observations suggest that it is possible to remove up to 75% of metal contaminants in mercury polluted soil samples by wetting them with 0.1 M EDTA, placing them in an experimental cell equipped with Ti electrodes, and then applying a 5 V electric field for 6 hours [140]. The pH remains nearly constant at 4 and conductivity showed values close to 10 mS cm^{-1} by the ionic species. The efficient removal of mercury contaminants observed under these conditions is attributed to electro-migration of the coordination complexes that form between the terminal hydroxyl groups in EDTA and divalent mercury (Hg^{+2}), which is probably strengthened by supramolecular interactions between unshared electrons at EDTA's tertiary amino nitrogens and Hg^{+2}. These interactions are particularly effective with the presence of potassium ions.

FUTURE PROSPECTIVES

To date mercury research has provided scientists and policy makers with a detailed understanding of mercury toxicology, biogeochemical cycling and past impacts on human exposure. The complexity of the global biogeochemical mercury cycle has led to repeated and ongoing paradigm shifts in numerous mercury-related disciplines [141]. Future research has to look further into health aspects of biota -human being, fauna and flora, on the methodologies of sampling and estimation of different forms of Hg and its transformation, the containment methods, on mercury toxicology and epidemiology, for marine mercury dynamics and ecology, for monitoring mercury emission from industrial sources. These studies must be taken at local, regional and global levels. Few other areas for continued research are: Arctic mercury-cyclestudies, mercury modeling, impact of regulations on combustion fuel (coal and petroleum), closer collaboration between observational mercury science and regulatory rules.

(A)

(B)

Figure 4. Localization of Queretaro in Mexico (A) satellite image from San Joaquin's Sierra Gorda, Queretaro, (B) where there is the Mine "El Rincón" with high concentration of Hg^{2+}, which was removed with electro-remediation process in presence of EDTA.

Conclusion

Hg toxicity has a long history and much is known about its biogeochemical cycle. Hg emissions from numerous sources are well accounted for. But the low concentrations and the complexity that characterize Hg toxicity and Hg cycling continue to impose major paradigm shifts. Thus continuous monitoring of Hg, in its different forms, its movement in our environment and health impacts are areas to be studied. This chapter describes the sources of Hg; physiochemical properties and toxicity of Hg, especially with human health prospective; estimation in air, water and soil; Hg remediation technologies and need of future studies in the field.

References

[1] http://www.rsc.org/periodic-table/element/80/mercury, accessed on 15.9. 2015.

[2] Mergler, D., Anderson, H. A., Chan, L. H. M., Mahaffey, K. R., Murray, M., Sakamoto, M. and Stern, A. H., (2007) Methylmercury exposure and health effects in humans: a worldwide concern. *Ambio* 36, 3-11.

[3] WHO, (1990) Methylmercury, International programme on chemical safety, World Health Organization, Geneva.

[4] Nriagu, J. O., (1989) A global assessment of natural sources of atmospheric trace metals. *Nature* 338, 47-49.

[5] Pacyna, E. G., Pacyna, J. M., Steenhuisen, F. and Wilson, S., (2006) Global anthropogenic mercury emission inventory for 2000. *Atmos. Environ.* 40, 4048-4063.

[6] Wood, J. M., Kennedy, F. S. and Rossen, C. G., (1968) Synthesis of methylmercury compounds by extracts of a methanogenic bacterium. *Nature* 220, 173-174.

[7] Kerin, E. J., Gilmour, C. C., Roden, E., Suzuki, M. T., Coates, J. D. and Mason, R. P., (2006) Mercury methylation by dissimilatory iron-reducing bacteria. *Appl. Environ. Microbiol.* 72, 7919-7921.

[8] Choi, S. C., Chase, T. and Bartha, R. (1994) Enzymatic catalysis of mercury methylation by *Desulfovibrio desulfuricans*. *Appl. Environ. Microbiol.* 60, 1342-1344.

[9] Choi, S.-C. and Bartha, R., (1993) Cobalamin-mediated mercury methylation by *Desulfovibrio desulfuricans*. *Appl. Environ. Microbiol.* 59, 290-295.

[10] Choi, S.-C., Chase, T. and Bartha, R., (1994) Metabolic pathways leading to mercury methylation in *Desulfovibrio desulfuricans LS*. *Appl. Environ. Microbiol.* 60, 4072-4077.

[11] Krishnamurthy, S., (1992) Biomethylation and environmental transport of metals. *Chem. Environ.* 69, 347-350.

[12] Craig, P. J. and Rapsomanikis, S., (1985) Methylation of tin and lead in the environment: oxidative methyl transfer as a model for environmental reactions. *Environ. Sci. Technol.* 19, 726-730.

[13] Falter, R. and Wilken, R. D., (1998) Isotope experiments for the determination of the abiotic mercury methylation potential of a River Rhine sediment. *Vom Wasser* 90, 217-231.

[14] Haris, H. H., Pickering, I. J. and George, G. N., (2003) The chemical form of mercury in fish. *Science* 201, 1203.

[15] Pongratz, R. and Heumann, K. G., (1998) Determination of concentration profiles of methyl mercury compounds in surface waters of polar and other remote oceans by GC-AFD. *Intl. J. Environ. Anal. Chem.* 71, 41-56.

[16] Strasdeit, H., (2008) Mercury-alkyl bond cleavage based on organomercurylyase. *Angew. Chem. Int. Ed.* 47, 828-830.

[17] Hinton, J., Veiga, M. (2001). Mercury contaminated sites: a review of remedial solu-tions. National Institute for Minamata Disease. In: *Proceedings of the NIMD* (National Institute for Minamata Disease).

[18] Bengtsson, G. (2008). Mercury sorption to sediments: dependence on grain size, dissolved organic carbon, and suspended bacteria. *Chemosphere*, 73, 526-531.

[19] Leopold, K., Foulkes, M., Worsfold, P. (2010). Methods for the determination and speciation of mercury in natural waters- a review. *Anal. Chim. Acta*, 663(2), 127-138.

[20] Nik, M. G. (2012). The estudy of mercury pollution distribution aroun a chlor-alkali petrochemical complex, Bandar Iman, southern Iran. *Environ. Earth Sci.*, 67 (5), 1485-1492.

[21] Pirrone, N., Keeler, G. J., Nriagu, J. O., (1996). Regional differences in worldwide emissions of mercury to the atmosphere. *Atmos. Environ.* 30, 2981-2987.

[22] Pacyna, E. G., Pacyna, J. M., (2002). Global emission of mercury from anthropogenic sources in 1995. *Water Air Soil Poll.,* 137, 149-165.

[23] Rytuba, J. J., (2000). Mercury mine drainage and processes that control its environmental impact. *Sci Total Environ.,* 260, 57-71.

[24] Rieuwerts, J. S., Farago, M., (1996). Mercury concentrations in a historic lead mining and smelting town in the Czech Republic: a pilot study. *Sci. Total Environ.* 188, 167-171.

[25] Lyons, W. B., Wayne, D. M., Warwick, J. J., Doyle, G. A., (1998). The Hg geochemistry of a geothermal stream, steamboat creek, Nevada: natural vs. anthropogenic influences. *Environ. Geol.* 34, 143-150.

[26] Schroeder, W. H., Yarwood, G., Niki, H., (1991). Transformation processes involving mercury species in the atmosphere-results from a literature survey. *Water Air Soil Poll.* 56, 653-666.

[27] Lindqvist, O., Rodlhe, H., (1985). Atmospheric mercury: a review. *Tellus* 37B, 136-159.

[28] Slemr, F., Langer, E., (1992). Increase in global atmospheric concentrations of mercury inferred from measurements over the Atlantic Ocean. *Nature* 355(6359), 434-437.

[29] Keeler, G. J., Glinsorn, G., Pirrone, N., (1995). Particulate mercury in the atmosphere: its significance, transport, transformation and sources. *Water Air Soil Poll.* 80 (1-4), 159-168.

[30] Siddiqi, Z. M., Song, X., Zhou, H., Lu, J., (2005) Studies on Mercury Associated with Airborne Particulate Matter in Downtown Toronto, XIII International Conference on Heavy Metals in the Environment (ICHMET), Rio de Janeiro, Brazil, *Paper # 061*, held on June 5-9, 2005.

[31] Siddiqi, Z. M., Lu, J., (2015) Measurement of mercury associated with airborne particle matter: use of two sampling devices, *Fresenius Environmental Bulletin*, 24(4), 1326-1332.

[32] Venkataraman, K., (2000). An analysis of mercury data in the Great Lakes area. *Civil Engineering*, The University of Toledo.

[33] Hermanson, M. H., (1993). Historical accumulation of atmospherically derived pollutant trace-metals in the Arctic as measured in dated sediment cores. *Water Sci. Technol.* 28(8-9), 33-41.

[34] Locotte, M., Mucci, A., Hillaire-Mariel, C., Pichet, P., Grondin, A., (1995). Anthropogenic mercury enrichment in remote lakes of Northern Quebec (Canada). *Water Sci. Technol.* 80(1-4), 467-476.

[35] Hermanson, M. H., (1998). Anthropogenic mercury deposition to arctic lake sediments, *Water, Air, and Soil Pollution* 101, 309-321.

[36] Tan, H., He, J. L., Liang, L., Lazoff, S., Sommer, J., Xiao, Z. F., Lindqvist, O., (2000). Atmospheric mercury deposition in Guizhou, China. *Sci Total Environ.* 259, 223-230.

[37] Dominguez, A., Gutierrez, M., Vazquez, F. A., (2001). Quantitative evaluation of atmospheric deposition flux of mercury in sediments within the City of Juarez, Mexico. *Water Air Soil Poll.* 132, 263-274.

[38] Guentzel, J. L., Landing, W. M., Gill, G. A., Pollman, C. D., (2001). Processes influencing rainfall deposition of mercury in Florida, *Environ. Sci. Technol.* 35, 863-873.

[39] Lacerda, L. D., Paraquetti, H. H. M., Rezende, C. E., Silva, L. F. F., Silva, E. V., Marins, R. V., Ribeiro, M. G., (2002). Mercury concentrations in bulk atmospheric deposition over the coast of Rio de Janeiro, Southeast, Brazil. *J. Brazilian Chem. Soc.* 13, 165-169.

[40] Landis, M. S., Vette, A. F., Keeler, G. J., (2002). Atmospheric mercury in the Lake Michigan basin: influence of the Chicago/Gary urban area. *Environ. Sci. Technol.* 36, 4508-4517.

[41] Bindler, R., (2003). Estimating the natural background atmospheric Deposition rate of mercury utilizing ombrotrophic bogs in southern Sweden. *Environ.l Sci. Technol.* 37, 40-46.

[42] Zhang, X., Siddiqi, Z., Song, X., Mandiwana, K. L., Yousaf, M., Lu, J., (2012) Atmospheric dry and wet deposition of mercury in Toronto, *Atmos. Environ..* 50, 60-65.

[43] Mason, R. P., Lawson, N. M., Sheu, G. R., (2000). Annual and seasonal trends in mercury deposition in Maryland. *Atmos. Environ.* 34, 1691-1701.

[44] Kolka, R. K., Nater, E. A., Grigal, D. F., Verry, E. S., (1999). Atmospheric inputs of mercury and organic carbon into a forested upland/bog watershed. *Water Air Soil Poll.* 113, 272-294.

[45] Heyvaert, A. C., Reuter, J. E., Slotton, D. G., Goldman, C. R., (2000). Paleolimnological reconstruction of historical atmospheric lead and mercury deposition at Lake Tahoe, Californiae Nevada. *Environ. Sci. Technol.* 34, 3588-3597.

[46] Smart, N. A., (1968). Use and residues of mercury compounds in agriculture. *Residue Reviews* 23, 1.

[47] Quemerais, B., Cossa, D., Rondeau, B., Gagnon, P., Fortin, B., (1999). Sources and fluxes of mercury in the St. Lawrence River. *Environ. Sci. Technol.* 33, 840-849.

[48] Carpi, A., (1997). TheSurface/Atmosphere Exchange of Elemental and Methyl Mercury over Background and Municipal Sewage Sludge Amended Soil. Cornell University, Ithaca, New York.

[49] Kerndorf, H., Schnitzer, M., (1980). Sorption of metals on humic acids. *Geochim. Cosmochim. Acta* 44, 1701-1708.

[50] De Oliveira, S. M. B., Melfi, A. J., Fostier, A. H., Forti, M. C., Favaro, D. I. T., Boulet, R., (2001). Soils as an important sink for mercury in the Amazon. *Water Air Soil Poll.* 26, 321-337.

[51] Wang, Q. C., Shen, W. G., Ma, Z. W., (2000) Estimation of mercury emission from coal combustion in China. *Environ. Sci. Technol.* 34, 2711-2713.

[52] Chang, M. B., Wu, H. T., Huang, C. K., (2000). Evaluation on speciation and removal efficiencies of mercury from municipal solid waste incinerators in Taiwan. *Sci. Total Environ.* 246, 165-173.

[53] Siddiqi, Z. M., Saleem, M., Basheer, C., (2015) Mercury focused preliminary assessment of water quality of logged water and sediments from water run-off canal system in Jubail Industrial City and comparative effect on mercury bio-accumulation in fish tissues, Environmental monitoring and assessment (Communicated).

[54] Beim, A. M., Grosheva, E. I., (1992). Ecological chemistry of mercury Contained in bleached kraft pulp mill effluents. *Water Air Soil Poll.* 65, 135-141.

[55] Moreira, J. C., Pivetta, F., (1997). Human and environmental contamination by mercury from industrial uses in Brazil. *Water Air Soil Poll.* 97, 241-246.

[56] Lippard, S. J. and Berg, J. M., (1994) *Principles of bioinorganic chemistry*, University Science Book, Princeton, NJ, US.

[57] Haraguchi, H. (2004) Metallomics as integrated biometal science. *J. Anal. At. Spectrom.* 19, 5-14.

[58] Szpunar, J., (2004) Metallomcs: a new frontier in analytical chemistry. *Anal. Bioanal. Chem.* 378, 54-56.

[59] Williams, R. J. P., (2001) Chemical selection of elements by cells. *Coord. Chem. Rev.* 216, 583-595.

[60] Boudou, A. and Ribeyre, F., (1997) Mercury in the food web: accumulation and transfer mechanisms. *Metal Ions Biol. Syst.* 34, 289-319.

[61] Magos, L. and Webb, M., (1980) The interactions of selenium with cadmium and mercury. *Crit. Rev. Toxicol.* 8, 1-42.

[62] Mason, R. P., Reinfelder, J. R. and Morel, F. M. M., (1996) Uptake, toxicity and trophic transfer of mercury in a coastal diatom. *Environ. Sci. Technol.* 30, 1835-1845.

[63] Watras, C. J. and Bloom, N. S., (1992) Mercury and methylmercury in individual zooplankton: implication of bioaccumulation. *Limnol. Oceanogr.* 37, 1313-1318.

[64] Wagemann, R., Trebacz, E., Boila, G. and Lockhart, W. L., (1998) Methylmercury and total mercury in tissues of arctic marine mammals. *Sci. Total Environ.* 218, 19-31.

[65] Kim, E. Y., Murakami, T., Saeki, K. and Tatsukawa, R., (1996) Mercury levels and its chemical form in tissues and organs of seabirds. *Arch. Environ. Contam. Toxicol.* 30, 259-266.

[66] Chen, R. W., Lacey, V. L. and Whanger, P. D., (1975) Effect of selenium on methylmercury binding to subcellular and soluble proteins in rat tissues. *Res. Comm. Chem. Pathol. Pharmacol.* 12, 297-308.

[67] Garcia, J. D., Yang, M. G., Wang, J. H. C. and Belo, P. S., (1974) Carbon-mercury bond cleavage in blood of rats fed methyl mercuric chloride. *Proc. Soc. Exp. Biol. Med.* 146, 66-70.

[68] Nordberg, G. F. and Skerfving, S., (1972) Metabolism, In: *Mercury in the environment* (Friberg, L. T. and Vostal, J. J., Eds.) pp. 29-91, CRC Press, Cleaveland, Ohio.
[69] Kerper, L. E., Ballatori, N. and Clarkson, T. W., (1992) Methylmercury transport across the blood-brain barrier by an amino acid carrier. *Am. J. Physiol.* 262, R761-R765.
[70] Rabenstein, D. L., (1978) The chemistry of methylmercury toxicology. *J. Chem. Edu.* 55, 292-296.
[71] Norseth, T. and Clarkson, T. W., (1970) Studies on the biotransformation of ^{203}Hg-labelled methyl mercury chloride in rats. *Arch. Environ. Health* 21, 717-727.
[72] Mehra, M. and Choi, B. H., (1981) Distribution and biotransformation of methyl mercuric chloride in different tissues of mice. *Acta Pharmacol. Toxicol.* 49, 28-37.
[73] Hansen, J. C., Reske-Nielsen, E., Thorlacius-Ussing, O., Rungby, J. and Dansher, G., (1989) Distribution of dietary mercury in a dog. Quantification and localization of total mercury in organs and central nervous system. *Sci. Total Environ.* 78, 23-43.
[74] Lind, B., Friberg, L. and Nylander, M., (1988) Primary studies on methylmercury biotransformation and clearance in the brain of primates II. Demethylation of mercury in brain. *J. Trace Elem. Exp. Med.* 1, 49-56.
[75] Fang, S. C., (1974) Induction of carbon-mercury cleavage enzymes in rat liver by dietary selenite. *Res. Comm. Chem. Pathol. Pharmacol.* 9, 579-582.
[76] Hirayama, K. and Yasutake, A., (1999) Effects of reactive oxygen modulators on *in vivo* demethylation of methylmercury. *J. Health Sci.* 45, 24-27.
[77] Rowland, I. R., Robinson, R. D. and Doherty, R. A., (1984) Effects of diet on mercury metabolism and excretion in mice given methylmercury: role of gut flora. *Arc. Environ. Health* 39, 401-408.
[78] Norseth, T. and Clarkson, T. W., (1971) Intestinal transport of ^{203}Hg-labelled methylmercury cysteine. *Arch. Environ. Health* 22, 568-577.
[79] Hirata, E. and Takahashi, H., (1981) Degradation of methyl mercury glutathione by pancreatic enzymes in bile. *Toxicol. App. Pharmacol.* 58, 483-491.
[80] Refsvik, T. and Norseth, T., (1975) Methylmercuric compounds in the bile. *Acta Pharmacol. Toxicol.* 36, 67-78.
[81] Reid, R. S. and Rabenstein, D. L., (1981) Nuclear magnetic resonance studies of the solution chemistry of metal complexes. XVII. Formation constants for the complexation of methylmercury by sufhydryl-containing amino acids and related molecules. *Can. J. Chem.* 59, 1505-1514.
[82] Rabenstein, D. L. and Evans, C. A., (1978) The mobility of methylmercury in biological systems. *Bioinorg. Chem.* 8, 107-114.
[83] Rabenstein, D. L., Isab, A. A. and Reid, R. S., (1982) A proton nuclear magnetic resonance study of the binding of methylmercury in human erythrocytes. *Biochim. Biophys. Acta* 720, 53-64.
[84] Rabenstein, D. L. and Reid, R. S., (1984) Nuclear magnetic resonance studies of the solution chemistry of metal complexes. 20. Ligand-exchange kinetics of methylmercury (II)-thiol complexes. *Inorg. Chem.* 23, 1246-1250.
[85] Prokof'ev, A. K., (1981) Chemical forms of mercury, cadmium and zinc in natural waters. *Russ. Chem. Rev.* 50, 32-48.
[86] Rabenstein, D. L. and Fairhurst, M. T., (1975) Nuclearmagneic resonance studies of the solution chemistry of metal complexes. XI. The binding of methylmercury by sulfhydryl-containing acids and by glutathione. *J. Am. Chem. Soc.* 97, 2086-2092.

[87] Davis, L. E., Kornfeld, M., Mooney, H. S., Fiedler, K. J., Haaland, K. Y., Orrison, W. W., Cernichiari, E. and Clarkson, T. W., (1994) Methylmercury poisoning: long-term clinical, radiological, toxicological, and pathological studies of an affected family. *Ann. Neurol.* 35, 680-688.

[88] Vogel, D. G., Margolis, R. L. and Mottet, N. K., (1985) The effects of methyl mercury binding to microtubules. *Toxicol. App. Pharmacol.* 80, 473-486.

[89] Abe, T., Haga, T. and Kurokawa, M., (1975) Blockage of axoplasmic transport and depolymerization of reassembled microtubules by methyl mercury. *Brain Res.* 86, 504-508.

[90] Hirayama, K., (1985) Effects of combined administration of thiol compounds and methylmercury chloride on mercury distribution in rats. *Biochem. Pharmacol.* 37, 2030-2032.

[91] Thomas, D. J. and Smith, J. C., (1982) Effects of coadministered low-molecular weight thiol compounds on short-term distribution of methylmercury in the rat. *Toxicol. Appl. Pharmacol.* 62, 104-110.

[92] Madsen, K. M., (1980) Mercury accumulation in kidney lysosomes or proteinuric rats. *Kid. Inter.* 18, 445-453.

[93] Zalups, R. K. and D. W., B. (1993) Intrarenal distribution of inorganic mercury and albumin after coadministration. *J. Toxicol. Environ. Health* 40, 77-103.

[94] Waku, K. and Nakazawa, Y., (1979) Toxic effect of several mercury compounds on SH- and non-SH enzymes. *Toxicol. Lett.* 4, 49-55.

[95] Grundt, I. K., Roux, F., Treich, I., Loriette, C., Raulin, J. and Fournier, E., (1982) Effects of methyl mercury and triethyllead on Na+K~ATPase and pyruvate dehydrogenase activities in Glioma C6 cells. *Acta Pharmacol. Toxicol.* 51, 6-11.

[96] Domanska-Janik, K. and Bourre, J. M., (1987) Effect of mercury on rabbit myelin CNP-ase *in vitro*. *Neurotoxicol.* 8, 23-32.

[97] Inoue, Y., Saijoh, K. and Sumino, K., (1988) Action of mercurials on activity of partially purified soluble protein kinase C from mice brain. *Pharmacol. Toxicol.* 62, 278-281.

[98] Ribarov, S. R. and Benov, L. C., (1981) Relationship between the hemolytic action of heavy metals and lipid peroxidation. *Biochim. Biophys. Acta* 640, 721-726.

[99] Nakada, S. and Imura, N., (1983) Susceptibility of lipids to mercurials. *J. Appl. Toxicol.* 3, 131-134.

[100] Walum, E. and Marchner, H., (1983) Effects of mercuric chloride on the membrane integrity of cultured cell lines. *Toxicol. Lett.* 18, 89-95.

[101] Ballatori, N., Shi, C. and Boyer, J. L., (1988) Altered plasma membrane ion permeability in mercury induced cell injury: Studies in hepatocytes of Elasmobranch *Raja erinacea*. *Toxicol. Appl. Pharmacol.* 95, 279-291.

[102] Rajanna, B. and Hobson, M., (1985) Influence of mercury on uptake of [^{3}H]dopamine and [^{3}H]norepinephrine by rat brain synaptosomes. *Toxicol. Lett.* 27, 7-14.

[103] Pfeifer, R. W. and Irons, R. D., (1985) Mechanism of sulfhydryl-dependent immunotoxicity, In: *Immunotoxicology and Immunopharmacology* (Dean, J. H., Luster, M. I., Munson, A. E. and Amos, H., Eds.) pp. 255-262, Raven Press, New York.

[104] Pařízek, J. and Ošádalová, I., (1967) The protective effect of small amounts of selenite in sublimate intoxication. *Experientia* 23, 142-143.

[105] Naganuma, A. and Imura, N., (1980) Bis(methylmercuric) selenide as a reaction product from methylmercury and selenite in rabbit blood. *Res. Comm. Chem. Pathol. Pharmacol.* 27, 163-173.

[106] Magos, L., Webb, M. and Hudson, A. R., (1979) Complex formation between selenium and methylmercury. *Chem. Biol. Interactions* 28, 359-362.

[107] Khan, M. A. K., Asaduzzaman, A. M., Schreckenbach, G., Wang, F., (2009) Synthesis, characterization and structures of methylmercury complexes with selenoamino acids. *Dalton Transac.* 29,5766-5772.

[108] Khan, M. A. K., Wang, F., (2010) Chemical Demethylation of Methylmercury by Selenoamino Acids. *Chem. Res. Toxicol.* 23(7), 1202-1206.

[109] Khan, M. A. K., Wang, F., (2009) Reversible Dissolution of Glutathione-Mediated $HgSe_xS_{1-x}$ Nanoparticles and Possible Significance in Hg-Se Antagonism. *Chem. Res. Toxicol.* 22(11), 1827-1832.

[110] Ganther, H. E., (1978) Modification of methyl mercury toxicity and metabolism by selenium and vitamin E: Possible mechanisms. *Environ. Health Perspect.* 25, 71-76.

[111] Beyrouty, P. and Chan, H. M., (2006) Co-consumption of selenium and vitamin E altered the reproductive and developmental toxicity of methylmercury in rats. *Neurotoxicol. Teratol.* 28, 49-58.

[112] Sheline, J. and Schmidt-Nielsen, B., (1977) Methyl mercury-selenium: Interaction in killifish, *Fundulusheteroclitus*, In: *Physiological Responses of Marine Biota to Pollutants* (Vemberg, F., Ed.) pp. 119-130, Symposium, Milford, CN.

[113] Holmes, J. and Lean, D., (2006) *Sci. Total Environ.*, 368, 306-319.

[114] Bulska, E., Baxter, D. C., Frech, W., (1991) *Anal. Chim. Acta*, 249, 545-554.

[115] Hintelmann, H., *Chemosphere*, (1999), 39, 1093-1105.

[116] Horvat, M., Bloom, N. S., Liang, L., (1993) *Anal. Chim. Acta*, 281,135-152.

[117] Erle, G., Jiang, G., He, B., Yin, Y., Shi, J., (2008) Speciation of mercury in coal using HPLC-CV-AFS system: Comparison of different extraction methods; *J. Anal. At. Spectrom.*, 23, 1397-1400.

[118] Application notes number 1067 of the Hydra IIAF, Hg Analyzer manufactured by Teledyne Leeman Labs.

[119] Chena, J., Chena, H., Jin, X., Chen, H., (2009) Determination of ultra-trace amount methyl-, phenyl- and inorganic mercury in environmental and biological samples by liquid chromatography with inductively coupled plasma mass spectrometry after cloud point extraction pre-concentration, *Talanta* 77, 1381-1387.

[120] Alp, O., Ertasx, N., (2009) Determination of inorganic and total mercury by flow injection vapor generation atomic absorption spectrometry using a W-coil atomizer, *J. Anal. At. Spectrom.*, 24, 93-96.

[121] Hashemi-Moghaddam, H., Saber-Tehrani, M. (2008) Sensitive mercury speciation by reversed-phase column high-performance liquid chromatography with UV-visible detection after solid-phase extraction using 6-mercaptopurine and dithizone; *Journal of AOAC International*, 1453(6).

[122] Cairns, W. R. L. Ranaldo, M., Hennebelle, R., Turetta, C., Capodaglioa, G., Ferraric, C. F., Dommergue, A., Cescona, P., Barbante, C., (2009) Speciation analysis of mercury in sea water from the lagoon of Venice by on-line pre-concentration HPLC–ICP-MS, *Analytica Chimica Acta*; 622, 62-69.

[123] Zhao, Y., Ma, X., Liu, S., Yao, J., (2009), Experiment on mechanism of simultaneous removal of Hg^0, SO_2 and NO from flue gas using $NaClO_2$ solution, *Environ. Technolo.* 30(3), 277-282.

[124] Lu, J. Y., Schroeder, W. H., Berg, T., Munthe, J., Schneeberger, D., Schaedlich, F., (1998) Field Intercomparison studies for evaluation and validation of the AES mini SampIR™ technique for sampling and analysis of total particulate mercury in the atmosphere, *Analytical Chemistry*, 70, 2403-2408.

[125] Risch, M. R., Prestbo, E. R., Hawkins, L., (2007) Measurement of Atmospheric Mercury Species with Manual Sampling and Analysis Methods in a Case Study in Indiana, *Water Air Soil Pollut.*, 184:285-297.

[126] He, S., Li, D., Zhu, C., Song, S., Wang, L., Long, L., Fan, C., (2008) Design of a gold nanoprobe for rapid and portable mercury detection with the naked eye, *Chem. Commun.*, 4885-4887.

[127] Hempel, M., Thoeming, J., (1999). Remediation Techniques for Hg-Contaminated Sites. In: *Mercury Contaminated Sites, Characterization, Risk Assessment and Remediation*. R. Ebinghaus, R. R. Turner, L. D., de Lacerda, O. Vasiliev and W. Salomons (Eds.), Springer Publishing, New York, NY, p. 113-130.

[128] Matsuyama, A., Iwasaki, H., Kigaki, K., Yabuta, H., Sano, T. and Akagi, H., (1999). Study on the Remediation Technology of Mercury-Compound Contaminated Soil by Low Temperature Thermal Treatment. In: *Mercury Contaminated Sites, Characterization, Risk Assessment and Remediation*. R. Ebinghaus, R. R. Turner, L. D. de Lacerda, O. Vasiliev and W. Salomons (Eds.), Springer Publishing, New York, NY, p. 421-440.

[129] Papp, C. S. E., Filipek, L. H., Smith, K. S., (1991). Selectrivity and Effectiveness of Extractants Used to Release Metals Associated with Organic Matter, *Applied Geochemistry*, 6, 349-353.

[130] Ebinghaus, R. Tripathi, R. M., Wallschlager, D. and Lindberg, S. E., (1999). Natural and Anthorpogenic Mercury Sources and Their Impacts on the Air-Surface Exchange of Mercury on Regional and Global Scales. In: *Mercury Contaminated Sites, Characterization, Risk Assessment and Remediation*. R. Ebinghaus, R. R. Turner, L. D. de Lacerda, O. Vasiliev and W. Salomons (Eds.), Springer Publishing, New York, NY, p. 1-50.

[131] Sexauer Gustin, M., Taylor, G. E. and Maxey, R. A., (1997). Effect of Temperature and Air Movement on the Flux of Elemental Mercury from Substrate to the Atmosphere, *J. Geophysical Research*, 102, 3891-3898.

[132] Domenico, P. A. and Schwartz, F. W., (1998). *Physical and Chemical Hydrogeology*. John Wiley and Sons, Inc., New York, NY, 506p.

[133] Xu, Y., Schwartz, F. W., (1994). Lead Immobilization by Hydroxyapatite in Aqueous Solutions. *J. Contam. Hydrol.*, 15(3), 187-206.

[134] Anonymous, (2000). Constructing a Solution to a Mercurial Problem, *Environ. Sci. Technol., Technol. Sol.* p. 251A.

[135] Palermo, M. R., (1998). Design Considerations for *In-Situ* Capping of Contaminated Sediments, *Wat. Sci. Tech.*, 37(6-7) 315-321.

[136] Rajeshwar, K., Ibanez, J. G., Swain, G. M., (1994). Electrochemistry and the environment, *J. Appl. Electrochem.* 24, 1077-1091.

[137] Huang, C. P., Cha, D., Chang, J. H., Qiang, Z., (2001). Electrochemical Process for in-situ Treatment of Contaminated Soils, Newark, Delaware.

[138] Acar, Y. B., Alshawabkeh, A. N., (1993). Principles of Electrokinetic Remediation. *Enviro. Sci. Technol.* 27(13), 2638-2647.

[139] Reddy, K. R., Chaparro, C., Saichek, R. E., (2003). Iodide-Enhanced Electrokinetic Remediation of Mercury-Contaminated Soils, *J. Environ. Eng. ASCE*, 129(12), 1137-1148.

[140] Robles-Gutiérrez, I., Solís-Valdéz, S., Hernández-Silva, G., Bustos, E., (2012). Elec-troremediation of Mercury Polluted Soil by Complexing Agents. Environmental In-fluences of Mercury Ore Processing: Case Studies Selected at Slovenian, Mexican, Hungarian Group Meeting in Idrija, Mateja Gosar, Tatjana Dizdarevič, Miloš Miler (Eds.) UNESCO, Slovenian National Committee of the International Geoscience Programme, Idrija, Slovenia.

[141] Sonke, G. E., Heimburger, L., Dommergue, A., (2013) Mercury biogeochemistry: Paradigm shifts, outstanding issues and research needs, *C. R. Geoscience,* 345, 213-224.

In: Heavy Metals
Editor: Deepak Pathania

ISBN: 978-1-63484-740-7

Chapter 2

OVERVIEW OF POLYMER COMPOSITES FOR HEAVY METAL REMOVAL

E. Vunain**, *A. K. Mishra*† and B. B. Mamba
Nanotechnology and Water sustainability Research Unit, College of Science, Engineering and Technology, University of South Africa, Florida Campus, Johannesburg, South Africa

ABSTRACT

Water availability in the form of portable and drinkable water is the most valuable resource to humans on earth. Water pollution by organic and inorganic pollutants is a serious environmental problem and unsafe water is classified amongst the top ten risks associated to human health. Unlike organic water pollutants which are biodegradable and photodegradable in nature, heavy metals are not biodegradable and their toxicity even at low concentrations constitutes a serious health problem. Therefore, the removal of heavy metal ions from wastewater is urgent and paramount due to their toxicities and environmental issues associated with wastewater containing heavy metals. This chapter highlights the use of polymer composites as sorbents to abstract specific toxic heavy metals from wastewater.

Keywords: water pollution, inorganic pollutants, polymer, composite

1. INTRODUCTION

1.1. General Background to Polymers

A polymer is a large molecule (macromolecule) that is built up from smaller molecules called monomers, covalently bonded in a polymer chain. Polymers are natural or synthetic

* Email: evunain@gmail.com.
† Email: mishrak@unisa.ac.za.

compounds of high molecular weight. Polymers range from natural biopolymers such as DNA and proteins to synthetic plastic such as polystyrene. Today, polymers have a range of applications in areas such as coatings, adhesives, packaging materials, foams, textiles, composites, and industrials fibers to electronic devices, biomedical and optical devices [1]. They also serve as precursors for a number of newly developed high-tech ceramics. In addition, polymers have been employed in wastewater treatment, toxic metal removal and enrichment of precious metals from hydrometallurgical liquids [2]. Polymers are classified as shown in Figure 1.

2. Composite Materials

Composite materials are made from two or more constituents' materials with different physical and chemical properties to obtain a combination of two or more simple materials with a matrix phase (or continuous phase) and a disperse phase. The matrix phase is the primary phase, having a continuous character while the disperse phase is embedded in the matrix in a discontinuous form. Examples of continuous phases are polymers, metals, ceramics, etc. and examples of the disperse phases are clay minerals, glass fibres, silica powder etc., [3].

The disperse phase is sometimes called reinforcing phase. The addition of a second phase is to create a synergy between the various constituents, such that the combined properties of the composites produce a synergistic effect far exceeding those of its constituents. Composites materials offers superior properties such as improved mechanical strength, thermal stability, chemical and solvent resistance, stiffness, low density or less expensive (economical) when compared to conventional materials.

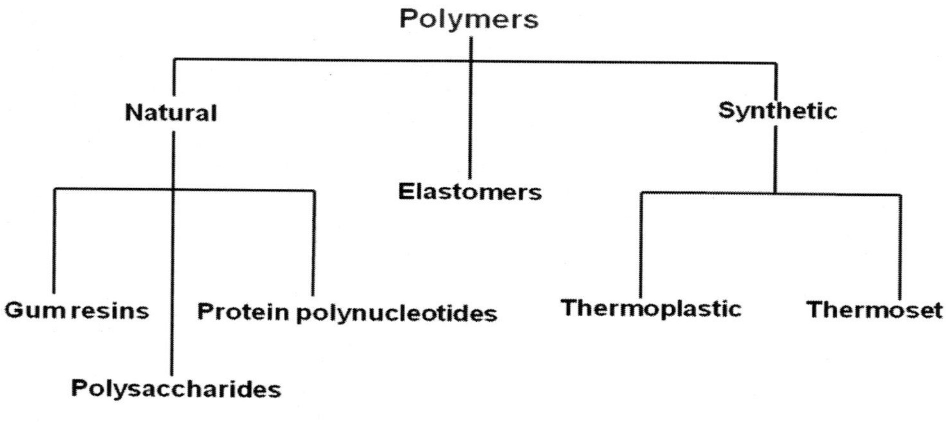

Figure 1. Polymer Classification.

2.1. Types of Composites

There are basically three types of composites, namely (a) metal composites (b) ceramic composites and (c) polymer composites.

Metal composites: Metal is the main phase with a nanomaterial as the reinforcement phase. An example is copper/carbon nanotube tube composite.

Ceramic composites: Ceramic is the main phase and composed of chemical compounds from the group of oxides, borides, nitrides and silicides. A metal is the reinforcement material and an example is alumina/copper nanocomposite.

Polymer composites: A polymer is the main phase and the reinforcement material is a nanomaterial. Example is polymer/clay nanocomposite.

3. Polymer Nanocomposites

Polymer nanocomposites are materials having nanoscopic inorganic particles (filler material) dispersed in a polymer matrix, where at least one of the dimension of the filler material is of the order of a nanometer (1 – 100 nm). Conventional fillers are materials in the form of particles (e.g., calcium carbonate), plate-shaped particles (e.g., mica) or fibre (e.g., glass fibres) [4]. The main purpose of combining nanoscale particles (nanoparticles) to polymers is to improve the performance properties of polymers and also to overcome the undesired properties of the particles [5, 6]. Nowadays, composites materials are being replaced by nanocomposites where nanoscale fillers and polymer materials offer multifunctional properties. These include increased strength and modulus, improved solvent and heat resistant, improved chemical resistance, decreased flammability and improved barrier properties [6]. It should be noted that a small amount of nanoparticles in the dispersed phase can significantly influenced the properties of the host polymer and as such a lot of attention has been focused on polymer nanocomposites by both industry and academia in the past decades. The properties of polymer nanocomposites depends on a range of factors such as the matrix material (which can exhibit nanoscale dimensions); loading, degree of dispersion, size, shape, orientation of the nanoscale second phase, processing methods and as well as interaction between the matrix and the second phase. The usefulness of polymer nanocomposites lies in their multifunctional properties which are not found with conventional polymers. In a nutshell, polymer nanocomposites have the ability to augment or replace traditional micron composites.

3.1. Different Types of Reinforcement Used in Fabrication of Polymer Composites

Many composites with enhanced properties have been reinforced with nano-fillers such as such as fibre, wood, clay, zeolites, nanoparticles etc. for heavy metal removal from wastewater [7, 8]. In the formation of these composites materials, the percentage of filler dispersed into polymeric matrix is directly proportional to the modulus [9]. It should be noted that poorly wetted filler particles can lead to the formation of void or air spaces around the particles thus resulting in reduce polymer strength [10]. When the material is subjected to severe stress, these air spaces could result to weak points which lead to reduced polymer strength. Vilakati et al. [11] had used lignocelluloses fiber-ethylene vinyl-acetate composites in the removal of Pb(II), Cr(III) and Cr(VI) from aqueous solution. In this study, adsorption equilibrium was reached after 3 h with a maximum lead adsorption of 80%. The adsorption of lead proved to be superior compared to Cr (III) and Cr(VI). Vunain et al. [12], used magnetite nanoparticles incorporated into ethylene vinyl-acetate (EVA) and polycaprolactone (PCL) in the removal of arsenic(III) from water. The composite was fabricated via the melt blending technique and a maximum sorption capacity of 2.83 mg/g was obtained at a temperature of 26°C and *p*H of 8.6. In this study, the process fits well with the Langmuir isotherm and As(III) removal obeyed the pseudo-second order kinetics. It is believed that the proper dispersion of magnetite nanoparticles into the polymer resulted in improved surface area for better adsorption of As(III) from solution. Furthermore, Dlamini et al. used clay/ethylene vinyl-acetate composite to remove Pb(II) from aqueous solution through adsorption [13].

4. Classification of Polymer Based Nanocomposites

Figure 2 shows the classification of polymer based nanocomposite materials.

Different dimensional morphology of nanocomposites can be obtained depending on the nanoscale filler. These inorganic components (filler) in the nanocomposites can either be three-dimensional framework such as zeolites, two dimensional layered materials such as metal oxides, clay and metal phosphate or even one or zero –dimensional materials [14-16]. The determination of which nanomaterials that can effectively be integrated into the nanocomposites is very important, so as to assess the potential value of the nanocomposites and improved properties of the nanocomposites.

For the purpose of this chapter, our focused is on polymeric nanocomposites and their application in removal of heavy metals from wastewater. This chapter will explore nanomaterials such as montmorillonite organoclays (MMT), metallic nanoparticles (e.g., Fe, Al, Au, Ag, Pd etc), metallic oxides (e.g., TiO_2, ZnO, Al_2O_3, $CaCO_3$ etc), carbon nanotubes (CNTs), carbon Nanofibers (CNFs), ceramic, non-metals oxides (e.g., SiO_2) and others like (Silicon carbide, SiC) used as fillers and dispersed into polymers for the formation of nanocomposites for removal of heavy metals from wastewater. In wastewater treatment, especially in the removal of heavy metals, the use of natural polymers such as chitosan, starch and its derivatives has gained a lot of attention due to their physico-chemical characteristics, low cost and the presence of reactive groups on the backbone chain [17].

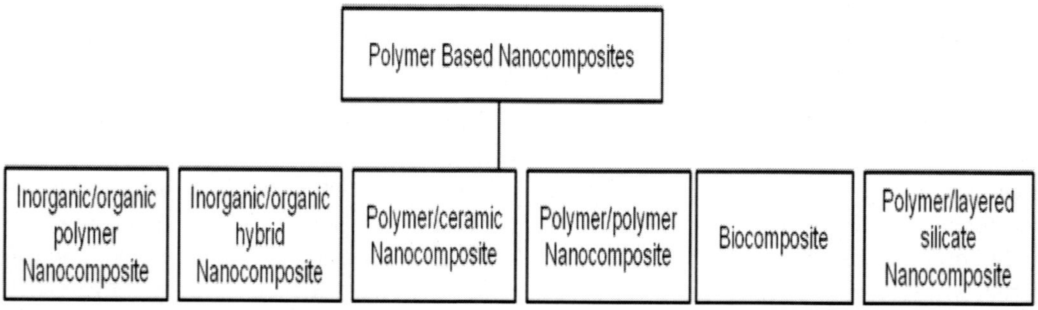

Figure 2. Classification polymer based nanocomposites materials.

5. Preparatory Methods for Polymer Based Nanocomposites

In the preparation of polymer based nanocomposites, three methods are commonly employed as shown in Figure 3 [3]. The first method is blending or direct mixing of the polymer and the inorganic fillers either as discrete phases (called the melt mixing) or in solution (called solution mixing). The second method is via sol-gel process which starts with a molecular precursor and then the formation of a metal oxide framework by hydrolysis and condensation. Finally, the third method is through in-situ polymerization of monomers in the presence of fillers.

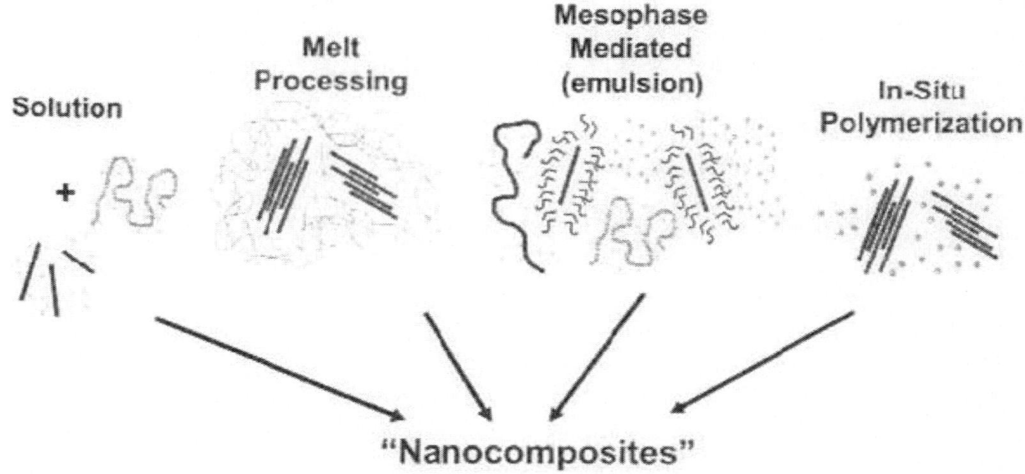

Figure 3. Methods for the synthesis of polymer based nanocomposites [3].

6. Heavy Metal Pollution

Heavy metals occur as natural constituents of the earth crust and soli. Although, there is no clear definition of a heavy metal, they are generally regarded as any dense metal or metalloid of environmental concern having atomic weights between 63.5 and 200.6 Dalton and specific gravity greater than 5 g/mL [18, 19]. Toxic heavy metals of concern in wastewater include lead, cadmium, copper, mercury, chromium, nickel arsenic, selenium, thallium, antimony, cobalt and zinc [20]. These metals are release to the environment by both natural and anthropogenic sources, especially industrial activities (painting, coating, tanneries, fertilizers etc.), mining, discharge of domestic waste, burning of fossil fuels such as coal, petrol, kerosene etc., and automobile exhausts (for lead) [21, 22]. Heavy metals have been used by humans for decades and when they leached into the underground waters, moving along pathways and eventually being deposited in aquifers or are washed away by run-off into surface waters, the result is water pollution [22]. Although several adverse health effects of heavy metal contamination have been known and documented, exposure to heavy metal contamination still continues more particularly in less developed countries [19].

6.1. Hazardous Effects of Heavy Metals

Heavy metal ions are non-biodegradable pollutants in water and have the ability to persist in natural ecosystems for longer periods and to accumulate in food chains and living tissues which greatly affects human health causing various diseases and disorders [23-25]. Furthermore, arbitrary disposal industrial and domestic waste (containing heavy metals) into aquatic ecosystems, threatens the inhabiting aquatic organisms [26]. Heavy metals in surface and groundwater supplies have been prioritized as the major inorganic contaminant in the environment due to their mobility in aquatic ecosystems and toxicities to higher life forms [27]. Water is the main portals of entry of these obnoxious elements into the human body and beyond the permissible limits; these elements cause direct toxicities to humans. In light of this therefore, the World Health Organization (WHO) and United State Environmental Protection Agency (USEPA) have defined maximum contamination level for some certain metal [26, 28]. The permissible limits for some heavy metals in drinkable water as stipulated by the United State Environmental Protection Agency (UNEPA) and World Health Organization (WHO) are listed on Table 1.

Cadmium toxicity can cause renal dysfunction, obstructive lung diseases such as lung cancer and hepatic injury. Cadmium may also produce bone defects (osteomalacia and osteoporosis) in both humans and animals [30]. High levels of lead exposure may result in toxicity in humans and may cause mental disorders, retardation, and semi-permanent brain damage. Lead can cause lesions in the central nervous system and apparently can damage the cells making up the blood-brain barrier that protects the brain [31]. Lead can also damage the kidney, liver and reproductive system of basic cellular processes. Toxic symptoms include; anemia, insomnia, headache, dizziness, and irritability, weakness of muscles, hallucination and renal damages [32]. High concentrations of mercury cause impairment of pulmonary function and kidney, chest pain and dyspnea [33]. Arsenic toxicity is cumulative, with long term exposure to arsenic contaminated water causing various types of cancers (such as skin, kidneys, bladder and lungs), damages blood vessels, central nervous system and causes reproductive disorders and high blood pressure [34].

Table 1. Permissible limits for some heavy metals in drinking water [28, 29]

Heavy Metal	USEPA maximum permissible limits of drinking water quality (mg/L)	WHO maximum permissible limits of drinking water quality (mg/L)
Cadmium	0.005	0.005
Mercury	0.002	0.001
Lead	0.015	0.05
Copper	1.3	1.0
selenium	0.05	0.01
Arsenic	0.05	0.05
Zinc	5.0	5.0
Chromium	0.1	–
Nickel	0.04	–

Considering the health risk associated with the exposure to these heavy metals, their removal from aqueous solution is one of the major problems in wastewater treatment because they are toxic even at very low concentration. Conventional remediation technologies such as lime precipitation, coagulation, flocculation, filtration, ion exchange, etc. for the removal of heavy metals are ineffective and/or expensive and are generally employed for the reduction of heavy metals to very low concentrations [35]. Therefore, it has become imperative to look for new treatment technologies for heavy metal removal/remediation to meet these requirements. The use of polymer composite materials in removal of heavy metals from wastewater is considered as an option in improving water quality for social and economic benefits.

7. Polymer Composites for Heavy Metal Removal

The following types of heavy metals have been removed by polymer-based composites:

7.1. Polymer/Clay Nanocomposites

Polymer/clay nanocomposites is a composite material in which the disperse phase is the silicate constituted of particles that have at least one dimension of the dispersed particles in the nanometer range (1-100 nm). Clays are natural, abundant, and much cheaper materials with high mechanical and chemical resistance and has been used as adsorbent to prevent or remediate environmental contamination by organic and inorganic pollutants due to their hydrophilic nature [36, 37]. The blending of polymers and inorganic clay minerals has produced a new kind of material which could be microcomposite and nanocomposite depending on the modification procedure with the clay mineral well dispersed in the polymer matrix or intercalated in the interlayer of the clay [38-41]. These characteristics of clay have favoured the adsorption of metal ions and other pollutants and have received enormous attention in recent years. Unlike other powder adsorbents, clay-polymer nanocomposites can be easily recovered after an adsorption process. Heavy metal removal by clay/polymer nanocomposites with clay such as Kaolinite, bentonite and montmorillonite has been studied. For example, Anirudhan et al. [42] have reported the removal of Cu(II), Zn(II), and Cd(II) from aqueous solution and industrial waste using amidoximated polyacrylonitrile/ organobentonite composite. In this study, a maximum adsorption of 99.8% Cu (II), 98.9% Zn (II) and 97.4% Cd (II) were removed with 2 g/L of adsorbent at *p*H of 6.0 for an initial metal concentration of 25 mg/L. A maximum monolayer capacity of 77.43, 65.40, 52.61 mg/g for Cu(II), Zn(II), and Cd(II) ions respectively was obtained at 30°C. Sölener et al. [43] reported the removal of Pb (II) ions from aqueous solution using clay/poly(methoxyethyl)acrylamide (PMEA) nanocomposites with a very high adsorption capacity of 81.02 mg/g and material showed no change in adsorption capacity within a pH range of 4.0 – 6.0. Also noted i this study was the fact that adsorbent showed good adsorption capacity for Pb (II) even at a lower *p*H of 3.0. Furthermore, clay-polymer nanocomposites hydrogels have been employed in removal of heavy metals from aqueous solution.

A hydrogel is a three-dimensional densed cross-linked polymer structure, with hydrophilic and hydrophobic portions in a defined proportion. In a typical example, Kasgöz et al. [26] used acrylamide (AAm)-2-acrylamido-2-methylpropane sulfonic acid (AMPS) sodium salt and clay nanocomposites as sorbent for the removal of Cu(II), Cd(II), and Pb(II) ions in competitive and non-competitive conditions. In this study, a 10% wt of clay was employed in the preparation of the nanocomposites to obtain high adsorption rates and capacities for the hydrogels. In another study, Unuabonah et al. [44] reported the use of polyvinyl alcohol-modified Kaolinite clay for the removal of Pb(II) and Cd(II) from aqueous solution.

The clay/polymer nanocomposites showed a good adsorption capacity of 56.18 mg/g and 41.67 mg/g for Pb (II) and Cd (II) ions respectively with a pHpzc value of 4.25. In this study, the proposed mechanism for the adsorption of the heavy metals from aqueous solution was through an inner-sphere complexation mechanism. Natural polymers have also been employed in the synthesis of efficient polymer nanocomposites as adsorbent for the removal of metal ions from wastewater [41]. For example, Shawky et al. [45], prepared Montmorillonite (Mt)-alginate nanocomposites using $CaCl_2$ as crosslinker for the removal of Pb(II) from aqueous solution. Based on alginate-montmorillonite beads packed columns, an efficient method for Pb(II) removal from aqueous solution was obtained with an adsorption capacity of 238.1 mg/g. In this study, it was noticed that the adsorption capacity of nanocomposites decreased with increased solution temperature.

7.2. Polymer/Nanoparticles Composites

Nanoparticles are particles with diameters below the micron dimension (below 0.1 μm (100 nm). They properties depends somehow directly on their size [46]. The addition of inorganic spherical nanoparticles to polymer allows the modification of the polymers physical properties as well as offering new features in the polymer matrix [46]. The introduction of the inorganic nanoparticles into the polymer matrix has proved to be an effective and low cost method to improve the performance of conventional polymers [47, 48]. Nanoparticles formed by metal or metal oxides are broadly used to remove heavy metal ions from wastewater. Nanosized metals or metal oxides include nanoparticles, ferric oxides, manganese oxides, magnesium oxides, titanium oxides, cerium oxides, copper oxides etc. which provide high surface area and specific affinity [49]. Thus, due to the synergistic action of the polymer and nanoparticle for heavy metal removal, polymer/nanoparticles composites have been widely employed wastewater treatment.

7.3. Polymer/Metallic Nanoparticles Composites

Khaydarov et al. in a recent study [50] developed a novel method of obtaining nanocarbon-conjugated polymer nanocomposites (NCPC) using nanocarbon colloids (NCC) and polyethyleneimine (PEI) for removal of metal ions from wastewater. In this approach, polymeric molecules operate as a sorbent and a carrier of metal ions, where the metal ions are bound to water soluble-polymers and the unbound ions pass through the membrane. However, in this study, NCPC had a bonding capacity of 4.0 – 5.7 mmol/g at *p*H 6 for most of the divalent ions such as Zn^{2+}, Cd^{2+}, Cu^{2+}, Hg^{2+}, Ni^{2+} and Cr^{2+} ions. The removal efficiency was higher than 99% with a life time of NCPC depending on the ration of polymeric molecules and carbon nanoparticles concentrations. This method has been based on complexation of metallic species using water-soluble polymer.

Nanoscale zero-valent iron (NZVI) has been employed in wastewater treatment to remove heavy metal ions very effectively due to its small size, large surface area and high reactivity as compared to normal iron powder [51, 52]. Fenglian et al. [53] and Manning et al. [54] have shown that nanoscale zero-valent iron has a high efficiency to remove one or more kinds of heavy metals in wastewater. In a recent, Liu et al. [55], used chitosan beads supported - nanoparticles for the removal of heavy metals from electroplating wastewater in permeable reactive barriers (PRBs). The main goal of the study was to use the polymer/nanoparticle composite to remediate electroplating wastewater composed four different heavy metals (Cr, Cu, Cd and Pb). Results showed a decreased in removal efficiency for Cr(VI) with increase of pH and initial Cr(VI) concentration whereas, the removal rates for Cu(II), Cd(II) and Pb(II) increased with increase *p*H as well as a decreased with an increase in their initial concentrations.

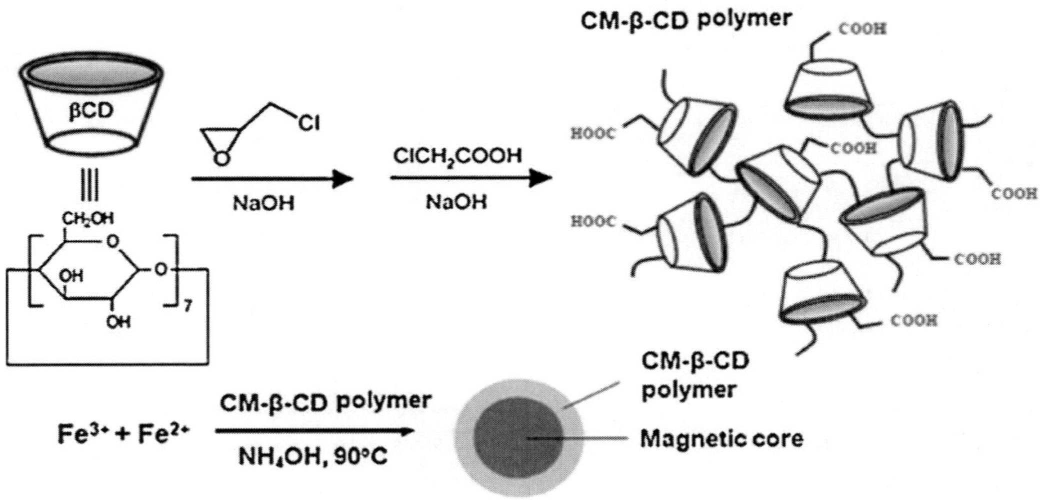

Figure 4. CM-β-CD polymer grafted on to Fe_3O_4 nanoparticles [59].

Only a few reports are available on the usage of chitosan/magnetite composites for removal of heavy metals from water. Huang et al. [56], reported the adsorption of chromium(VI) from aqueous solution using cross-linked magnetic chitosan beads. The use of hydrogel (2-acrylamido-2-methyl-1-propansulfonic acid, AMPS) cross-linked chitosan for the removal of Pb(II) and Ni(II) from aqueous solution have been reported by Tran et al. [57]. In another reported study, magnetic chitosan nanocomposites was employed for the removal of Pb(II), Cu(II), and Cd(II) from wastewater by external magnetic field [58]. It was found that a decrease in magnetic saturation occurred due to an increased mass of gluteraldehyde and chitosan on the surface of the magnetic nanoparticles. It was also possible to recycle and reuse the magnetic nanocomposites because the magnetic chitosan could be made to release the metal ions when subjected to ultrasound radiation in weakly deionized water.

Table 2. Application of nanoparticles/polymer nanocomposites in removal of pollutants from aqueous solutions

Sr. No.	Composite	Targeted Heavy metal	Type of removal	Reference
1	Polyacrylic acid/Superparamagnetic magnetite nanoparticles	Cd(II), Pb(II), Ni(II) and Cu(II)	Adsorption	Mahdavian et al. [60]
2	Polymer/modified magnetic nanoparticles (magnetic nanoparticles modified with 3-aminopropyltrietho-xysilane and copolymers of acrylic acid (AA) and crotonic acid (CA)	Cd(II), Zn(II), Pb(II) and Cu(II)	Adsorption	Ge et al. [61]
3	Magnetic/alginate beads	Pb(II)	Adsorption	Bée et al. [62]
4	Chitosan/coated magnetic nanoparticles modified with α-ketoglutaric acid	Cu(II)	Adsorption	Zhou et al. [63]
5	Magnetic alginate beads based on maghemite nanoparticles	Pb(II)	Adsorption	Idris et al. [64]
6	Chitosan/magnetic nanoparticles	Cu(II)	Adsorption	Chang et al. [65]
7	Maghemite/homopolymers	Cd(II), Hg(II), Pb(II) and Ag(I)	Adsorption	Madrakian et al. [66]
8	Magnetic alginate beads containing Cyanex 272	Co(II)	Adsorption	Ngomsik et al. [67]
9	Iron oxide-alumina mixed nanocomposites fiber	Cu(II), Pb(II), Ni(II) and Hg(II)	Adsorption	Mahapatra et al. [68]
10	Poly(methyl methacrylate) (PMMA) and poly(ethyleneimine) (PEI)/core shell nanoparticles	Cu(II)	Adsorption	Wu et al. [69]
11	Poly(γ-glutamic acid)/Fe_3O_4 nanoparticles	Cr(III), Cu(II), Pb(II) and Ni(II)	Adsorption	Jing et al. [70]
12	Polyvinylacohol (PVA)/chitosan magnetic composite	Co(II)	Adsorption	Zhu et al. [71]
13	Chitosan/ Fe^o nanoparticles	Cr(III)	Adsorption	Geng et al. [72]
14	Fe_3O_4–SiO_2/poly(1,2-diaminobenzene) composite	As(III), Cu(II), and Cr(III)	Adsorption	Zhang et al. [73]

Fe_3O_4/cyclodextrin polymer nanocomposites have been used in the removal of heavy metals from wastewater. In a typical example, carboxylated- β-cyclodextrin (CM-β-CD) polymer was synthesized and grafted onto Fe_3O_4 nanoparticles surface, as shown in Figure 4. The Figure shows carboxyl groups and hydroxyl groups crossed linked on the β-CD polymer to improve on metal complexing ability and selectivity of cyclodextrin. The polymer grafted magnetic nanoparticles (CDpoly-MNPs) were used as easily separable, recyclable and highly selective nanoadsorbent for removal of Pb(II), Cd(II) and Ni(II) ions from wastewater [59]. In this study, the adsorption capacities of Fe_3O_4 nanoparticles were enhanced due to the modification with CM-β-CD polymer with a strong affinity for metal ions as a result of the two crosslinked ligands. Parameters such as temperature, solution *p*H and ionic strength strongly affected the adsorption capability of the polymer with adsorption equilibrium attained within 45 min. A maximum uptake of 64.5, 27.7 and 13.2 mg/g for Pb(II), Cd(II) and Ni(II) respectively, was obtained at 25°C as well as the equilibrium data fitting well with Langmuir model.

Table 2 show a list of some polymer/nanoparticles composites as adsorbents for removal of some heavy metal metals from wastewater.

7.4. Polymer/Metal Oxide Nanoparticles Composites

Metal oxides such as Fe(III), Zr(IV), Ti(IV) and Al(III) are environmentally benign and exhibit amphoteric sorption behaviours around neutral pH. This implies that they can selectively bind both transition metal cations (e.g., Cu^{2+}) and anionic ligands (e.g., arsenate or $HAsO_4^{2-}$) [74]. The application of iron oxide based nanomaterials for removal of heavy metals has attracted a lot of attention because of their important features like small size, high surface area and magnetic property [75-77]. In addition, metal oxides have been employed in removal of heavy metals from wastewater because they offer minimal environmental impact with no secondary pollution formed during its application. An example is the use of hydrated Fe(III) oxides or HFO particles to selectively sorb dissolved heavy metal like copper, zinc or metalloids like arsenic oxyacids or axyanions [78].

Lim and co-workers [79] synthesized polypyrrole(PPy)/silica nanocomposite as adsorbent for the removal of Hg^{2+}, Ag^{+} and Pb^{2+} from wastewater. The composites were prepared by a modified silica-templated oxidation/polymerization of pyrrole in the presence of $FeCl_3$. In this study, the adsorbent showed high adsorption capacities for Hg^{2+}, Ag^{+} and Pb^{2+}, as a result of the complexation reaction between the secondary amine of the pyrrole and the heavy metal ion. In another example for the use of polymer/metal oxides in removal of heavy metals, a composite of polyamide 6 (PA6)/chitosan @Fe_xO_y have been used in the removal of chromium (VI) ion from wastewater [80]. According to this work, a simple and effective route based on electrospinning and pyrolytic reaction was employed for the fabrication of iron oxides loaded nano-fibrous membranes and then applied as adsorbent for Cr(VI) removal from aqueous solution. Furthermore, an effective, simple and low cost adsorbent for the removal of cadmium from wastewater using acrylamide-titanium nanocomposites (TiO_2-AM) as adsorbent have been used for the removal of Cd(II) from wastewater [81].

7.5. Polymer/Carbon Nanotubes (CNTs) Nanocomposites

A carbon nanotube (CNT) consist of one (single-walled, SWCNT) or more (multiwalled, MWCNT) graphene sheets rolled into a cylinder [82] consisting of several tens of graphic shells with adjacent shell separation of ~ 0.34 nm, diameters of ~ 1 nm and large length/diameter ratio [83]. The structures of SWCNT and MWCNT are shown in Figure 5.

Much attention have been focused on carbon nanotubes (CNTs) for the past decades due to their peculiar properties such as chemical, electrical, optical and mechanical properties [84]. Most recently, carbon nanotubes have been widely used as adsorbent for the removal of metallic pollutants from wastewater due to their small size, high surface to weight ratios, hollow/layered structures and hydrophobic surfaces [85-87].

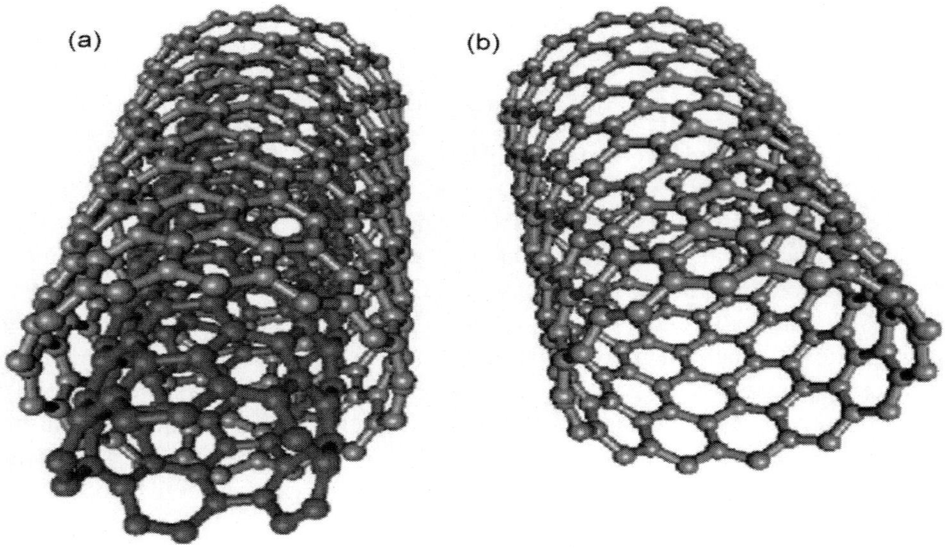

Figure 5. (Super) structure representations of (a) MWCNT and (b) SWCNT [31].

Significant work have been done to reveal the unique structural, electrical, mechanical and electrochemical properties shown by carbon nanotubes and the modification of CNTs with polymers has been of great interest since the synergistic effects of the properties are exceptional in various applications [88-90]. Modification of CNTs with biopolymers has attracted a lot of interest because of their hydrophilicity, excellent dispersion in aqueous solution and excellent adsorption properties [91-92]. One of the common polymers employed for the CNTs modification is chitosan (a polysaccharide biopolymer) because it is economically processed from chitin, hydrophilic, non-toxic, biocompatible, ease of chemical derivatization and biodegradable [93, 94]. In addition, chitosan has $-NH_2$ and –OH groups with good complexing ability that can chelate heavy metal ions from various wastewaters [95-97].

As typical examples, Kim et al. [98], demonstrated that chitosan functional gel included with multiwall carbon nanotube and substituted polyaniline as adsorbent exhibited good adsorption capability and high adsorption efficiency for Cr(VI) ion. The [Functional Gel(chitosan, multiwall carbon nanotube, poly(acrylic acid) and poly(4-aminodiphenylamine, FG (CS–MWNT–PAA–PADPA/FG] composites was synthesized by free radical polymerization and cross-linking reactions. The maximum Cr(VI) adsorption efficiency of ~100% was achieved in a period of 1 h owing the synergistic effects from individual components and adsorption of Cr(VI) was significantly influenced by *p*H. Results from this study showed a complexation reaction between adsorbent and Chromium. Sheng and co-workers [99], used polyaniline (PANI)-grafted multiwalled carbon nanotubes (MWCNTs/PANI), prepared by a plasma-induced grafting technique for the sequestration of a radionuclide 64Cu(II) ion in aqueous solution. The sorption of Cu(II) on MWCNTs/PANI increase with pH from 3.0 – 7.0 and maintained a higher level of adsorption at a pH above 7.0. At lower *p*H, the sorption was dominated by inner-sphere surface complexation and precipitation at high *p*H.

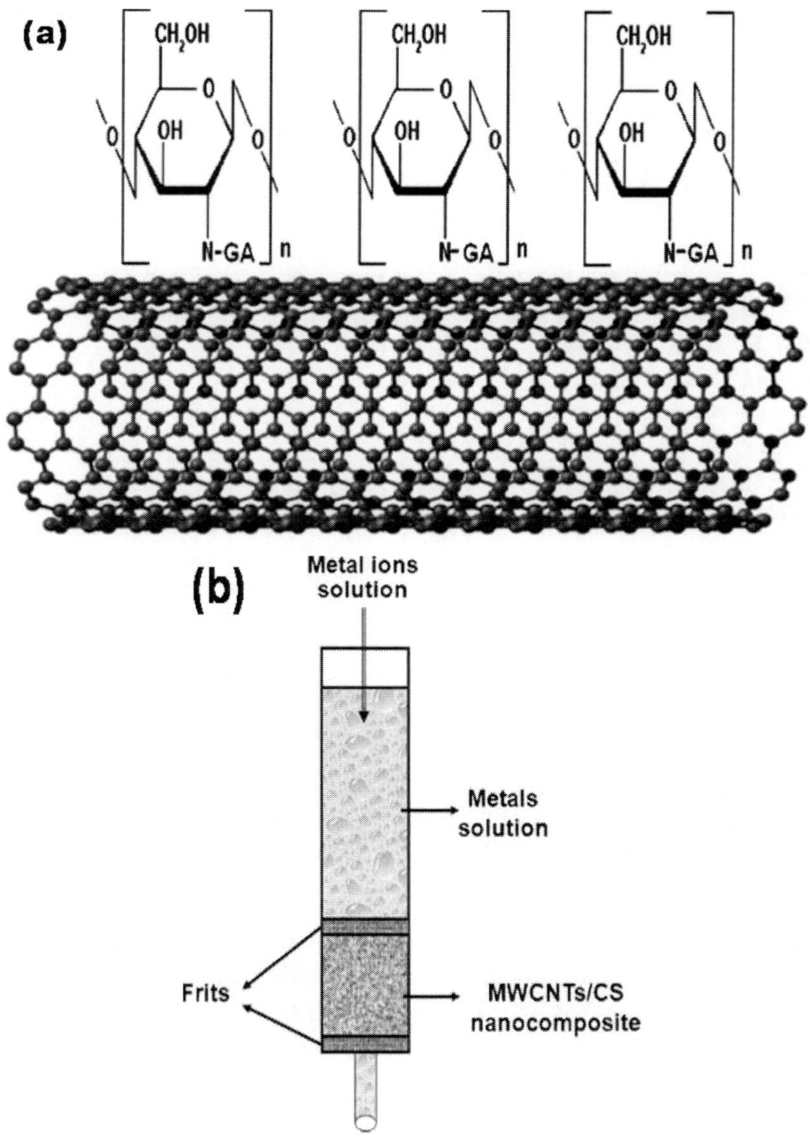

Figure 6. (a) Physical structure of MWCNTs/chitosan (b) MWCNTs/chitosan packed inside a glass column [101].

In another study, a nanocomposite made of from chitosan/Sulfydryl-functionalized graphene oxide (CS/GO-SH) was synthesized via covalent modification and electrostatic self-assembly. The (CS/GO-SH) composite as adsorbent was used in the removal of Cu(II), Pb(II) and Cd(II) in a single and multi-metal ion system [100]. Results revealed that the composite (CS/GO-SH) had potential applications in adsorption of materials due to its specific surface area and superior chemical characteristics. Furthermore, Salam et al. [101] prepared a nanocomposite by modifying multi-walled carbon nanotubes with chitosan for the removal of heavy metals from aqueous solution. The MWCNTs/chitosan nanocomposites were characterized and packed inside a glass column for the removal of Cu(II), Zn(II) Cd(II) and

Ni(II) ions from aqueous solution [Figures 6(a) and (b)]. Results showed that MWCNTs/chitosan nanocomposites could effectively remove most of the metal ions with a competition amongst ions in the following order Cu(II) > Cd(II) ≈ Zn(II) > Ni(II). This shows that the composites material could be applied in different environmental applications.

Table 3. Application of CNTs/polymer nanocomposites in removal of pollutants from aqueous solutions

No.	Composite	Targeted Heavy metal	Type of removal	Reference
1	Chitosan/poly(vinyl) alcohol/amino-functionalized MWCNTs	Cu(II)	Adsorption	Salehi et al. [103]
2	Dodecyl-benzene-sulfonic acid (DBSA) doped polyaniline/multi-walled carbon nanotubes (DP/MWCNTs) nanocomposites	Cr(VI)	Adsorption	Kumar et al. [104]
3	Polyacrylamide functionalized graphene	Pb(II)	Adsorption	Xu et al. [105]
4	β-cyclodextrin/magnetic graphene oxide	Cr(VI)	Adsorption	Hui et al. [106]
5	Graphene/Fe_3O_4@polyrrole nanocomposites	Cr(VI)	Adsorption	Yao et al. [107]
6	Amine-functionalized MWCNTs/polysulfone composites membrane	Cr(VI) and Cd(II)	Adsorption	Shah and Murthy [108]
7	Polydopamine/poly ethyleneimine graphene oxide composite nanosheets	Cu(II), Cd(II), Pb(II) and Hg(II)	Adsorption	Dong et al. [109]
8	Polyaniline/reduced graphene oxide nanocomposite	Hg(II)	Adsorption	[110]

In another case, Yang et al. [102], used Polyacrylamide (PAAM) grafted multi-walled carbon nanotubes (MWCNTs/PAAM) and humic acid (HA) (natural organic macromolecular compound) as additive to improved the removal of lead from aqueous solution. In this study, the presence of humic acid led to an increased in lead adsorption at lower *p*H. This was attributed to the modification of the surface of the adsorbent and partial complexation of Pb(II) with the adsorbed HA [102]. Table 3 shows a range of heavy metals that have been studied as targeted pollutants on CNTs with various surface chemistry and physical structures.

7.6. Polymer/Nanofibers Nanocomposites

Electronspun nanofibers can be fabricated into mats comprising of oriented nanofibers or deposited on other substrates to produce a nanofibrous composites media used in removal of heavy metals from wastewater [111-113]. Nanofibers mats offer many advantages such as high porosity, high gas permeability, and high specific surface area per unit mass, which leads to their high adsorption capacities (see Figure 7) [114].

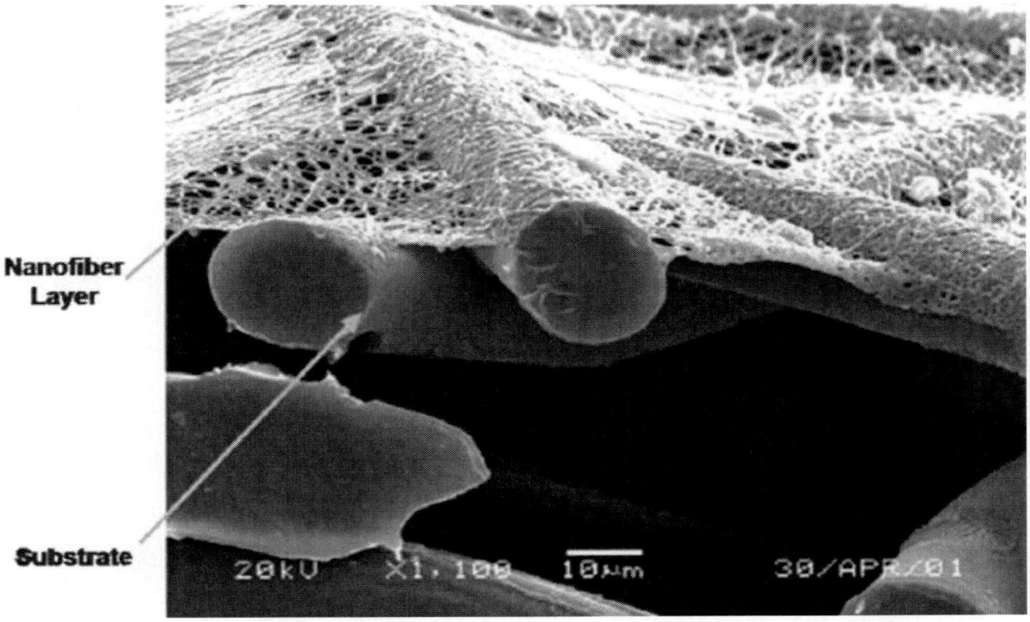

Figure 7. Scanning electron microscopy of Electrospun nanofibers on a polyester substrate [115].

For example, Haider and Park [116], fabricated chitosan nanofibers and the mats, which was employed in adsorption of Cu(II) and Pb(II) from aqueous solution due to the good erosion stability in water and high affinity for the heavy metals. Results from this study showed that the adsorption data for Cu(II) ions on chitosan electrospun nanofiber were ~ 6 and ~ 11 times higher than the reported values of chitosan microsphere and the plain chitosan. Xiao et al. [113], had reported the use of water stable Electrospun Polyacrylic acid-based nanofibrous mats for the removal of Cu(II) ions in aqueous solution. Furthermore, aminated polyacrylonitrile (APAN) nanofiber mat with surface modification of diethylenetriamine (EDTA) have been employed for the removal of Ag(I), Cu(II), Pb(II) and Fe(II) ions from aqueous solution [114]. The adsorbent showed the highest adsorption capacity for Ag(I) and Cu(II) ions. The adsorbent was eluted with nitric acid and results showed that APAN nanofiber mats could be reused several times.

Conclusion and Perspectives

The rapid growth of industrialization and population has resulted in increased environmental problems such as water, air and land pollution. It is reported that water pollution particularly heavy metal pollution causes harmful effects to both humans and aquatic systems. This is a cause of concern. Extensive research and financial resources have been put by academia and industry on the treatment options to minimize the impact of heavy metals on the environment. Besides, all these efforts, much still need to be done in remediating heavy metal pollution in the environment and as such low cost adsorbents are being developed for its removal from the environment. The chapter focuses on the use of eco-friendly and cost-effective polymer based nanocomposites as adsorbents with enhanced performance for the removal of heavy metals compared to conventional methods with are ineffective and expensive. Thus, we have highlighted the use nanofillers such as nanoparticles, nanoclays, metal oxides etc. embedded into polymers forming composites as efficient and economical technology for removal of heavy metal from wastewater solution.

The future of composites in water treatment is quite bright but it needs the collaborative efforts of both academia and industry to materialize a dream of fast, economical and feasible water treatment technology. Extensive research in this area will complement existing water treatment technologies to ensure clean and safe drinking water at a lower cost.

Acknowledgments

Dr. Ephraim Vunain acknowledges the university of South Africa Research board for the award of a Postdoctoral Research Fellowship.

References

[1] Zoveidavianpoor, M., Gharibi, M., 2015. Application of polymers for coating of proppant in hydraulic fracturing of subterraneous formations: A comprehensive review. *J. Natural Gas Sci. Eng*. 24, 197-209.

[2] Reddy, K.H., Reddy, A.R., 2003. Removal of heavy metal ions using the chelating polymers derived by the condensation of poly(3-Hydroxy- 4-acetylphenyl methacrylate) with different diamines. *J. Applied Polymer Science,* 88, 414-421.

[3] Singh, N.B., Rai, S., Agarwal, S., 2014. Polymer Nanocomposites and Cr(VI) Removal from Water. *Nanosci Technol*. 1(1), 1-10.

[4] Pavlidou, S., Papaspyrides, C.D., 2008. A review on polymer–layered silicate nanocomposites. *Progr. Polym. Sci*. 33, 1119-1198.

[5] Paul, D.R., Robenson, L.M., 2008. Polymer nanotechnology: Nanocomposites. *J. Polymer* 49, 3187-3204.

[6] Lagashetty, A., Venkataraman, A., 2005. Polymer nanocomposites. Resonance, general article, 49-57.

[7] Saba, N., Md Tahir, P., Jawaid, M., 2014. A Review on potentiality of nano filler/natural fiber filled polymer hybrid composites. *Polymers*, 6, 2247-2273.

[8] Kadirvelu, K., Thamaraiselvi, k., Namasivayam, C., 2001. Removal of heavy metals from industrial wastewaters by adsorption onto activated carbon prepared from an agricultural solid waste. *Bioresour. Technology* 76, 63-65.

[9] Zou, D.Q., Yoshida, H., 2010. Size effect of silica nanoparticles on thermal decomposition of PMMA. *J. Therm. Anal. Calorim*. 99, 21-26.

[10] Dorigato, A., D'Amato, M., Pegoretti, A., 2012. Thermo-mechanical properties of high density polyethylene – fumed silica nanocomposites: effect of filler surface area and treatment. *J. Polym. Res*. 19, 9889-9900.

[11] Vilakati, G.D., Mishra, A.K., Mishra, S.B., Mamba, B.B., Thwala, J.M., 2012. Metal ion adsorption behavior of lignocellulosic fiber–ethylene vinyl acetate composites. *Polym. Eng. Sci*. 52 (4), 760-767.

[12] Vunain, E., Mishra, A.K., Krause, R.W., 2013. Fabrication, Characterization and Application of Polymer Nanocomposites for Arsenic(III) Removal from Water. *J. Inorg. Organomet. Polym.* 23, 293-305.

[13] Dlamini, D.S., Mishra, A.K., Mamba, B.B., 2012. Kinetic and equilibrium studies of the removal of Pb^{2+} from aqueous solutions using Na_2SO_4-EVA/Cloisite® 20A composite. *Mater. Chem. Phys.* 133, 369-375.

[14] Liu, Y.J., DeGroot, D.C., Schindler, J.L., Kannewurf, C.R., Kanatzidis, M.G., 1993. Intercalation of water-soluble polymers in V_2O_5 xerogel. *Adv. Mater.* 5(5), 369-372.

[15] Kanatzidi, M.G., Wu, C.J., Marcy, H.O., Kannewurf, C.R., 1989. Conductive-polymer bronzes. Intercalated polyaniline in vanadium oxide xerogels. *J Am Chem Soc.* 111(11) 4139-4141.

[16] Bissessur, R., Schindler, J.L., Kannewurf, C.R., Kanatzidis, M., 1994. Nanoscale Composites Formed by Encapsulation of Polymers in MoS_2. From Conjugated Polymers to Plastics. Detection of Metal to Insulator Transition. *Mol. Cryst. Liq. Cryst.* 245(1), 249-254.

[17] Crini, G., 2005. Recent developments in polysaccharide-based materials used as adsorbents in wastewater treatment. *Progress Polym. Sci.* 30, 38-70.

[18] Hernández-Morales, V., Nava, R., Acosta-Silva, Y.J., Macías-Sánchez, S.A., Pérez-Bueno, J.J., Pawelec, B., 2012. Adsorption of lead (II) on SBA-15 mesoporous molecular sieve functionalized with –NH2 groups. *Micropor. Mesopor. Mater.* 160, 133-142.

[19] Järup, L., 2003. Hazards of heavy metal contamination. *British Med. Bull.* 68, 167-182.

[20] Akpor, O.B., Ohiobor, G.O., Olaolu, T.D., 2014. Heavy metal pollutants in wastewater effluents: Sources, effects and remediation. *Adv. Biosci. Bioeng,* 2(4), 37-43.

[21] Duruibe, J.O., Ogwuegbu, M.O.C., Egwurugwu, J.N., 2007. Heavy metal pollution and human biotoxic effects. *Inter. J. Phys. Sci.* 2 (5), 112-118.

[22] Gavrilescu, M., 2004. Removal of heavy metals from the environment by biosorption. *Eng. Life Sci.* 4(3) 219-232.

[23] Jadia, C.D., Fulekar, M.H., 2009. Phytoremediation of heavy metals: Recent techniques. *African J. Biotechnol.* 8, 921-928.

[24] Nomanbhay, S.M., Palanisamy, K., 2005. Removal of heavy metals from industrial wastewater using chitosan coated oil palm shell charcoal. *Electronic J. Biotechnol.* 8, 43-53.

[25] Yao-Wen, Q., 2015. Bioaccumulation of heavy metals both in wild and mariculture food chains in Daya Bay, South China. Estuarine, *Coastal Shelf Sci.* 163, 7-14.

[26] Kumar, K.S., Hans-Uwe, D., Eun–Ji, W., Jae–Seong, L., Kyung–Hoon, S., 2015. Microalgae- A promising tool for heavy metal remediation. *Ecotoxicol. Environ. Safety,* 113, 329-352.

[27] Khan, M.A., Khan Rao, R.A., Ajmal, M., 2008. Heavy metal pollution and its control through nonconventional adsorbents (1998-2007): a review. *J. Int. Environ. Appl. Sci.* 3(2) 101-141.

[28] Hajiaghababaei, L., Badiei, A., Ganjali, M.R., Heydari, S., Khaniani, Y., Ziarani, G.M., 2011. Highly efficient removal and preconcentration of lead and cadmium cations from water and wastewater samples using ethylenediamine functionalized SBA-15. *Desalination* 266, 182-187.

[29] Kumar, M., Puri, A., 2012. A review on permissible limits of drinking water. *Indian J. Occup. Environ. Med.* 16(1), 40-44.

[30] Sari, A., Mendil, D., Tuzen, M., Soylak, M., 2008. Biosorption of Cd(II) and Cr(III) from aqueous solution by moss (Hylocomium splendens) biomass: Equilibrium, kinetic and thermodynamic studies. *Chem. Eng.* 144, 1-9.

[31] Anwar, J., Shafique, U., Waheed-uz Zaman, M.S., Memoona, M., 2009. Adsorption study of cadmium(II) and lead(II) on radish peels. *J. Sci. Res. XXIX*, 29-34.

[32] Naseem, R., Tahir, S.S., 2001. Removal of Pb(II) from aqueous/acidic solutions by using bentonite as an adsorbent. *Water Res.* 35, 3982-3986.

[33] Zabihi, M., Ahmadpour, A., Asl, A.H., 2009. Removal of mercury from water by carbonaceous sorbents derived from walnut shell. *J. Hazard. Mater.* 167, 230-236.

[34] Halem, V.D., Bakker, S.A., Amy, G.L., Van Dijk, J.C., 2009. Arsenic in drinking water: a worldwide water quality concern for water supply companies. *Drink. Water Eng.* Sci. 2, 29-34.

[35] Wilde, E.W., Benemann, J.R., 1993. Bioremoval of heavy metals by the use of microalgae. *Biotechnol. Adv.* 11, 4, 781-812.

[36] Kasgöz, H., Durmus, A., Kasgöz, A., 2008. Enhanced swelling and adsorption properties of AAm-AMPSNa/clay hydrogel nanocomposites for heavy metal ion removal. *Polym. Adv. Technol.* 19, 213-220.

[37] Manohar, D.M., Kkrishnan, K.A., Anirudhan, T.S., 2002. Removal of mercury(II) from aqueous solutions and chlor-alkali industry wastewater using 2-mercaptobenzimidazole-clay. *Water Rev.* 36(6) 1609-1619.

[38] Kilaris, P., Papaspyrides, C.D., 2010. Polymer/layered silicate (clay) nanocomposites: An overview of flame retardancy. *Prog. Polym. Sci.* 35, 902-958.

[39] Pavlidou, S., Papaspyrides, C.D., 2008. A review on polymer–layered silicate nanocomposites. *Prog. Polym. Sci.* 32, 1119-1198.

[40] Scocchi, G., Posocco, P., Fermeglia, M., Pricl, S., 2007. Polymer-clay nanocomposites: A multiscale molecular modeling approach. *J. Phys. Chem.* B. 2143-2151.

[41] Unuabonah, E.I., Taubert, A., 2014. Clay–polymer nanocomposites (CPNs): Adsorbents of the future for water treatment. *Appl. Clay Sci.* 99, 83-92.

[42] Anirudhan, T.S., Ramachandran, M., 2008. Synthesis and characterization of amidoximated polyacrylonitrile/organobentonite composite for Cu(II), Zn(II), and Cd(II) adsorption from aqueous solutions and industry wastewaters. *Ind. Eng. Chem. Res.* 47, 6175-6184.

[43] Sölener, M., Tunali, S., Ozcan, A.S., Ozcan, A., Gedikbey, T., 2008. Adsorption characteristics of lead ions onto the clay/poly(methoxyethyl) acrylamide (PMEA) composite from aqueous solutions. *Desalination* 223, 308-322.

[44] Unuabonah, E.I., Olu-Owolabi, B.I., Adebowale, K.O., Yang, L.Z., 2008. Removal of lead and cadmium ions from aqueous solution by polyvinyl alcohol-modified kaolinite clay: A novel nano clay adsorbent. *Adsorpt. Sci. Technol.* 26(6) 383-405.

[45] Shawky, H.A., 2011. Improvement of water quality using alginate/montmorillonite composite beads. *J. Applied Polym. Sci.* 119(4) 2371-2378.

[46] Hanemann, T., Szabó, V., 2010. Polymer-nanoparticle composites: From synthesis to modern applications. *Materials,* 3, 3468-3517.

[47] Mukherjee, M., Datta, A., Chakravorty, D., 1994. Electrical resistivity of nanocrystalline PbS grown in a polymer matrix. *Appl. Phy. Lett.* 64, 1159-1161.

[48] Ramos, J., Millan, A., Palacio, F., 2000. Production of magnetic nanoparticles in a polyvinylpyridine matrix. *Polymer,* 41, 8461-8464.

[49] Wang, X., Guo, Y., Yang, L., Han, M., Zhao, J., Cheng, X., 2012. Nanomaterials as sorbents to remove heavy metal ions in wastewater treatment. *J. Environ. Anal. Toxicol.* 2(7), 1-7.

[50] Khaydarov, R.A., Khaydarov, R.R., Gapurova, O., 2010. Water purification from metal ions using carbon nanoparticle-conjugated polymer nanocomposites. *Water Res.,* 44, 1927-1933.

[51] Geng, B., Jin, Z., Li, T., Qi, X., 2009. Kinetics of hexavalent chromium removal from water by chitosan-Fe0 nanoparticles. *Chemosphere* 75, 825-830.

[52] Kanel, S.R., Grenche, J.M., Choi, H., 2006. Arsenic(V) removal from groundwater using nano scale zero-valent iron as a colloidal reactive barrier material. *Environ. Sci. Technol.* 40, 2045-2050.

[53] Fenglian, F., Jun, M., Liping, X., Bing, T., Weijiang, H., Suya, L., 2013. Chromium removal using resin supported nanoscale zero-valent iron. *J. Environ. Management* 128, 822-827.

[54] Manning, B.A., Kiser, J.R., Kwon, H., Kanel, S.R., 2007. Spectroscopic investigation of Cr(III)- and Cr(VI)-treated nanoscale zerovalent iron. *Environ. Sci. Technol.* 41, 586-592.

[55] Liu, T., Yang, X., Wang, Z.L., Yan, X., 2013. Enhanced chitosan beads-supported Fe0-nanoparticles for removal of heavy metals from electroplating wastewater in permeable reactive barriers. *Water Res.* 47, 6691-6700.

[56] Huang, G.L., Zhang, H.Y., Jeffrey, X.S., Tim, A.G.L., 2009. Adsorption of chromium(VI) from aqueous solutions using cross-linked magnetic chitosan beads. *Industrial Eng. Chem. Res.*, 48, 2646-2651.

[57] Tran, H.V., Tran, L.D., Nguyen, T.N., Preparation of chitosan/magnetite composite beads and their application for removal of Pb(II) and Ni(II) from aqueous solution. *Mater. Sci. Eng.* C, 30, 304-310.

[58] Liu, X.W., Hu, Q.Y., Fang, Z., Zhang, X.J., Zhang, B.B., 2009. Magnetic chitosan nanocomposites: A useful recyclable look for heavy metal ion removal. *Langmuir,* 25, 3-8.

[59] Badruddoza, A.Z.M., Shawon, Z.B.Z., Daniel, T.W.J., Hidajat, K., Uddin, M.S., 2013. Fe_3O_4/cyclodextrin polymer nanocomposites for selective heavy metals removal from industrial wastewater. *Carbohyd. Polymer,* 91, 322-332.

[60] Mahdavian, A.R., Mirrahimi, M. Al-S., 2010. Efficient separation of heavy metal cations by anchoring polyacrylic acid on superparamagnetic magnetite nanoparticles through surface modification. *Chem. Eng. J.* 159, 264-271.

[61] Ge, F., Li, M.-M., Ye, H., Zhao, B.X., 2012. Effective removal of heavy metal ions Cd^{2+}, Zn^{2+}, Pb^{2+}, Cu^{2+} from aqueous solution by polymer-modified magnetic nanoparticles. *J. Hazard. Mater.* 211-212, 366-372.

[62] Bée, A., Talbot, D., Abramson, S., Dupuis, V., 2011. Magnetic alginate beads for Pb(II) ions removal from wastewater. *J. Colloid Interf. Sci.* 362, 486-496.

[63] Zhou, Y.–T., Nie, H.–L., Branford-White, C., He, Z.–Y., Zhu, L.–M., 2009. Removal of Cu^{2+} from aqueous solution by chitosan-coated magnetic nanoparticles modified with α-ketoglutaric acid. *J. Colloid. Interf. Sci.* 330, 29-37.

[64] Idris, A., Ismail, N.S.M., Hassan, N., Misran, E., Ngomsik, A.–F., 2012. Synthesis of magnetic alginate beads based on maghemite nanoparticles for Pb(II) removal in aqueous solution. *J. Ind. Eng. Chem.* 18, 1582-1589.

[65] Chang, Y.–C., Chen, D.-H., 2005. Preparation and adsorption properties of monodisperse chitosan-bound Fe_3O_4 magnetic nanoparticles for removal of Cu(II) ions. *J. Colloid Interf. Sci.* 283, 446-451.

[66] Madrakian, T., Afkhami, A., Zadpour, B., Ahmadi, M., 2015. New synthetic mercaptoethylamino homopolymer-modified maghemite nanoparticles for effective removal of some heavy metal ions from aqueous solution. *J. Ind. Eng. Chem.* 21, 1160-1166.

[67] Ngomsik, A.-F., Bée, A., Siauguea, J.-M., Talbot, D., Cabuil, V., Cote, G., 2009. Co(II) removal by magnetic alginate beads containing Cyanex 272®. *J. Hazard. Mater.* 166, 1043-1049.

[68] Mahapatra, A., Mishra, B.G., Hota, G., 2013. Electrospun Fe_2O_3–Al_2O_3 nanocomposite fibers as efficient adsorbent for removal of heavy metal ions from aqueous solution. *J. Hazard. Mater.* 258-259, 116-123.

[69] Wu, A., Jia, J., Luan, S., 2011. Amphiphilic PMMA/PEI core–shell nanoparticles as polymeric adsorbents to remove heavy metal pollutants. *Colloids Surf. A: Physicochem. Aspects* 384, 180-185.

[70] Jing, C., Zhaoxiang, Z., Hong, X., Zhong, Y., Rizhi, C., 2013. Fabrication of poly(γ-glutamic acid)-coated Fe_3O_4 magnetic nanoparticles and their application in heavy metal removal. *Chinese J. Chem. Eng.* 21(11), 1244-1250.

[71] Zhu, Y., Hu, J., Wang, J., 2014. Removal of Co2+ from radioactive wastewater by polyvinyl alcohol (PVA)/chitosan magnetic composite. *Progr. Nuclear Energy* 71, 172-178.

[72] Geng, B., Jin, Z., Li, T., Qi, X., 2009. Preparation of chitosan-stabilized Fe0 nanoparticles for removal of hexavalent chromium in water. *Sci. Total Environ.* 407, 4994-5000.

[73] Zhang, F., Lan, J., Zhao, Z., Yang, Y., Tan, R., Song, W., 2012. Removal of heavy metal ions from aqueous solution using Fe3O4–SiO2-poly(1,2-diaminobenzene) core–shell sub-micron particles. *J. Colloid Interf. Sci.* 387, 205-212.

[74] Puttamraju, P., SenGupta, A.K., 2006. Evidence of tunable on-off sorption behaviours of metal oxide nanoparticles: Role of ion exchanger support. *Ind. Eng. Chem. Res.* 45, 7737-7742.

[75] Warner, C.L., Chouyyok, W., Mackie, K.E., Neiner, D., Saraf, L.V., Droubay, T.C., Warner, M.G., Addleman, R.S., 2012. Manganese doping of magnetic iron oxide nanoparticles: tailoring surface reactivity for a regenerable heavy metal sorbent. *Langmuir* 28(8), 3931-3937.

[76] Karatapanis, A.E., Petrakis, D.E., Stalikas, C.D., 2012. A layered magnetic iron/iron oxide nanoscavenger for the analytical enrichment of ng-L−1 concentration levels of heavy metals from water. *Analytica Chimica Acta,* 726, 22-27.

[77] Liu, Z., Wang, H., Liu, C., Jiang, Y., Yu, G., Mu, X., Wang, X., 2012. Magnetic cellulose–chitosan hydrogels prepared from ionic liquids as reusable adsorbent for removal of heavy metal ions. *Chem. Commun.* 48 (59), 7350-7352.

[78] Cumbal, L., Greenleaf, J., Leun, D., SenGupta, A. K., 2003. Polymer supported inorganic nanoparticles: characterization and environmental applications. *Reactive Funct. Polym.* 54, 167-180.

[79] Lim, C.W., Song, K., Kim, S.H., 2012. Synthesis of PPy/silica nanocomposites with cratered surfaces and their application in heavy metal extraction. *J. Ind. Eng. Chem.* 18, 24-28.

[80] Lia, C.-J., Zhang, S.-S., Wanga, J.-N., Liu, T.-Y., 2014. Preparation of polyamides 6 (PA6)/Chitosan@FexOy composite nanofibers by electrospinning and pyrolysis and their Cr(VI)-removal performance. *Catal. Today* 224, 94-103.

[81] Sharma, A., Lee, B.-K., 2014. Cd(II) removal and recovery enhancement by using acrylamide–titanium nanocomposite as an adsorbent. *Appl. Surf. Sci.* 313, 624-632.

[82] Iijima, S., 1991. Helical microtubules of graphitic carbon. *Nature* 354, 56-58.

[83] Popov, V. N., 2004. Carbon nanotubes: properties and application. *Mater. Sci. Eng. R* 43, 61-102.

[84] Vul A. Y., Sokolov, V. I., 2009. Nanocarbon Studies in Russia: From Fullerenes to Nanotubes and Nanodiamonds. *Nanotechnol. Russia*, 4 (7–8), 397–414.

[85] K. Yang, K., Xing, B., Adsorption of organic compounds by carbon nanomaterials in aqueous phase: Polanyi theory and its application. *Chem. Rev.* 110, 5989-6008.

[86] Shao, D., Chen, C., Wang, X., 2012. Application of polyaniline and multiwalled carbon nanotube magnetic composites for removal of Pb(II). *Chem. Eng. J.* 185–186, 144-150.

[87] Li, J., Chen, S., Sheng, G., Hu, J., Tan, X., Wang, X., 2011. Effect of surfactants on Pb(II) adsorption from aqueous solutions using oxidized multiwall carbon nanotubes. *Chem. Eng. J.* 166, 551-558.

[88] Salam, A. M., Burk, R., 2009. Solid phase extraction of polyhalogenated pollutants from freshwater using chemically modified multi-walled carbon nanotubes and their determination by gas chromatography. *J. Sep. Sci.* 32, 1060-1068.

[89] Ge, J.J., Zhang, D., Li, Q., Hou, H., Graham, M.J., Dai, L., Harris, F.W., Cheng, S.Z.D., 2005. Multiwalled Carbon Nanotubes with Chemically Grafted Polyetherimides. *J. Am. Chem. Soc.* 127 (28), 9984-9985.

[90] Zhang, D., Shi, L., Fang, J., Li, X., Dai, K., 2005. Preparation and modification of carbon nanotubes. *Mater. Lett.* 59, 4044-4047.

[91] Chang, P.R., Zheng, P., Liu, B., Anderson, D.P., Yu, J., Ma, X., 2011. Characterization of magnetic soluble starch-functionalized carbon nanotubes and its application for the adsorption of the dyes. *J. Hazard. Mater.* 186, 2144-2150.

[92] Li, C., Yang, K., Zhang, Y., Tang, H., Yan, F., Tan, L., Xie, Q., Yao, S., 2011. Highly biocompatible multi-walled carbon nanotube–chitosan nanoparticle hybrids as protein carriers. *Acta Biomater.* 7, 3070-3077.

[93] Rinaudo, M., 2006. Chitin and chitosan: Properties and applications. *Prog. Polym. Sci.* 31, 603-632.

[94] Mourya, V.K., Inamdar, N.N., 2008. Chitosan-modifications and applications: Opportunities galore. *React. Funct. Polym.* 68, 1013-1061.

[95] Rhazi, M., Desbrieres, J., Tolaimate, A., Rinaudo, M., Vottero, P., Alagui, A., 2002. Contribution to the study of the complexation of copper by chitosan and oligomers. *Polymer* 43, 1267-1276.

[96] Kandile, N.G., Nasr, A.S., 2009. Environment friendly modified chitosan hydrogels as a matrix for adsorption of metal ions, synthesis and characterization. *Carbohydr. Polym.* 78, 753-759.

[97] Findon, A., McKay, G., Blair, H.S., 1993. Transport studies for the sorption of copper ions by chitosan. *J. Environ. Sci. Health, Part A, Environ. Sci. Eng.* 28 (11), 173-185.

[98] Kim, M.K., Sundaram, K.S., Iyengar, G.A., Lee, K.-P., 2015. A novel chitosan functional gel included with multiwall carbon nanotube and substituted polyaniline as adsorbent for efficient removal of chromium ion. *Chem. Eng. J.* 267, 51-64.

[99] Sheng, G., Li, Y., Dong, H., Shao, D., 2012. Environmental condition effects on radionuclide 64Cu(II) sequestration to a novel composite: polyaniline grafted multiwalled carbon nanotubes. *J. Radioanal. Nucl. Chem.* 293, 797-806.

[100] Li, X., Zhou, H., Wu, W., Wei, S., Xu, Y., Kuang, Y., 2015. Studies of heavy metal ion adsorption on Chitosan/sulfydrylfunctionalized graphene oxide composites. *J. Colloid Interf. Sci.* 448, 389-397.

[101] Salam, M.A., Makki, M.S.I., Abdelaal, M.Y.A., 2011. Preparation and characterization of multi-walled carbon nanotubes/chitosan nanocomposites and its application for the removal of heavy metals from aqueous solution. *J. Alloys Compounds* 509, 2582-2587.

[102] Yang, S., Hu, J., Chen, C., Shao, D., Wang, X., 2011. Mutual effects of Pb(II) and humic acid adsorption on multiwalled carbon nanotubes/polyacrylamide composites from aqueous Solutions. *Environ. Sci. Technol.* 45, 3621-3627.

[103] Salehi, E., Madaeni, S.S., Rajabi, L., Vatanpour, V., Derakhshan, A.A., Zinadini, S., Ghorabi, S., Monfared, H.A., 2012. Novel chitosan/poly(vinyl) alcohol thin adsorptive membranes modified with amino functionalized multi-walled carbon nanotubes for Cu(II) removal from water: Preparation, characterization, adsorption kinetics and thermodynamics. *Sep. Purif. Technol.* 89, 309-319.

[104] Kumar, R., Ansari, M.O., Barakat, M.A., 2013. DBSA doped polyaniline/multi-walled carbon nanotubes composite for high efficiency removal of Cr(VI) from aqueous solution. *Chem. Eng. J.* 228, 748-755.

[105] Xu, Z., Zhang, Y., Qian, X., Shi, J., Chen, L., Li, B., Niu, J., Liu, L., 2014. One step synthesis of polyacrylamide functionalized graphene and its application in Pb(II) removal. *Appl. Surf. Sci.* 316, 308-314.

[106] Hui, W., Yun-guo, L., Guang-ming, Z., Xin-jiang, H., Xi, H., Ting-ting, L., Hua-ying, L., Ya-qin, W., Lu-hua, J., 2014. Grafting of β-cyclodextrin to magnetic graphene oxide via ethylenediamine and application for Cr(VI) removal. *Carbohyd. Polym.* 113, 166-173.

[107] Yao, W., Ni, T., Chen, S., Li, H., Lu, Y., 2014. Graphene/Fe3O4@polypyrrole nanocomposites as a synergistic adsorbent for Cr(VI) ion removal. *Composites Sci. Technol.* 99, 15-22.

[108] Shah, P., Murthy, C.N., 2013. Studies on the porosity control of MWCNT/polysulfone composite membrane and its effect on metal removal. *J. Membrane Sci.* 437, 90-98.

[109] Dong, Z., Zhang, F., Wang, D., Liu, X., Jin, J., (2015). Polydopamine-mediated surface-functionalization of graphene oxide for heavy metal ions removal. *J. Solid State Chem.* 224, 88-93.

[110] Li, R., Liu, L., Yang, F., 2013. Preparation of polyaniline/reduced graphene oxide nanocomposite and its application in adsorption of aqueous Hg(II). *Chem. Eng. J.* 229, 460-468.

[111] Sang, Y., Gu, Q., Sun, T., Li, F., Liang, C., 2008. Filtration by a novel nanofiber membrane and alumina adsorption to remove copper(II) from groundwater. *J. Hazard. Mater.* 153, 860-866.

[112] Hota, G., Kumar, B.R., Ng, W., Ramakrishna, S., 2008. Fabrication and characterization of a boehmite nanoparticle impregnated electrospun fiber membrane for removal of metal ions. *J. Mater. Sci.* 43, 212-217.

[113] Xiao, S., Shen, M., Ma, H., Guo, R., Zhu, M., Wang, S., Shi, X., 2010. Fabrication of water-stable electrospun polyacrylic acid-based nanofibrous Mats for removal of copper (II) ions in aqueous solution. *J. Appl. Polym. Sci.* 116, 2409-2417.

[114] Kampalanonwat, P., Supaphol, P., 2014. The Study of Competitive Adsorption of Heavy Metal Ions from Aqueous Solution by Aminated Polyacrylonitrile Nanofiber Mats. *Energy Procedia* 56, 142-151.

[115] Yoon, K., Hsiao, B.S., Chu, B., 2008. Functional nanofibers for environmental applications. *J. Mater. Chem.* 18, 5326-5334.

[116] Haider, S., Park, S.-Y., 2009. Preparation of the electrospun chitosan nanofibers and their applications to the adsorption of Cu(II) and Pb(II) ions from an aqueous solution. *J. Membr. Sci.* 328, 90-96.

In: Heavy Metals
Editor: Deepak Pathania

ISBN: 978-1-63484-740-7

Chapter 3

MINING AND MINERAL PROCESSING AS HEAVY METALS POLLUTION SOURCE

Marinela Panayotova*
University of Mining and Geology, Department of Chemistry,
Sofia, Bulgaria

ABSTRACT

Mining, mineral beneficiation and extractive metallurgy (pyro-, hydro- and electrometallurgy) are discussed as sources of heavy metals to all environmental compartments. Examples of elevated heavy metal concentrations in the environment, caused my metals' extraction from ores, as well as the resulting impact on flora, fauna, humans in different countries are presented. Possible prevention/mitigation measures for each stage of metals' recovery are briefly commented.

Keywords: heavy metals, mining, mineral beneficiation, extractive metallurgy, wastewater

1. INTRODUCTION

The general process of metals' production from ores passes through different stages - mining, mineral beneficiation and extractive metallurgy (pyrometallurgy, hydrometallurgy, and electrometallurgy). Each of these stages can be a source of heavy metals[1] release to all environment compartments (air, soil, surface water, sediment, ground water) - Table 1.

* Corresponding Author address: Bulgaria, Sofia 1700, Boyan Kamenov Str. 1, University of Mining and Geology; E-mail: marichim@mgu.bg.

[1] In this chapter any metal (cadmium - Cd, chromium - Cr, cobalt - Co, copper - Cu, lead - Pb, mercury - Hg, nickel - Ni, zinc - Zn) or metalloid (antimony Sb, arsenic - As, selenium - Se) of environmental concern is referred to as heavy metal.

Table 1. Sources of heavy metals, pathways and impacted environment compartments at extracting metals from ores

Activity	Wastes	Pathways
Open pit and underground mining	Waste rock, overburden soils, mining water, atmospheric emissions	Movement of mine/waste water; Run-off, infiltration through waste piles and through soil/vadoze zone; Surface water - movement and re-suspension of sediments; Groundwater; Air - aerosols movement; Uptake by biota
Mineral beneficiation	Tailings, wastewater, atmospheric emissions	
Extractive metallurgy: - Pyrometallurgy - Hydrometallurgy - Electrometallurgy	 Slag, wastewater, flue dusts Leached ores, wastewater, tailings Anodic sludge, wastewater, tailings	

Nonferrous mining activities are some of the largest sources of heavy metals emissions to the environment. Studies estimated that cumulatively about 1.62 Mt Pb and 3.32 Mt Zn were emitted into the ambient environment during the mining, processing and smelting of Pb and Zn in China up to 2007. Among these three types of activities, mineral processing contributes the most to the total emission of 50.67% lead and 45.51% zinc [1].

2. Mining as Source of Heavy Metals Pollution

Ores mining activities leave huge amounts of waste rock (heaped or damped), bearing different heavy metals - depending on the ore type. For example: Deep mining of Pb and Cu deposits, conducted for centuries in the district of Linares (Jaén, Spain), generated large mine waste accumulations, bearing (in mg/kg): Pb - 19 913, Cu - 418, manganese (Mn) - 4149, As - 43, Zn - 190, and iron (Fe) - 26200 [2]. Average concentrations of Cd, Cu, Pb and Zn found in tailings of an abandoned metalliferous mine in Korea are 8.57, 481, 4450 and 753 mg/kg, respectively [3]. The total concentrations of some heavy metals determined in dump material, rich in chalcopyrite, of an abandoned mine in Galicia (NW Spain) are as follows (in mg/kg): Fe - 4315 ÷ 31 578, Cu - 273 ÷ 5241, Mn - 294 ÷ 2105, Zn - 73 ÷ 894, Cr - 0.01 ÷ 30. Total As concentrations in the range of 300–67000 mg/kg are found in spoil heap samples from different abandoned mines in Spain [4]. Arsenic concentrations in the range 175–2300 mg/kg are determined in an tungsten–tin abandoned mining area, where the exploitation ceased more than three decades ago [5]. Mine tailings in Liwu mud-retaining dam (China) are with low pH (2.43) and rich in some metals with mean concentrations, in mg/L: Fe - 337.17, Mn - 66.28, Cu - 14.68, Zn - 80.41, Pb - 1.47 resulting in a total discharge to the environment, in kg/day: Fe - 4967.55, Mn - 994.20, Cu - 220.20, Zn - 1206.15 [6]. Heavy metals, available in the fine particles of the waste material can be dispersed around by wind blowing. Heavy metals in the waste rock can be mobilized through natural weathering process as well as by leaching, facilitated by precipitations and water runoff or percolation.

2.1. Air Pollution

Mining activities can "contribute" in different way to air pollution - by evaporation, wind erosion, wind transport. According to UNEP, the artisanal and small-scale gold mining (ASGM) and coal burning are the major sources of anthropogenic Hg emissions to air. Annual emissions from ASGM are estimated at 727 t, making this the largest sector accounting for more than 35% of total Hg anthropogenic emissions [7].

Wind erosion of mine wastes leads to creation of fine particulate dust emissions, often bearing Pb, Cd, As, Hg. There are many evidences for such type of air pollution:

Measurements of short-term atmospheric contaminant emissions in mining and smelting communities in Australia gave 24-h maximum suspended particulate (TSP) for lead-in-air and arsenic-in-air 12.8 $\mu g/m^3$ and 2973 ng/m^3, respectively in Mount Isa, Queensland and 22.57 $\mu g/m^3$ and 250 ng/m^3, respectively - in Port Pirie, South Australia. Estimated atmospheric emissions from these sites for the period 2011- 2012 were 33000 kg As and 120000 kg Pb - from Mount Isa and 1300 kg As and 46000 kg Pb - from Port Pirie [8].

Contaminated dust and aerosol from mining operations can be transported more rapidly (compared to the other major transport pathways - water, soil and biota) and over relatively long distances. Further, atmospheric deposition contributes to accumulation of contaminants in soil, water and biota. For example, it has been found that abandoned mine wastes are the source of the annual levels of total bulk deposition (soluble and insoluble fractions) registered in the Rio Tinto Mining District (Iberian Pyrite Belt) ranging between 18 and 43 g/m^2 depending on the distance of the sampling station from the mine waste deposits. Levels of deposited Zn and Cu were in a range of 9–62 mg/m^2 not only in the insoluble but also in the soluble fraction. Toxic trace elements such as As, Sb, Ba, Pb, Cd, Sn and Bi showed greater deposition fluxes in the sites exposed to dust-bearing winds downwind of the mining area [9].

Interactions between transported particles and repository media may affect speciation, mineralogy and bioavailability of contaminants and enhance the negative effects to surrounding soils, plants and humans. Air transport of pollutants, associated with mining and metallurgical activities, was found as main reason for availability of many heavy metals (As, Cd, Cu, Pb, Sb, Sn, Zn) in indoor dust in several Andean cities and villages. The highest concern is related to lead pollution (600 $\mu g/g$) [10]. Further study gave concentrations of heavy metals in dust samples in a mining district of Oruro, Bolivia [11] – Table 2.

Table 2. Median concentrations of trace elements measured in dust samples at mining district of Oruro, Bolivia. Percentiles 5% and 95% are shown in parentheses

Pollutant	As	Cd	Pb	Sb	Sn	Cu	Zn
Concentration, ppm	88.60 (18.54-69.51)	10.40 (5.10–38.15)	560.60 (87.86-198.00)	108.70 (27.08–2770.28)	63.40 (25.96–309.12)	80.15 (31.70–325.50)	414.54 (167.80–1705.50)

2.2. Water Pollution

Mining impacted water (MIW) is a major problem for metal-mining operations. MIW composition is formed as a result of water contact with surrounding solids (waste rock, tailings, mine or mineral surfaces) in their pathways. Broadly, this term includes acid rock drainage (ARD), neutral mine drainage (NMD), and saline drainage (SD). The approximate thresholds between the three types of mine drainage are: above pH 6 - NMD and SD, below pH 6 - ARD. Usually, the drainage is considered saline at sulfate (SO_4^{2-}) concentration $\geq$ 1000 mg/L. Typical ore-deposit types which are usually associated with ARD include the following deposits: volcanogenic massive, high sulfidation epithermal, porphyry copper and skarn. Coal deposits also often generate ARD. Typical deposit types associated with SD include low-sulfide gold-quartz vein and Mississippi-Valley type deposits. NMD can be formed in wide variety of ore deposits, depending on the type of alteration and sulfide content, including most types listed for ARD and SD. Usually, most of the problems related to water pollution due to mining activities, are caused by ARD (often named also acid mine drainage - AMD). The acidic water results from the pyrite oxidation at contacting with atmospheric oxygen and water [12]. The overall reaction generally is written as:

$$FeS_2 + 7/2O_2 + H_2O \Leftrightarrow Fe^{2+} + 2SO_4^{2-} + 2H^+ \quad (1)$$

Sulfuric acid is generated. Pyrite can be directly oxidized abiotically or biotically (mediated through microorganisms). Indirect oxidation, i.e., pyrite dissolution, followed by oxidation can also take place.

Oxygen dissolved in water can also oxidize pyrite but due to oxygen limited solubility in water, this process is much less prominent. Aqueous ferric iron can also oxidize pyrite:

$$FeS_2 + 14Fe^{3+} + 8H_2O \Leftrightarrow 15Fe^{2+} + 2SO_4^{2-} + 16H^+ \quad (2)$$

This reaction is faster (2 to 3 orders of magnitude) than reaction (1) and generates higher acidity per mole of pyrite oxidized. However, it proceeds in acidic conditions under which significant amounts of dissolved ferric iron are available. Therefore, pyrite oxidation starts through reaction (1) at nearly neutral or higher pH, then, when conditions have become sufficiently acidic, reaction (2) takes place. Ferric iron is generated through oxidation of ferrous iron by atmospheric oxygen as follows:

$$2Fe^{2+} + ½O_2 + 2H^+ \Leftrightarrow 2Fe^{3+} + H_2O \quad (3)$$

Reaction (3) may be catalyzed by bacteria (primarily members of the *Acidithiobacillus genus*) which are obligate aerobes. Therefore, some amount of oxygen is needed for this process to be effective, although the oxygen requirement is less than for abiotic oxidation.

Ferrous iron is removed from solution under slightly acidic to alkaline conditions through oxidation and subsequent hydrolysis. Relatively insoluble iron hydroxide is formed:

$$Fe^{2+} + ¼O_2 + 2½H_2O \Leftrightarrow Fe(OH)_3 + 2H^+ \quad (4)$$

In the general case, when conditions are not acidic (i.e., pH > 4.5), reactions (1) and (4) proceed together and pyrite oxidation generates higher acidity compared to reaction (1):

$$FeS_2 + 15/4O_2 + 7/2H_2O \Leftrightarrow Fe(OH)_3 + 2SO_4^{2-} + 4H^+ \quad (5)$$

In addition, bacteria, such as *A. ferrooxidans* and *A. thiooxidans* play an important role in sulfide oxidation and in the formation of ARD.

Most important chemical and physical factors, governing the rate of ARD formation through sulfides oxidation, are: type of sulfide mineral, its crystallinity, morphology and contacting surface area; pH and oxidation-reduction potential values of water, contacting with the rocky material; temperature; type (oxygen from air, Fe^{3+}) and availability of oxidants.

Conditions for growing of bacteria of the *Acidithiobacillus* species are the important biological factors. These include temperature, nutrients (i.e., nitrogen, phosphorus), oxygen and carbon sources (for example, CO_2 from air) availability.

Formed ARD leaches out of minerals different major and trace metals. Usually Fe and aluminium (Al) are the major dissolved metals, with concentrations in the range of 1000 ÷ 10000 mg/L in ARD. Trace metals, such as Pb, Zn, Cu, Cd, manganese (Mn), Co, and Ni are also leached and are found in ARD in elevated concentrations, usually in the range of 100 ÷ 1000 mg/L. The solubility of metals vary with pH. Cationic species, such as Cu, Pb, Zn, Cd, Ni, and Co, are more soluble at low pH. Elements that form anionic species, such as As, Sb, Se, and Cr are more soluble at high pH. Other factors that impact the heavy metals' mobilization in MIW are redox conditions, availability of iron- and aluminum-oxide precipitates that can act as effective sorbents for a variety of metals, availability of other adsorbents, availability of different inorganic and organic ligands, biota intervention, etc. Table 3 supplies a general guide to predict heavy metals behavior in surface water environments [13]. In mine discharges with less acidic character, trace metal concentrations are lower due to formation of secondary mineral phases and increased sorption of heavy metals onto different minerals, they are contacting with. However, anion-forming metals and metalloids (As, Se, and Sb) remain in solution in neutral MIW and as the pH increases amphoteric metals (for example, Zn in absence of carbonates) tend to remobilize.

Mine waste pile runoff and mine drainage are the major sources of heavy metals entering the MIW from open pit mines. Metals from underground mines enter surface water through adits, mine tunnels, and naturally-forming streams. Subsurface inputs also contribute substantially to metal loading in mining-impacted streams. Formation of MIW continues long after the mine closure. The situation is worse, when the mineralization in the region is sulfidic. Then weathering and already formed ARD continues to expose fresh sulfide surface to oxygen from air and water (from precipitation, infiltration, seepage, etc.). Further formation of ARD and further acidification of contacting water is observed, so further the power of this water to elute/dissolve heavy metals from the solid materials (mine waste, ore bodies, already precipitated or adsorbed heavy metals) is increased.

The transport of heavy metals by MIW is realized by runoff and overland flow with eventual discharge into surface water and transport via surface water, infiltration through the mine waste facility and into the soil/vadose zone followed by transport in groundwater, uptake by biota, and physical movement of mine waters as part of mine water management. During its transport MIW is modified by physical (climate conditions, temperature, precipitation events, and water movement), chemical (pH, redox conditions, chemical composition of drainage, neutralization and secondary mineral formation reactions, complexation, sorption) and biological (microbial growth and fade) factors.

Some examples of MIW composition, formed in different mining areas worldwide, are presented in this chapter. Average metal loading in MIW from mine waste pile runoff and mine drainage discharge in Tar Creek (Oklahoma, USA), which drains an abandoned Zn/Pb mining area, as well as in the receiving stream, is presented in Table 4 [4].

Table 3. Generalized relative mobility of heavy metals and uranium under different environmental conditions

Environment condition	Very mobile	Mobile	Mobile to some extent	Scarcely mobile to immobile
Oxidizing, at pH < 3	Cd, Co, Cu, Ni, Zn	As, Hg, Se	Ag, Cr, Pb, Sb, Tl	-
Oxidizing, at pH > 5 to nearly neutral (pH = 6.5 to 7.5), no Fe substrates	Cd, Zn	Se, U	As, Co, Cr, Cu, Hg, Ni, Sb, Tl	Ag
Oxidizing, pH > 5 to nearly neutral, with abundant Fe substrates	-	Cd, Zn	Co, Hg, Ni, Sb, Se, Tl	Ag, As, Cr, Cu, Pb, U
Reducing, at pH > 5 to nearly neutral, no hydrogen sulfide	-	Cd, Cu, Pb, Zn	As, Co, Cr, Hg, Tl	Ag, Sb, Se, U
Reducing, at pH > 5 to nearly neutral, with hydrogen sulfide	-	-	Tl	Ag, As, Cd, Co, Cr, Cu, Hg, Ni, Pb, Sb, Se, U

Table 4. Concentrations of heavy metals in mine waste pile runoff and mine drainage discharge in Tar Creek samples from Zn/Pb mining area

Parameter	pH	Fe, mg/L	Mn, mg/L	Zn, mg/L	As, μg/L	Cd, μg/L	Cu, μg/L	Ni, μg/L	Pb, μg/L
Waste pile runoff	7.3	1.6	0.26	41	0.36	190	1.5	110	14
Mine drainage	5.9	39	1.5	6.2	16	0.88	0.32	190	2.6
Stream	7.14	1.33	0.855	3.97	0.841	4.31	0.783	9.60	2.39

Mine waste pile runoff contributed 70% of Cd, while mine drainage contributed 90% of Pb, and both sources contributed to Zn loading. Subsurface inputs accounted for 40% of flow and 40–70% of metal loading along the studied 0.9-km stretch of Tar Creek. Streambed iron oxide aggregate material contained elevated amounts of Zn (up to 27000 μg/g), Pb (up to 550 μg/g) and Cd (up to 200 μg/g). Rock waste dumps and reject wastes from Sarcheshmeh, Iran (one of the largest porphyry Cu deposits in the world) are the main sources of ARD that contain heavy metals, such as, in mg/L: Cu (150), Cd (0.31), Co (2.13), Mn (73.7), Ni (1.74), and Zn (20.3) [15]. Mine tailings from an abandoned mine, Cabezo Rajao, in SE Spain showed the following mean concentrations of Pb, Zn, Cd, Cu and As: 8.3 g/kg, 12.5 g/kg, 40.9 mg/kg, 332.1 mg/kg, 314.7 mg/kg respectively. The heavy metals concentrations in runoff water were correspondingly, in mg/L: 1.5, 50.3, 13.6, 17.2, and 1.7 [16]. In addition to above mentioned heavy metals, mine discharges can be polluted with Se. In mining wastewaters, selenium can be found in concentrations ranging from 3 μg/L to above 12 mg/L;

in uranium mine discharge - concentration of 1.6 mg/L is reported, and concentrations in the range of 0.2-33 mg/L were reported for gold (Au) mines wastewater [17].

Acidic mine-shaft water from abandoned Pb-Zn mining area in Sardinia exhibited high concentrations of heavy metals [18]. Its composition is formed by water contact with tailings stored in mine. In time the water table reached the topographical level of the ''Fais Shafts'' mouth and entered the groundwater spring, flooding into ''Irvi'' stream. Thus, Rio Irvi (a tributary of Rio Piscinas) and Rio Piscinas became acidic and contaminated by dissolved heavy metals - Table 5. Heavy metals, moved from contamination sources by polluted surface waters, were bound to the solid-phase of rivers' sediments. Metals are accumulated in river sediments and pose an environmental risk at events leading to certain decrease in water pH values. In addition, authors pointed out that the contamination level of some metals (Cd, Co, Fe, Mn, Ni, Pb and Zn) did not change significantly during a 10 years period.

Studies showed that due to mining activities the surface water of Linglong gold mining area (Shandong Province, China) is seriously polluted by heavy metals: Hg (up to 827.2 μg/L), Zn (up to 3720 μg/L), Cd (up to 194.5 μg/L), Pb (up to 433.9 μg/L), Cu (up to 14605 μg/L), Cr (up to 75.06 μg/L), and As (up to 187.05 μg/L) [19].

Accidents with waste holding dams can cause serious problems. On 25 April 1998 a dam, holding mine tailings from Los Frailes Cu and Zn mine near Aznalcóllar, Seville province, burst and 4–5 million cubic meters of acidic tailings, containing heavy metals such as Zn, Pb, As, Co, Sb, Co, thallium (Tl), bismuth (Bi), Cd, silver (Ag), Hg and Se were released in the surroundings, reached the nearby Agrio River, and then its affluent the Guadiamar River. The Guadiamar is the main water source for the Doñana National Park, a UNESCO and one of the largest national parks in Europe.

Table 5. Concentrations of heavy metals in mine-shaft water, rivers impacted my mining activity, mine tailings and river sediments

Parameter	Cd	Co	Cu	Fe	Mn	Ni	Pb	Zn	Cr	As	Se
Tailings, mg/kg	71	15	84	64500	3600	25	4100	7300	-	-	-
Sediments, mg/kg	130	19	490	76800	2900	28	12400	20800	-	-	-
Mine shaft	1100	2900	4.2	240	98	4600	570	780	6.8	170	31
Rio Roia Cani[1]	650-990	7.1-17	1.2-8.8	0.02-0.08	0.2-06	57-110	270-710	45-76	0.94-1.3	<0.7	2.8-3.8
Rio Irvi[2]	1000-1100	2700	4.2-7.7	140-200	88-93	4100-4300	280-780	710-730	1.9-3.3	44-110	23
Rio Piscinas[3]	170-1700	1.9-1400	5.4-22	49-54	0.05-59	16-3200	150-770	12-620	1.4-1.9	5.2-8.6	1.8-22

[1] – Impacted by waste tailings generated during mineral processing; [2] - Impacted by ARD from the mine shaft; [3] – Collects waters from Rio Roia Cani (the low range values) and Rio Irvi (the high range values); Fe, Mn and Zn concentrations in mg/L; Cd, Co, Cu, Pb, Ni, As concentrations in μg/L.

The ARD waters coming from the tailings sites of the Lousal abandoned mine, Portugal showed values of pH ranging from 1.9 to 2.9 and concentrations of 9249 to 20700 mg/L SO_4^{2-}, 959 to 4830 mg/L Fe and 136 to 624 mg/L Al. The acid effluents and mixed stream waters also carried high contents of Cu, Pb, Zn, Cd, and As, generally exceeding the Fresh Water Aquatic Life Acute Criteria. High concentrations of Cu, Pb, Zn, As, Cd and Hg have been determined in the stream sediments downstream of the mine tailings sites, in mg/kg: As - 817, Cd - 6.7, Cu - 1568, Pb - 1059, Sb - 82.4, and Zn - 4373 [20].

Data on elevated As concentrations due to gold mining were summarized [21]: Arsenic concentrations in ground water in Obuasi (Ghana) were in the range of 1÷175 μg/L; In Victoria (Australia) As concentrations in ground water were in the range of 1÷12 μg/L, in surface water – in the range of 1÷220 μg/L, and in drinking water – in the range of 1÷73 μg/L. Polluted with As groundwater (with the maximum concentration of 0.45 mg/L As) was found in the San Antonio-El Triunfo mining district (Mexico) where Au and Ag extraction through the cyanide leaching was carried out [22]. Arsenic concentrations in the range of 33÷911 μg/L were measured in the groundwater samples impacted by the borate mines in Bigadic district, Turkey [23].

Rivers, affected by ARD and polluted with heavy metals can cause further pollution. They transport and disperse heavy metals under dissolved or particulate form. It has been found that Huelva estuary is heavily contaminated by metals and metalloids, due to the pollution of Tinto and Odiel rivers caused by the long-term mining activities in Iberian Pyrite Belt. It has been estimated that the Tinto and Odiel rivers transport dissolved metals to the estuary: Fe - 7900 t/yr, Al - 5800 t/yr, Zn - 3500 t/yr, Cu - 1700 t/yr, Mn - 1600 t/yr and minor quantities of other metals and metalloids [24].

The following concentrations of heavy metals (in mg/kg) were measured in the surface sediments of the Bistriţa River (Romania), which is impacted by manganese ores and uranium mining activities: 39 ÷ 99 for Cr, 11.2 ÷ 38.5 for Co, 16 ÷ 48 for Ni, 17 ÷ 451 for Cu, 50 ÷ 1117 for Zn, 0.11 ÷ 2.38 for Cd, 17 ÷ 139 for Pb and 8.2 ÷170 for As [25].

Polluted rivers pollute flooded areas. For example, increased concentrations of heavy metals have been found in floodplain sediments of the River Severn, UK, in mg/kg: Pb – 33.2÷291.1, Zn – 107.6÷231.9, Cu – 14.4÷24.5, Co – 19.9÷31.8, and Cd – 1.6÷2.7 [26]. Mining for Pb, Cu and Zn was operating up to the early 20th century in the upper reaches of the river. Floodplain areas in the Lower Severn receive sediment transported from upstream.

When physico-chemical conditions change, heavy metals available in water-body sediments and/or floodplain sediments can re-mobilize and thus present an environmental hazard, becoming secondary sources of diffuse pollution.

High concentrations of Hg (up to 2580 μg/kg; mean: 70 μg/kg) were measured in surface water sediments at elevated total Hg concentrations determined in waters (up to 223 ng/L; mean: 52 ng/L) impacted by historical use of mercury (Hg) for the extraction of gold in an abandoned mine in Randfonetin (South Africa) [27]. The Krugersdorp Game Reserve watersheds which are downstream from the mine drainage are impacted - average Hg content of 223 ng/L was measured in a borehole, thus suggesting a groundwater pollution.

2.3. Soil Pollution

Uncontrolled pilling of waste rock, scattering of milled mineral material, improper mine-site remediation measures, flooding by polluted water, irrigation with polluted water, wind transport of fine material, bearing heavy metals, and acidic precipitations (which usually enable heavy metals mobilisation) can be pointed out as main factors which facilitate soil pollution as a result of mining activities. Some examples of soil pollution with heavy metals, due to mining activities, are described below.

It has been found that topsoil from old heaps of mining waste rock left by former Zn–Pb mining in western Małopolska (Poland) and located in agricultural land contain high total heavy metal concentrations, in mg/kg: 5 ÷ 522 Cd, 94 ÷ 23006 Pb, and 394 ÷ 70435 Zn [28]. The amount of total and EDTA-extractable (i.e., bio-available) heavy metals in soil generally decreased when increasing the distance from the heaps, but still remained high in agricultural soil sampled 10 m from the foot of the heaps thus showing that remnants of the historical Zn–Pb ore mining are "hot spots" of soil contamination and may pose an environmental problem.

Studies on the soil contamination in the vicinity of the Dabaoshan Mine, Guangdong Province, China showed that the main pollution sources, affecting soils, are mine tailings and acid mine drainage. Average concentration of Cu, Zn, Cd, and Pb in the soil were 1486, 2516, 6.42, and 429 mg/kg, respectively. These metals were continuously dispersed downstream from the tailings and waste waters, and therefore their concentrations in the paddy soils were 567, 1140, 2.48, and 191 mg/kg, respectively. The amounts of metals that were bound to Fe-Mn oxides and organic matter were relatively higher compared to those bound to carbonates. At certain acidification, these metals could be transformed from an inert state to an active state, thus posing environmental risk [29].

Jishui River in Jiangxi Province of China, has been polluted by non-ferrous heavy metal mining activities. Long-term irrigation using Jishui River water has caused severe heavy metal pollution of soil. It has been found that Cd and Cu were two primary pollutants in the soils with concentrations of 0.52–2.55 mg/kg and 27.87–426.15 mg/kg, respectively. The mean concentrations of As, Mn, Pb and Zn in the soils were 33.99 mg/kg, 468.70 mg/kg, 125.32 mg/kg and 171.48 mg/kg, respectively [30].

It has been found that total heavy metal contents in soils in an abandoned Pb/Zn-mining area in Spain varied from 2 to 95 mg Cd/kg, 0.2 to 24.8 mg Hg/kg, 46 to 6100 mg Pb/kg, and 340 to 52000 mg Zn/kg. Total and EDTA-extractable metal concentrations were higher in mine impacted zone (MZ) than in tailings pond impacted zone (TP) [31].

Carbonate-rich mine tailings are exposed to severe erosion and dispersion in Mediterranean climates. In Jalta mining area (Tunisia), the dispersion of mine tailings' particles (that bear Pb, Zn and Cd) results in the contamination of the surrounding agricultural soils. These soils have high concentrations of Pb (17229 mg/kg), Zn (4813 mg/kg) and Cd (42 mg/kg). Aeolian and hydrological erosion were the main ways for mobilization of heavy metal rich particles from mining waste [32].

High concentrations of total and bio-available As have been determined in soils in Salamanca, Spain where mining activities were important during the 1940–1987 period [33]. The As contents in ores ranged between 0.1% and 0.9%. About 10 Mt tailings were left, distributed along the area. The soils in the studied area showed a total As concentration from 5.5 mg/kg to 150 mg/kg, water-soluble As concentration from 0.004 mg/kg to 0.107 mg/kg, and bio-available As concentration from 0.045 mg/kg to 0.760 mg/kg. The vertical distribution of As through soil profiles pointed a deposition of As on the top-soils due to the wind or water transport of mine tailings. Leaching tests pointed at As potential mobility threatening water by continuous leaching.

Increased concentrations of Hg in soils (2.4 ÷4164 mg/kg) have been determined in San Joaquin region, Mexico as a result of cinnabar mining [34]. Soil pollution due to mining activities in several countries has been summarized – Table 6 [35].

Table 6. Heavy metal concentrations in soils due to mining activities - mean values, mg/kg

Country (No. of examined mining areas)	As	Cd	Cr	Cu	Ni	Pb	Zn	Hg
China (72)	195.5	11.0	84.28	211.9	106.6	641.3	1163	3.82
Iran (3)	146.2	1.49		88.40		1002	363.4	3.13
Spain (16)	191.9	6.59	63.20	120.8	28.35	881.8	465.8	52.9
S. Korea (70)	70.08	1.99		79.09	22.00	111.1	183.2	1.12
Vietnam (3)	3144	135	1501	271.4	2254	30635	41094	
India (5)	18.62	3.82	1509	63.49	1069	304.7	338.8	

Knowledge on total concentrations of heavy metals in soils may not be enough to understand the mobility and bioavailability of the elements. Usually sequential extraction procedures are applied to differentiate amounts of heavy metals bound to different phases of the corresponding substrate and thus having different mobility and bioavailability and posing different environmental risk.

A seven-step sequential extraction procedure was used to investigate the mobility and retention behaviour of Al, Fe, Mn, Cu, Zn, Pb, Cr, Co, Ni, Mo, Cd, Bi, Sn, W, Ag, As and U in specific physical–chemical and mineral phases in mine tailings and soils in the surroundings of the abandoned Ervedosa mine [36]. Beyond the tailings, the highest recorded concentrations for most elements are in soils situated in mineralized areas or area impacted by tailings. It has been found that some metal cations (Mn, Cd, Cu, Zn, Co, Cr, Ni) showed important enrichment in the most mobilizable and bio-available (water-soluble and exchangeable) fractions due likely to the acidic conditions in the area. In contrast, oxy-anions such as Mo and As showed lower mobility because of adsorption to Fe oxy-hydroxides. The residual fraction comprised largest proportions of Sn and Al and to a lesser extent Zn, Pb, Ni, Cr, Bi, W, and Ag, which were also present at low concentrations in the bioavailable fractions. The elements in secondary mineral phases (mainly Fe, Mn, Cu, Zn, Cd, Pb, W, Bi, Mo, Cr, Ni, Co, As and U) as well as in organic matter and sulfides are temporarily withheld, suggesting that they may be released to the environment by changes in physico-chemical conditions.

It is known that pH as an important factor controlling the distribution of the studied elements. Type and content of soil organic matter could play a significant role in determining the concentration, profile distribution, and mobilization of heavy metals in soil.

2.4. Pollution Impact on Flora, Fauna, Humans

Air, water and soil polluted with heavy metals, arising from mining activity, cause further pollution and/or anomalies in the flora and fauna of the different environmental compartments, and (often) as a final result – in humans. Some examples are presented below.

Negative impacts have been found in the diatom communities growing in stream waters polluted by heavy metals in the Lousal (Portugal) abandoned mine area [20].

The uptake of heavy metals, As, and Sb by aquatic plants (fluvial horsetail, platyphyllous cattail, etc.) growing in ponds impacted by metal mining industry in the Kemerovo region, Russia, was studied. Cu, Pb, Cd, Zn, As, and Sb are the major pollutants in these plant habitats. The data obtained demonstrate high pollutant concentrations in sediments, water, and plants, especially in their roots. Submerged hornwort and pond scum appeared to be hyper-accumulators. Analysis of fresh platyphyllous cattail acid extracts has demonstrated that As and Sb present in plants are serious threat to human and animal health, as they are capable of entering food chains in large amounts [37].

Accumulation of Se in lakes downstream of a uranium mine operation in northern Saskatchewan, Canada has been studied. Selenium concentrations in sediment and biota were elevated even though water concentrations were low. The determined pattern of selenium accumulation was: periphyton < plankton and filterer invertebrates < detritivore and predator invertebrates < small bodied (forage) fish and predatory fish. Bio-magnification of Se resulted in an approximately 1.5-6 fold increase in the Se content between plankton, invertebrates and forage fish. No bio-magnification was observed between forage fish and predatory fish. The study concluded that the Se released into these aquatic systems has the potential to bio-accumulate and reach levels that could impair fish reproduction [38].

Blood and liver samples representing fish of three species (largescale stoneroller, Campostoma oligolepis; longear sunfish, Lepomis megalotis; and northern hog sucker, Hypentelium nigricans) were collected from 16 sites impacted by lead–zinc mining and ore beneficiation in southeastern Missouri. Blood and liver Pb concentrations were highly correlated in all three species and were significantly greater at sites < 10 km downstream of active lead–zinc mines and mills and in a historical lead–zinc mining area than at reference sites. Blood and liver Cd concentrations were highest in all three species at sites near mines. Concentrations of Co and Ni in blood and liver were significantly higher at a site near to an active mine than at a reference site. The findings indicate that Pb and other metals, released to streams from active lead–zinc mines are available for uptake by aquatic organisms [39].

High As concentrations were found in brown rice collected from the Lianhuashan mining area - the total As was in the range of 0.03 ÷1.04 mg/kg, which is significantly greater than those measured in grains of rice grown in uncontaminated soils [40]. It was found that the paddy fields, distributed in areas of mining activities in northern Hunan Province, China were heavily polluted by Cu, Zn and Cd, due to long-term irrigation with the nearby stream water contaminated by mining wastewater. Soil enzyme activities and microbial biomass were significantly inhibited by the heavy metal pollution. It has been found that the rice grain produced at the fields had a high risk of Cd contamination [41].

High Cd and Pb concentrations were found in surface water and in paddy soil around the Chatian mercury mining deposit in western Hunan Province, China. As a consequence, both metals pollution occurs in brown rice [42].

A field survey, conducted to study the metal and As contamination in soils and vegetables in non-ferrous mining and smelting area Baiyin, China showed that soils were contaminated with Cd, Cu and As. It was found that the vegetables, grown in the area, possess a health hazard for human consumption. The leafy vegetables were more heavily contaminated than non-leafy vegetables. Chinese cabbage was the most severely contaminated, with concentrations of Cd exceeding the maximum permissible levels (0.05

mg/kg) by 4.5 times. Bio-accumulation factor shows that Cd can entry to food chain plants [43].

Heavy metal concentrations in wheat, grown on contaminated soils in Jalta mining area (Tunisia), are high (Pb contents of 1824 and 54 mg/kg and Zn contents of 1546 and 636 mg/kg in the roots and aerial parts, respectively). These concentrations exceed the thresholds for plant toxicity [32].

Significantly elevated lead levels in blood (PbB) of cattle and sheep raised in an ancient mining area were found, when compared to the controls. In 91.4% of cattle and 13.5% of sheep sampled in the mining area, PbB levels corresponded to subclinical exposure. Two young cattle had PbB levels indicative of clinical poisoning. Observed PbB levels suggest that a potential risk to human consumers of beef from the areas may exist [44].

Blood, liver and muscle Pb levels in sheep from the ancient mining area of Sierra Madrona and Alcudia Valley (Spain) were studied. Blood Pb concentration in the mining area (mean: 6.7 μg/dl in ewes and 10.9 μg/dl in rams) was above background levels (>6 μg/dl) in 73.3% of animals. Liver Pb concentration in 68% of sheep from the mining area (mean: 6.16 μg/g dry weight, d.w.) exceeded the minimum level associated with toxic exposure (5 μg/g d.w.) and 87.5% of liver samples were above European Union Maximum Residue Levels (MRL) established for offal destined for human consumption (0.5 μg/g w.w.~1.4 μg/g d.w.). None of the muscle samples in ewes exceeded the EU MRL (0.1 μg/g w.w.~0.34 μg/g d.w.) established for meat. The results suggest a potential health effect for sheep exposed to Pb pollution in this area and implications for food safety [45].

Partial remediation actions at a former gold mine in Southern France led to a mosaic of contaminated and rehabilitated zones. Arsenic levels were measured in the soil and in small mammal species (the wood mouse (*Apodemus sylvaticus*), the Algerian mouse (*Mus spretus*). Arsenic concentrations in the soil were highly spatially heterogeneous (ranging from 29 to 18900 μg/g). Despite the decrease in arsenic concentrations in the remediated soils, both wood mice and Algerian mice experienced higher oral exposure to As in remediated zones than in the control area. A weak but significant correlation between As concentration and body condition was observed, and weak relationships between the liver/kidney/lung mass and As levels were detected, suggesting histological alterations [46].

Health hazard posed by As in contaminated soil and water in the vicinity of closed or abandoned metal mines was studied by comparing the urinary level of As in 2 groups: first – individuals living near closed metal mines (the exposed group) and second - non-exposed group. It has been found that mean concentrations for urinary As of people in areas with high As level in soil and water were significantly higher than that for the control areas [47].

Findings on the environmental arsenic (As) contamination and the consequent health effects in a community located near historic gold mining activities in the Mangalur green stone belt of Karnataka, India have been summarized. Arsenic contents in water, soil, hair and nail were measured. Of 59 tube-well water samples, 79% had As above 10 μg/L (maximum 303 μg/L). Of 12 topsoil samples, six were found to contain As greater than 2000 mg/kg. All hair and nail samples collected from 171 residents contained elevated As. Arsenical skin lesions were observed among 58.6% of a total 181 screened individuals. Based on the time-course of As-like symptoms reported by the community as well as the presence of overt arsenicosis, it was hypothesized that the primary route of exposure in the study area was via contaminated groundwater; however, the identified high As content in residential soil could also be a significant source of As exposure via ingestion [48].

An assessment has been made of the correlations of the metallic concentrations in household dust and in children hair of children grown in a mining district of Oruro, Bolivia. Significant correlations between non-essential metallic elements (As, Cd, Pb, Sb and Sn) in dust and hair have been found. Children who played with dirt had higher dust-hair correlations for Pb, Sb, and Cu and children who put hands or toys in their mouths had higher dust-hair correlations of Cd [11].

2.5. Prevention, Remediation, Treatment

Different measures may be taken to prevent or mitigate the negative impact of mining waste rock and tailings piles or dumps on the environment. In order to prevent weathering, wind erosion and wind transport of fine particles, the waste material has to be compacted, stabilized and/or covered with non-acid generating material and topsoil and re-vegetated. Sometimes flooding is also applied, but care has to be taken to avoid risk of partial flooding and mine openings have to be properly sealed. Suitable pollutant-tolerant species are used for re-vegetation and accumulation and extraction of heavy metals.

Different methods are applied to minimize formation of ARD, to neutralize it or to intercept its transportation. Waste material is compacted or covered in order to decrease the air and water permittivity, mixed with alkaline material. The already generated ARD is treated on the site or diverted to treatment/storage areas using ditches/channels. The number of tested technologies for recovering some metals from the ARD is increasing.

Below are some examples of the above described ways for preventing or mitigating the negative impact of mining on the environment. The attention is drawn to some unexpected results, as well as to positive and negative features of some of the widely applied methods.

The agglomeration process, realized with the use of a conventional binder such as Portland cement, immobilizes the reactive components of the mine tailings and results in the formation of strong pellets which remain impervious to weathering and leaching. Fly ash was evaluated as a binder, either alone or in combination with cement, in the mitigation of AMD caused by the reactive tailings. Results showed that the use of fly ash significantly reduced the requirement of cement and improved crushing strength of the agglomerates. The leaching (simulated toxicity characteristic leaching procedure-TCLP) tests performed on agglomerates prepared using fly ash/cement led to the much lower level of extractable ions concentration than from the agglomerates prepared using either fly ash or cement alone [49].

A restoration of mining tailing were carried out in the Mining District of Cartagena-La Unión in 1982, by means of the sealing of the mining wastes with a layer of 0.5 m of soil, to allow the colonization by the vegetation. A study was carried out to evaluate the metal mobilization on the restored site [50]. Results showed that after 30 years the sealing did not prevent the ascent of heavy metals to the soil surface. Heavy metals concentrations reached levels above the levels established by legislation. The lowest values in the soil profile were found in the soil surface - the total heavy metal concentrations were 12 mg/kg, 4616 mg/kg, 67 mg/kg, and 3635 mg/kg, for Cd, Pb, As and Zn correspondingly. Some vegetation showed excessive concentrations of Cd or Zn in tissues. This shows that even after 30 years some heavy metals, especially Cd and Zn still remain highly mobile. Consequently, the sealing of the mining tailing with a soil cover is not enough and before the sealing the installation of physical barriers on mining waste should be employed to break the bottom-up flux of metallic salts. P. fugax and E. hirsutum have been pointed as provisional hyper-accumulators of Cu and therefore may be suitable candidate species for phytoremediation to clean-up soils from Cu contamination due to mining activities [51].

Organic amendments are used in mining areas to encourage establishment of vegetation that may minimize heavy metal contamination through runoff and erosion. However, increased risk due to heavy metal leaching in the presence of organic amendments should be considered. A study found that composted yard waste, composted cattle manure, and cattle manure aged for one month increased heavy metal leaching from mine tailings when compared to an non-amended control. Aged and composted cattle manure significantly increased Zn concentration in the leachate. All organic amendments increased Cd leachate concentrations. Aged cattle manure greatly increased Pb concentrations in the leachate [52].

Selenium can be removed from aqueous solution, including MIW, by chemical reduction. Zero-valent iron (ZVI) can be used as reductant, acting also as a catalyst [53]. The technology recommended by USEPA (Best Demonstrated Available Technology) is ferrihydrite precipitation with concurrent adsorption of selenium onto the ferrihydrite surface [54].

Wetlands are demonstrated as a suitable biological passive treatment for selenium [55].

An enhanced coagulation process with ferric ions and coarse calcite (38–74 mm) followed by conventional filtration was proposed for removing arsenic from high-arsenic water in a mine drainage system [56].

Studies have shown that constructed wetlands have considerable potential to remove As from polluted mine waters. Major As removal mechanisms in the wetlands include precipitation, co-precipitation and sorption. Microorganisms such as sulfate-reducing bacteria, iron-oxidizing bacteria, arsenite-oxidizing bacteria and arsenate-reducing bacteria can mediate the removal mostly by oxidation/reduction reactions, followed by precipitation, co-precipitation or sorption depending on the environmental conditions [57].

The generation of ARD is practically unavoidable and it is difficult to inhibit the formation of ARD at its source. Methods for treating ARD are divided into active and passive processes [58]. Active treatment involves addition of alkaline chemicals (such as limestone, lime, hydrated lime, caustic soda, soda ash) while passive treatment involves developing natural chemical and biological systems that are self-operating, like constructed wetlands, open limestone-based channels, anoxic limestone drains.

A comprehensive technology is proposed for recovery of Fe, Cu, Zn and Mn from AMD and prevention of metal pollution by mining activities. Initially ferrous ions were oxidized by H_2O_2, and then a four-step fractional precipitation was applied by selective addition of $Ca(OH)_2$ and Na_2S solutions. The Fe, Cu, Zn and Mn contents of obtained sludges were 45.91%, 11.58%, 31.06% and 7.95% correspondingly, and the recovery efficiencies of Fe, Cu, Zn and Mn from AMD reached 99.51%, 86.09%, 87.87% and 79.71%, respectively. The metals concentrations in the treated effluent were below the required limits. Fe oxide red was obtained by roasting the Fe sludge, resulting in a Fe_2O_3 content of 85.18%. Cu and Zn rough concentrates were produced by a flotation process; the Cu and Zn contents of these concentrates were 35.72% and 55.13% correspondingly, and the recovery efficiencies of the Cu and Zn were 72.66% and 76.18%, respectively. The Mn sludge obtained can be used in cement mixes to replace 45% of ordinary Portland cement [6].

Chitorem SC-20 (raw crushed crab shells containing 40% (w/w) $CaCO_3$, 30% protein, 20% chitin, 7% moisture, and 3% ash) was found effective at neutralizing the strong acidity of MIW, even at loads as low as 1 g/L the equilibrium pH was neutral. At a load of 2 g/L, SC-

20 showed a final pH of 7.94 with almost complete (> 99.8%) removal of Fe (120 mg/L), Pb (1.1 mg/L) and Zn (79 mg/L), along with partial removal of Cd (96% of 1.3 mg/L), Co (54% of 0.78 mg/L), Cu (42% of 72 mg/L), and Mn (64% of 52 mg/L) from MIW [59].

Basic oxygen slag was found to have a significant potential as a remediating agent in treating AMD and may be used as a replacement reagent for lime [60].

An open limestone channel (OLC), constructed along a creek surrounded by upstream tailings deposits, in an abandoned sulfide-mining site, which generates acidic and heavy metals-rich drainage water during the occasional precipitation, proved to be efficient means to counteract the ARD. Results showed that the values of pH and Ca were increased downstream of the channel, whereas the EC, TS, and concentrations of K, Mg, SO_4^{2-}, Al, Mn, Fe, Ni, Cu, Zn, As, Cd, and Sb decreased towards the mouth of the creek. The OLC reduced the input of heavy metals into the Mar Menor lagoon by one order of magnitude [61].

The main disadvantage with limestone beds is the regular maintenance required to ensure maximum life and effectiveness. Usually 1-2 years are enough for the silted material in the channel to prevent the contact of AMD with the bed rock. For this reason removal of the silt layer should be performed on a yearly basis. A sediment trap could be set up across the channel to collect the residues and prevent them from being transported downstream. Another option is placing a permeable geotextile cover over the limestone, which could be shaken and raised up to eliminate the sediment.

Using wetlands represents a low cost way to cope with ARD. The wetlands absorb and bind heavy metals and make them slowly concentrated in the sedimentary deposits to become part of the geological cycle [62]. Chemical and physicochemical processes involved in the removal of heavy metals in the wetlands are adsorption, oxidation, hydrolysis, precipitation, co-precipitation with Fe. More than 50% of the heavy metals can be easily adsorbed onto particulate matter in the wetland and thus be removed. Lead and Cu in general tend to be adsorbed most strongly and Zn, Ni and Cd are usually held weakly which implies that they are likely to be more liable and bio-available. The adsorption of metals varies with the fluctuation of pH in the outflow water. Oxidation and hydrolysis of metals also occurs. Thus Fe, Al and Mn are removed, since they can form insoluble compounds through hydrolysis and/or oxidation that occur in wetlands. Iron removal depends on pH, oxidation–reduction potential and the presence of various anions. Precipitation and co-precipitation of heavy metals depends on the solubility product K_{sp} of the metal involved, pH of the wetland water and concentration of metal ions and relevant anions. Heavy metals may form carbonates when the bicarbonate concentration in water is high. Although carbonates are less stable than sulfides, they can have a significant role in initial trapping of metals. Wetlands with appropriate substrate promote the growth of sulfate reducing bacteria in anaerobic conditions. In ARD which is rich in sulfates, these bacteria generate hydrogen sulfide (H_2S). Heavy metals react with H_2S and form highly insoluble metal sulfides. Biological removal is perhaps the most important pathway for decreasing the heavy metal concentrations in the wetlands. Probably the most widely recognized biological processes for metal removal in wetlands is plant uptake.

Positive features of active treatment systems are that they can cope with high flow rates, are flexible and usually require smaller area. Disadvantages include high operating costs associated with alkaline chemicals, necessity of constant monitoring and maintenance, of skilled manpower and production of voluminous sludge that should be handled.

The advantage of passive treatment systems is that they require little maintenance. Both natural and artificially constructed wetlands achieve an efficient treatment at low investment costs, low operating costs and no external energy input. The disadvantage of anaerobic wetlands is the requirement for large area of land for effective treatment. Although aerobic wetlands are able to remove various metals from effluents, their main disadvantage lies in the fact that they cannot handle typical AMD effluents and they require vast surface areas for their operation. A further drawback is their inability to reach pH levels greater than 8.

3. MINERAL PROCESSING AS SOURCE OF HEAVY METALS

3.1. Generated Waste

Minerals beneficiation generates different types of waste in liquid and solid phase. Solid waste represents rocky materials and non-extracted minerals. Mechanical and physical beneficiation technologies give rise to suspensions from grinding and gravitational enrichment. This waste is deposited in tailings ponds. Flotation (beneficiation, based mainly on physico-chemical processes) has as waste mainly liquid phase consisting of water, dissolved and suspended compounds and non-consumed during the flotation reagents. It is typically passed to tailings ponds. Characteristics of liquid and solid beneficiation wastes vary considerably, depending on the processed ore and applied technology. Usually, the wastewater is with high content of suspended solids. Table 7 summarizes generally the basic inorganic pollutants in wastewater from flotation of ores of nonferrous metals. Part of this water is treated and recycled to the factory, other part enters the tailings pond. Often the wastewater is polluted with organic reagents, which by formation of stabile complexes, keep the heavy metals mobile.

Table 7. Major inorganic pollutants in wastewater from flotation of ores of nonferrous metals

Type of flotation	Major inorganic pollutants
Flotation of copper ores	As, Sb, Cd, Cu, Co, Mn, Fe, Mo, Pb, Hg, Ni, Se, Zn, Ag, CN^-
Flotation of waste from copper metallurgy	As, Cd, Cu, Fe, Pb, Hg, Ni, Se, Zn, CN^-
Flotation of lead ores	Pb, Zn, Hg, Cd, Cu, Cr, Mn, Fe, CN^-
Flotation of zinc ores	Pb, Zn, Hg, Cd, Cu, Cr, Mn, Fe, CN^-

Some natural physico-chemical processes take place in the tailings pond and lead to the formation of slightly soluble compounds, including precipitates of heavy metals. Supernatant liquid can be returned back to the mills or flotation circle or can be discharged in natural water bodies, provided that it meets the standards for water quality. Concentration of heavy metals in tailings pond water and sediment can vary considerably, depending on the treated ores, local climatological and physico-chemical conditions, technologies applied to treat the wastewater before its disposal to the tailings pond or directly in the pond, manner of usage of the pond (only for flotation waste or water is mixed with MIW, and especially - ARD). Below are given some examples for the heavy metals concentrations determined in tailings ponds.

Total concentrations of heavy metals in three tailings dams (sediments) from activities processing Ag, Pb, Cu and Zn ores in Queretaro, Mexico were determined to be in the ranges, in mg/kg: As (183–14660), Cd (45–308), Pb (327–1754), Cu (149–459) and Zn (448–505). Generally, higher heavy metals concentrations were found in the tailings dam, where the tailings are acid generating (mainly sulfidic) [63].

Mean concentrations determined in two tailings ponds located in the Sierra Almagrera mining district (Pb/Zn mining), Spain - the El Arteal deposit (SA) and the Jaraviras deposit (JAR) are as follows: in SA pond, in mg/L: Ag (26.6), As (278.4), Ba (5800), Cu (59), Pb (2879.3), Sb (169.8) and Zn (2179.2). In the JAR pond, in mg/L: Ag (15.8), As (302.2), Cr (28.5), Pb (1830), Sb (69.1) and Zn (1557.5) [64]. Studies showed that tailings ponds contain high concentrations of water-soluble Zn (1463 mg Zn/kg of tailing pond material) and bio-available Pb and Zn (2449 mg Zn/kg and 5298 mg Pb/kg) [65].

Concentrations of typical heavy metals in the clear water of a tailings pond usually decrease considerably due to metals immobilization in the sediments - Table 8 [66].

Table 8. Concentration of heavy metals in beneficiation waste water (BWW) and in the clarified water (CW) of the tailings pond of a copper-ores beneficiation factory

Metal, mg/L, Sample	Cu	Pb	Zn	Cd	As
BWW	24.83	1.088	1.717	0.072	0.080
CW	0.017	0.026	0.013	0.0017	0.074

While the tailings ponds usually reduce the concentrations of typical heavy metals, this is not always the case with As, Se and Sb - Table 9 [67].

Table 9. As and Sb concentrations, in mg/L, in water samples of tailings ponds of Giant Mine, Yellowknife, Northwest Territories, Canada

Sampled pond, Year	September 2005		July 2006	
	As	Sb	As	Sb
North West pond	13.9	0.813	26.8	1.62
North pond	4.73	0.655	2.64	0.926

3.2. Water, Air and Soil Pollution

Events leading to certain acidification or dilution of water in the tailings pond (for example, entrance of AMD from a nearby mine) can lead to heavy metals mobilization from the pond sediments. Studies, carried out with tailings sediment from Dexing Cu mine indicated that the maximum leaching ratios of Zn, Cu, Fe and Mn at pH 2.0 were 5.4%, 5.8%, 11.1% and 34.1%, respectively. At prevailing quantity of AMD entering from the nearby opencast mine, the tailings water may pose an environmental problem [68].

Rupture or overflow of tailings dam can be with serious environmental consequences.

Fish kill resulted from the contamination of the coastal water with cyanide and Hg as a results of overflow and spill of the mine tailings pond of a gold mining company in Rapu-rapu Island in the Philippines into Albay Gulf during a storm in October 2005. Next year studies on heavy metals concentrations in water, sediments, fish and shellfish samples collected in coastal area showed that all water samples yielded Hg, As, Cd and Pb concentrations less than the local quality standards for the water use for fishery and aquatic life. Some water samples showed Cu concentrations exceeding the standards. Mercury and Cd were not detected in the sediment samples. Arsenic, Pb and Cu were detected above the limit of quantification in sediments from most of the sites. The different species of fish, small shrimps, shells and cuttlefish collected in Sorsogon and Rapu-rapu Island showed varying concentrations of Hg which ranged from less than the method detection limit (MDL) of 0.02 mg/kg wet weight to 0.25 mg/kg wet weight. Lead concentrations in the fish samples ranged from less than the MDL of 0.04 mg/kg wet weight to 0.21 mg/kg wet weight. The Hg and Pb concentrations in the fish investigated were below the internationally accepted regulation limits of 0.5 mg/kg methyl mercury and 0.3 mg/kg Pb. Cadmium was not detected in most biota. However, the Cu and As contamination in water and sediments in Rapu-rapu Island show that the present and past mining activities in the island have caused an elevation of the metal contamination in some areas in Albay Gulf [69].

Infiltration of polluted water from evaporation ponds of mineral processing facilities can pose environmental problems. At a heavy mineral separation plant situated in the West Coast of South Africa, process waters, with a low pH and high SO_4^{2-} and trace element contents, are neutralized and disposed of in evaporation ponds. Because of the permeability of the ponds, the waste water infiltrates to soils and groundwater. Liquid–solid partitioning coefficients suggested that Se, As, Cu and Zn were the most mobile elements in the soils. Results showed that U species were likely to remain in solution as carbonate complexes in the aquifer. Modeling showed that Cd is also likely to remain in solution. The study identified the evaporation and seepage (used to drain evaporation pond effluent) ponds and mineral separation plant as the main sources of SO_4^{2-}, Cu, Zn, Ni, Th and U distribution in the aquifer. The study demonstrated that the process waters are likely to contaminate distant water systems and enter food web - Table 10 [70].

The following average concentrations of heavy metals, in μg/L, were determined in a stream near the amalgamation tailings pond in Talawaan – Tatelu Gold Mining, North Sulawesi, Indonesia: As - 10.00, Cd < 0.01, Cr - 5.00, Cu < 5.00, Ni < 1.00, Pb - 23.00, Zn < 5, and Hg < 0.05 [71].

Table 10. Average concentrations of some heavy metals, Th and U in the aquifer located around the evaporation pond of a heavy mineral sand mine in the West Coast of South Africa

Pollutant	As	Cu	Ni	Se	Zn	Th	U
Concentration range, μg/L	21.0÷ 87.0	14.5÷ 83.4	9.0÷ 330.0	34.6÷ 177.0	40.0÷ 350.0	0.01÷ 0.05	0.6÷ 34.0

Impact of tailings pond can cause pollution of nearby soil. Maximum concentrations of As in tailings in Zloty Stok, Poland was 8500–17 200 mg/kg. Alluvial soils in the above-mentioned region contained up to 6800 mg/kg of As, which is related to long-lasting ore processing activities [72]. Contamination of soils with dust fallout from the tailings dam at the Rosh Pinah area, Namibia has been reported [73]. The amount of suspended dust particles in the air (TSP) monitored on the surface of the tailings dam ranged between 8 and 84 mg/m^3.

The air contained (median, in μg/m^3): Zn (792), Pb (412), Mn (163), Cu (42), As (5) and Cd (2). Sampling of topsoil together with reference sampling of subsurface soil revealed that the shape of the dust contamination halo is elongated northwestward the tailings pond according to the prevailing direction of winds during summer. Dispersion modeling of TSP in the air revealed that while the whole surface of the tailings dam was dry in the past due to infiltration and evaporation of water, the average yearly concentrations of TPS in the Rosh Pinah town amounted to 100–300 μg/m^3. Bio-accessible lead exceeded the tolerable daily intake for children. The topsoil metal and As contents are in good agreement with their contents in the grass (*Stipagrostis geminofolia L*). The contents of Pb in 52% of collected samples of grass exceed permissible values for dry fodder.

The environmental dust metal and metalloid hazards (As, Cd, Pb and Ni) associated with bulk mineral transport, loading and unloading port operations has been studied in children's playgrounds in the inner city of Townsville, Queensland [74]. The mean increase in Pb on post-play hand wipes was 965 μg/m^2/day at mean pre-play loadings of 95 μg/m^2/day. Maximum loading values after a 10-minute play period were 3012 μg/m^2, while the Australian standard is 400 μg/m^2. Maximum daily Ni post-play hand loadings of 404 μg/m^2 were determined, while the German Federal Immission Control Act 2002 annual benchmark is 15 μg/m^2/day. Repeat sampling over the 5-day study period showed that hands and surfaces were re-contaminated daily from the deposition of metal-rich atmospheric dusts. Lead isotopic composition analysis of dust wipes showed that Pb in surface dust was similar to Pb in Mount Isa type ores, which are exported through the Port of Townsville.

3.3. Prevention, Treatment, Mitigation

Heavy metals from flotation wastewater, before entering the tailings pond, can be removed in different ways:

- Chemical precipitation - by forming practically insoluble oxides and/or hydroxides or carbonates of heavy metals by means of alkalization of the wastewater to a pH of approximately 9.5-10.0. This should be done with environmentally safe and relatively inexpensive reagents (lime, fly ash, limestone). At treatment with lime milk As forms insoluble calcium arsenate. The precipitation with above-mentioned regents is still the most widely used method for the removal of heavy metals from mineral processing wastewater. Addition of coagulants is commonly used to improve the aggregation of colloidal dispersed precipitates and to increase the efficiency of subsequent separation of the solid from the liquid phase. The last operation is the neutralization of the treated water to a pH of 7-8.5. Less commonly is used precipitation of the heavy metals as sulfides, precipitation of As in the form of scorodite - $FeAsO_4 \cdot 2H_2O$. Although chemical precipitation is a widely used treatment method, it has a number of disadvantages such as the generation of bulky sludge, which must be further treated to meet the criteria for disposal, and loss of precious metals.
- Adsorption of metals from wastewater on different materials: iron hydroxide, aluminium oxide, natural and modified zeolites, organic polymers, or on solid waste

from a variety of processes (fly ash, sludge from aluminium production and organic materials).

- Extraction with solvents, followed by electro-deposition may be used. Due to high costs, the method is most suitable for small volumes of water bearing large amounts of dissolved metal.
- Ion exchange by use of ion exchange resins, cementation, membrane processes are also applied for metals extraction from the wastewater. At the application of those methods, the treated water is sufficiently purified, in order to return back into the flotation process.
- Destruction of stable complex-forming reagents is achieved by anodic oxidation, and often in the electrochemical reactor simultaneously with the anodic oxidation useful metals are separated on the cathode.

The following measures can be implemented to reduce pollution, including release of heavy metals, from tailings ponds:

- Submission of additional flocculants and/or alkalizing agents (lime) to accelerate the flocculation and sedimentation;
- Accumulation of fine waterproof material at the bottom and walls of the tailings pond to prevent or reduce the infiltration of contaminated water to the underground or surface waters;
- Installation of drainage system to pond to reduce losses by filtration;
- Installation of systems for early detection of leaks.

4. Extractive Metallurgy as Source of Heavy Metals

Extractive metallurgy is the industry and science of obtaining metals from ores. It uses chemical methods, often at high temperature and pressure, for treating ores to recover metals in a pure form. Generally, extractive metallurgy is divided into three areas: pyro-, hydro-, and electrometallurgy [75].

Pyrometallurgy comprises dry methods usually conducted at high temperature (oxidation, reduction, chlorination, melting). It often involves melting of the minerals and the separation of the valuable components in the liquid state. Typically ores of iron, copper, and lead are treated by this technology. Hydrometallurgy comprises wet methods, usually carried out at room temperature or near the boiling point of water. It includes ores' leaching, precipitation of a metal or its compounds from aqueous solutions, methods of isolation and purification (ion exchange, solvent extraction, sorption). Typically ores of gold, uranium, copper, aluminium, nickel are treated by this technology. Hydrometallurgical methods are used also to recover zinc and expensive and minor metals (vanadium, niobium, tantalum, zirconium, hafnium, beryllium, tungsten, and the rare earths). Electrometallurgy comprises all processes based on the use of electric current for metal recovery or refining (in aqueous solution or in a fused salt). Typically aluminium, copper, and zinc are metals produced by this technology. The electrometallurgy is used also in production of Ni, Co, Pb, Sn, Sb, Mn, Cr, Cd, Au, Ag, Te, Be, Na, Li, Bi, In, Ga.

Above-mentioned areas cannot be considered isolated one from the other, because a combination of these processes is usually used in the production of a single metal. Here are some examples: Bauxite (the most important aluminium ore) is always treated by a hydrometallurgical method but the final aluminium production step is achieved by electrometallurgical technique. Uranium ores are most often treated by hydrometallurgical technique while the metal production is achieved by pyrometallurgical methods. Copper ores are treated by pyro- or hydrometallurgical methods, but electrometallurgy is used for producing refined metal. One metal can be extracted by different routes - for example, nearly 80% of the zinc produced annually is by the hydrometallurgy–electrometallurgy route while the remaining 20% is produced by pyrometallurgical technique. Practically, all areas of the extractive metallurgy can be to some extent sources of heavy metals pollution.

4.1. Pollution Caused by Pyrometallurgy

Gases, aerosols, dust, slag, and wastewater are the waste products of the pyrometallurgic processing of ores. In dusts, fumes and smoke, emitted by pyrometallurgical plants, the following harmful metals, metalloids and metal compounds may be found: As, Cd, Hg (inorganic and organic), Mn, Mo, Pb, Sb, Se, Te, U, V, Zn, Zr, chromium trioxide, ferrovanadium, iron oxide, magnesium oxide, titanium dioxide, divanadium pentoxide, zinc oxide. Some figures are presented below.

Primary nonferrous metallurgy is one of the most significant anthropogenic Hg emission sources. It was calculated that in 2010, the Hg emission from primary Zn, Pb and Cu smelting was 7490, 6050 and 1100 kg, respectively. Calculations showed that during 1949–2010 the cumulative Hg emission from China's nonferrous metal smelting was 323000 kg, of which 44.6% from Pb smelting, 32.8% from Zn smelting, and 22.6% from copper smelting [76].

Steelmaking generates dusts in amount of 15 - 20 kg per a tone of produced steel, and the concentration of heavy metals in the dust can be in the range, presented in Table 11 [77].

Table 11. Heavy metals in dust from carbon and stainless steel production, in wt %

Steel, Element	Cd	Cr	Ni	Pb	Zn
Carbon	0.90	< d.l. - 0.81	< d.l. - 0.1	0.2 - 4.05	17.05 - 33.00
Stainless	0.48	9.99 - 18.16	1.4 - 6.7	0.29 - 1.4	0.96 - 8.1

d.l. - detection limit.

Umicore (Olen, Belgium) treats particulate emissions, in a way that the resulting As concentration in the exhaust gas is around 0.47 mg/m^3, which is below the limit specified in the emission legislation (1 mg/m^3). However, the absolute amount of As emitted into the atmosphere is significant (295.6 kg/year) [78].

Data on mass release of heavy metals from some primary zinc plants in Europe are presented in Table 12 [79].

Concentrations of As of 10.59 ng/m^3 were measured in PM_{10} in 2005 in the area impacted by Cu-smelter emission plumes, city of Huelva (SE Spain). The $PM_{2.5}$ arsenic fraction made up to 85% of the total PM_{10} arsenic fraction. The elevated As concentration was synchronic with increased concentrations of other metals such as Cu, Zn, Se and Bi, and was around 3–5 times higher than at other sites in Spain [80].

Studies in the >100 km^2 area surrounding the smelter in Torreon (NE Mexico) showed that in outdoor ambient air concentrations of Cd, Pb and Zn ranged from 0.7 to 12.6, 65 to 705 and 30 to 790 ng/m^3, respectively. Atmospheric Pb outdoor deposition averaged from 30 to 1350 mg/(m^{2}d). Metals concentrations in outdoor dust ranged from 4.2 to 56.2 mg/g (median 12.6 mg/g) for Cd, 150 to 14365 mg/g (median 880 mg/g) for Pb, and 509 to 16483 mg/g (2451 mg/g) for Zn. Concentrations were greatest closest to the smelter, decreased with distance from the smelter, the highest levels corresponded to the prevailing wind direction [81].

Table 12. Mass release of metals from some primary zinc plants in Europe, in g/t product

Process	Product	Production, t	Dust	Zn	Pb	Cd	As
Roasting, purification, electrolysis (with casting)	Zn or Zn alloys	130000-145000	3-17	1-15	< 0.3	< 0.05	< 0.05 as oxide
Roasting, purification, electrolysis (all process stages from concentrate with low Fe content)	Zn + Cd	238850	3	2	0.007	0.0004	< 0.0001
Waelz process	Waelz oxide	29000 - 60000	14-73	4-21	0.3-2.0	0.13-0.42	< 0.1

The impact of airborne pollution resulting from smelting activities on the soils of the Almalyk mining area (NE Uzbekistan) was studied. Enriched contents, compared to background values, were found in the upper soil layers near the metallurgical complex: Zn ≤ 3010 mg/kg, Pb ≤ 630 mg/kg, Cd ≤ 30 mg/kg. The morphology and internal microstructure of metal-bearing particles suggested that these particles were a result of inefficient flue gas cleaning technique of the smelter. Significant correlations were found between the dissolved metal content and the microbiological soil ecosystem health parameters, suggesting that the contaminated soil has significant negative impact on the functioning of the microbiological community [82].

Soil pollution due to the Cu mining and smelting activities in the Copper belt province (Zambia) was studied in a 32-km^2 zone around the smelter. The spatial distribution of the major contaminants indicated the highest contamination NW of the smelter stack, corresponding to the prevailing wind direction. The highest pollutant concentrations in the topsoil were, in mg/kg: As (255), Cu (27410), Co (606), Pb (480) and Zn (450) [83].

The impact of a large scale Pb/Zn smelter in Hunan province, China on soil pollution was studied. It was found that the soil located 4 km from the smelter was severely contaminated, with maximum concentrations of Hg, Pb, Zn, Cd, Cu and As as high as 2.89, 1200, 3350, 41.1, 157 and 93 mg/kg, respectively. Concentrations of soil metal and As decreased with distance in the dominant wind direction [84].

Generally, pyrometallurgy generates liquid streams, contaminated with heavy metals, metalloids and organic contaminants - Table 13.

Table 13. Main pollutants in wastewater from metallurgical production of base metals

Metallurgy	Major pollutants
Iron and steel production - agglomeration plant	Suspended particles (including such that bear As, Cr, Cu, Hg, Ni, Pb, Zn), resistant organic compounds, PAH, PCB, sulfur compounds, F^-, Cl^-
Iron and steel production - pelletization plant	Cd, Co, Cr, Cu, Hg, Pb, Fe, Mn, Zn, Ni, As, N, P
Iron and steel production - blast furnace	CN^-, SO_4^{2-}, NH_4^+, S, Fe, Mn, Zn, Pb
Iron and steel production - granulation of blast furnace slag	Pb, Cr, Cu, Zn, Cd, Ni, Fe
Steel production and casting	Pb, Cr, Cu, Zn, Cd, Fe, Ni, mineral oil
Steel production - electric arc furnace	Fe, Zn, Ni, Cu, mineral oil
Copper production - smelting and casting of copper cathodes	As, Cd, Cu, Fe, Pb, Hg, Se, Zn
Copper production - refining in an anode furnace	Cl^-, Al, As, Cu, Zn
Lead production - fuming process	Cu, Pb, Fe, Ni, Zn, Hg
Lead production - granulation of the slag	Cu, Pb, Fe, Ni, Zn, Hg
Zinc production	Zn, Cd, Fe, Al, Mn, Cu, As, Pb

Typical concentrations of heavy metals in gas-cleaning effluents from Zn-Cd and Pb-Sn nonferrous metallurgy are as follows, in mg/L: As - 5÷95, Cd - 1÷95, Pb - 1÷13, Hg - 0.1 ÷9, Se - 0.1÷50, and Zn - 0.1÷2.5 [79].

Wastewater and gas-cleaning effluents are treated with different technologies (neutralization, alkalization, precipitation as hydroxides, sulfides and other low soluble compounds, adsorption, membrane processes). Most often neutralization/alkalization is applied, since not only typical heavy metals, but also As is precipitated as calcium arsenate. Especially for As, the other often used treatment procedure is to bind it in insoluble scorodite - $FeAsO_4 \cdot 2H_2O$. However, recent studies showed that the dissolution of calcium arsenates in residues disposed on an inactive smelter site has caused high levels of As pollution in the adjoining down-gradient 6-km perched aquifer, reaching up to 158 mg/L of dissolved As, and releasing a total of ca. 7.5 tons of As in a year [85]. Furthermore, the latest studies demonstrate that scorodite should not be considered an environmental stable repository for As attenuation when dumped outside because natural rainfall and the resulting runoff drive As dispersion in the environment [86].

Surface and ground water can be polluted due to heavy metals leaching from metallurgical solid waste. Pyrometallurgy produces slags in large amounts. Only a fraction is used in cement manufacturing and in road construction. The main amount of the slag is taken to the dam. Leaching of accumulated pyrometallurgical slag can be source of water pollution with heavy metals and metalloids. Slag from production of several, most technologically important metals (iron and iron alloys, copper, zinc and lead) is considered in this chapter.

Production of ferrous metals generates four types of solid waste: slag from blast furnaces, dust/sludge from the control of air pollution in blast furnaces, slag from furnaces for steel production and dust/sludge from the control of air pollution in furnaces for steel production. Blast furnace slag comprises oxides of silicon, aluminium, calcium and magnesium, together with other trace elements. The slag from the furnace for steel production consists of ferrite and calcium silicates, combined with molten oxides of iron, aluminium, manganese, calcium and magnesium. Dust/sludge from the control of air pollution in blast furnaces contains mainly iron, calcium, silicon, magnesium, aluminium, and manganese. Dust/sludge from the control of air pollution in the furnaces for steel production contains mainly iron and small amounts of calcium, silicon and other metals.

Data from tests made by the Environmental Protection Agency (EPA) of the United States show [87]: a) The main contaminants that can leach from the slag from blast furnaces are Mn, Fe, Pb, As, Al, Ag, Hg, Sb; b) The main contaminants which can leach from the slag from the furnace for steel production are Mn, F, As, Pb, Ag, Fe, Mo, Ba. EPA and industrial test data show that some pollutants from the slag from blast furnaces and furnaces for the steel production can leach in relatively high concentrations. Taking into account only those elements that are relatively mobile in groundwater (given the existing slag management methods and expected pH levels of leachate), As and Hg can be referred to as contaminants which are the main potential threat at leaching of blast furnace slag. Fluorine (F), As and Mo are the greatest potential threat at leaching slag from furnaces for steel.

EPA analyzes show that: a) The main contaminants that can leach from the dust/sludge from the control of air pollution in blast furnaces are Mn, Pb, As, Al, Fe, Zn, F, Se, Tl, Ag, Hg, Cu, Sb, Cd, Cr, Ba, Ni; b) The main contaminants that can leach from the dust/sludge from the control of air pollution in furnaces for steel production are Zn, Mn, Cd, Fe, Mo, Pb, Se, Hg. Test data show that As, F, Se, Hg, Cd and Cr can be referred to as contaminants which are the main potential threat at leaching of dust/sludge from the control of air pollution in blast furnaces. Cadmium, Mo, Se and Hg are the greatest potential threat at leaching the dust/sludge from the control of air pollution in furnaces for steel production.

Surface waters can be polluted by discharge of groundwater contaminated by leachate from the above-mentioned slag/sludge/dust or by direct runoff of dissolved or suspended contaminants at heavy rain/storms. Pollutant concentrations and pH values of leachate show that there is potential for migration of leached contaminants in surface water.

The operations of the primary production of copper include, in general, melting, converting, fire refining in an anode furnace, and electrolytic refining. The products of each of these operations are, respectively, a copper matte, a converter copper with a purity of 98%, copper anodes and refined copper. Smelters produce two separate molten streams: copper-iron sulfide - matte and slag. Slag melt is usually poured hot, still molten and cooled by air or water (pelletized) prior to disposal. Slags from some smelters are with higher Cu content compared to primary ores. They are sent back for beneficiation and then concentrate returns in the metallurgy. The waste from the slag processing is the second type of waste from primary Cu processing. Converter slag is disposed. Anode furnace generates slag which is also disposed. The mixed slag is composed mainly of iron silicate, calcium oxide, aluminium oxide, and contains small amounts of Cu, Pb, Zn and other metals. Data from tests made by the EPA [87] indicate that the potential environmental pollutants due to leaching by water percolated through the slag are Pb, Cu, As, Mo, Cd, Hg, Fe, Ba and Ni. EPA tests, considered together with the existing slag management practices and the neutral pH of leachate, showed that As, Mo, Cd, and to a lesser extent, Ba and Cr can be mobile in groundwater. Leachate from waste from the slag processing typically contains Mo, As, Cr, Pb, Ag, Ni, Hg. EPA studies show that As, Hg and Mo are the main possible contaminants of groundwater.

Lead production includes the following generalized steps: sintering, melting in blast furnaces (generating slag for disposal), purification in a slag furnace (generating a slag which is sent back in the process of sintering) and refining (generating slag for disposal).

Slag from lead production is composed mainly of oxides of iron, silicon, aluminium and calcium. Other metals may also be present in lesser amounts: Sb, As, Be, Cd, Cr, Co, Cu, Pb, Mn, Hg, Mo, Ag and Zn. EPA and industrial test data indicate that the elements which can be leached from the lead slag in dangerous concentrations are Pb, Cd, As, Zn, Fe, Co, Cu, Mn, Hg, Se, Ag, Sb. In general, concentrations of As, Pb and Hg are the highest [87]. Given the existing slag management practices and that leachate is practically neutral, only As, Cd, Se, Co and Hg are expected to be mobile in ground waters, if leachate flows into them.

The slag, removed from the furnace for zinc production, is rocky solid material composed mainly of iron, silicon and non-reacted coke. It is subjected to a series of operations "shredding/separation" and results in four material flows: a fine, rich in Zn fraction; recovered coke; processed slag; and ferrosilicon. The fine fraction and the coke are sent back in the process. EPA studies indicate that the elements which can be leached from the processed zinc slag into leachate in dangerous concentrations are Pb, Mn, Zn, Cu, Cd, Ni, As, Se, Fe [87]. EPA data indicate that the elements which can be leached from ferrosilicon in dangerous concentrations are Pb, Mn, Cu, Ni, Zn, Se and Fe. Described results of analyzes do not prove that the slag from Zn production or the ferrosilicon pose a significant risk, but rather show that there could be a risk at incorrect management. Given the existing management practices of the treated slag and ferrosilicon, as well the fact that leachate is practically neutral, only Cd, As and Se are expected to be mobilized in groundwater from treated slag and only Se to mobilize from ferrosilicon. Leaching of pollutants from processed slag from Zn production or from accumulated ferrosilicon and their transport to groundwater depends entirely on the extent to which precipitation can percolate the accumulated slag and/or ferrosilicon. Leached Cd, As and Se from the treated zinc slag and ferrosilicon can be mobile in groundwater.

The potential for the contaminants from the above-mentioned smelting activities to reach usable aquifer and to be transported to points of abstraction is determined by various specific factors, such as the availability of engineering means for the protection of groundwater, groundwater level, precipitation, etc. Surface waters can be polluted by discharge of groundwater contaminated by leachate from the slag or by direct runoff of dissolved or suspended ingredients at heavy precipitation/storms.

A few concrete examples of water and soil pollution, caused by heavy metals mobilization from metallurgical slag are described briefly below.

The weathering of slag dumps in the Sierra Almagrera area (SE Spain) near the Mediterranean shoreline had been identified as reason contaminating the soils and groundwater [88]. The following concentrations of heavy metals in groundwater were determined: 0.64 mg/L Cu, 40 mg/L Fe, 0.6 mg/L Mn, 7.6 mg/L Zn, 5.1 mg/L Pb and 19 μg/L As.

A survey, conducted in the area of the former Capital Iron & Steel Factory, Beijing, China, showed that surface soil was contaminated with Cu, Pb, Cd and Zn mainly due to pollutants mobilization from steel-smelting activities [89].

Significantly increased concentrations of heavy metals in the surface soil in the region of former zinc smelter in Kosovska Mitrovica, Kosovo are reported. It has been found that in the investigated area, the average content of Pb in soil is 20-fold higher than the European median, Cd 11-fold, Hg 5.5-fold, As 4.6-fold, Zn 4.2, and Cu 3.2-fold higher [90].

The leaching study of a slag from a primary lead smelter, operated in Santo Amaro (Brazil) between 1970 and 1990, showed that in an acidic environment ($pH < 2.8$) the solubilization of Pb was enhanced significantly [91].

Run-off draining and leaching of the roaster complex solid waste and scorodite cemented substrate at the Prohibition Mill site, New Zealand contained 30 mg/L As. The wetland water, which drains most site run-off, was heavily contaminated with As (up to 52 mg/L) [92].

A study throughout the Kola region of Russia observed a large number of lakes that are heavily contaminated by Ni and Cu. The increased heavy metal concentrations in water are related to local discharges from metallurgical and mining industry, as well as to indirect leaching of elements by acid precipitation. Fish in these lakes contain high concentrations of Ni and Cu and display frequent pathologies, mostly associated with the kidneys [93].

Elevated levels of Se (maximum 200 mg/kg) and Te (maximum 11 mg/kg) have been reported in topsoils from around a long-established Ni refinery at Clydach, the Lower Swansea Valley, UK. The study showed that the elevated Se and Te concentrations resulted from contamination produced during the early period of refinery operation (1902–1935) and that the mobile pollutants have remained in the soil for 80 years [94].

4.1.1. Pollution Impact on Flora, Fauna, Humans

Polluted soil, water and air can cause further negative impact. Below are some examples:

Elevated heavy metal concentrations were found in corn plants around Pb-Zn smelter in Hezhang County, western Guizhou, China - total Pb (0.80–1.5 mg/kg) and total Cd (0.05–0.76 mg/kg). A conclusion was drawn that the soil-to-crop transfer of heavy metals might pose a potential health risk to the local residents [95].

The soil samples from the arable lands surrounding the smelting sites in Huixian, Northwest China contained high levels of Cd, Pb, and Zn, which were 6.59–237, 3.27–218 and 0.375–2.38 times respectively above the background values in the study areas. The main causes to the dispersion of pollution were proposed to be the continuous irrigation with polluted water and wind transport of dust from smelters. The levels of Cu, Cd, Pb, and Zn in aerial parts of typical plants in the area ranged from 0.048 to 19.9, 0.122 to 8.17, 11.1 to 48.6, and 0.646 to 85.8 mg/kg, respectively. The accumulation ability changed in the order: Cd > Zn > Cu > Pb and a high level was detected in maize (Cd), ainsliaea (Cd), goose grass herb (Zn), and poplar (Zn). Lead and Cd were potentially risky to local inhabitants [96].

Considerable contamination by Hg and Cu were determined in wild mushroom species collected in a polluted from historical polymetallic ores mining and smelting area in eastern Slovakia. The highest levels of Hg, up to 50 mg/kg dry matter, were found in Boletus reticulatus and Lycoperdon perlatum, and of Cd, up to 20 mg/kg dry matter - in Xerocomus chrysenteron and Lycoperdon perlatum. The latter species also had extremely high Pb and Cu concentrations - in hundreds of milligrams per kilogram dry matter [97].

Studies showed that around Cu–Ni and Cu smelters the pollution negative impact on bird is accelerated when the potentially bioavailable Cu concentrations in forest litter exceed 1000 µg/g. Total bird densities, biomasses and species diversities decreased in the vicinity of smelters. Bird density and diversity generally decreased towards the pollution sources [98].

Lead poisoning combined with Cd in sheep and horses living on farmland in the vicinity of non-ferrous metal smelters in Baiyin of Gansu province in China was diagnosed [99]. The concentrations of Pb, Cd, Cu and Zn in soils, water, forages, were much higher than those in the control area. The contents of Pb and Cd were 9.3 and 690 times in forages, respectively, 10 and 35.6 times in feed higher than reference values. The concentrations of Pb and Cd in blood, hair and tissues of the affected sheep and horses were significantly higher than reference values and control animals. The study stipulated that the disease of sheep and horses in this region was caused by Pb poisoning combined with Cd. Levels of contamination in animal food and grain pose a significant potential risk to human health through food chain.

Concentrations of some heavy metals (Cd, Pb) were determined in environmental samples and bovine tissues collected from the areas around a lead–zinc smelter in Guizhou, China. Cadmium in soil (10 mg/kg) and feed (6.6 mg/kg) from the polluted areas was 10 times higher than the Chinese Standards, resulting in higher Cd in bovine kidney (38 mg/kg) and liver (2.5 mg/kg). Lead in feed (132 mg/kg) from the polluted area was much higher than unpolluted areas, causing higher Pb levels in bovine tissues [100].

Impact of mining and smelting of polymetallic ores in the Bolivian Altiplano (Highlands) region on air pollution with $PM_{2.5}$ aerosols was shown. Correlation to children's hair metal contents, collected in five schools from different districts, showed a direct exposure to smelting activity - fingerprinted by a unique trace elements pattern (Ag, As, Cu, Pb, Sb) [101].

4.1.2. Prevention, Remediation, Treatment

Different techniques can be applied to prevent diffuse emissions from pyrometallurgy, such as: Furnace sealing (or the use of sealed furnaces) combined with process control; Use of sealed charging systems; Regular inspection and preventative maintenance of the collector hood, the ducts, and the filter system.

The following systems have to be widely used for the removal of particles such as dust, metals and fumes: electrostatic precipitators, wet electrostatic precipitators, cyclones, fabric or bag filters, ceramic and metal mesh filters, wet scrubbing systems, wet scrubbers, dry and semi-dry scrubbers.

The material collected as dust by the off-gas-cleaning systems can be agglomerated and sent back to the smelter or supplied as a raw material for further metal recovery in other facilities. For example, the zinc-rich dust from a converter or an electric slag cleaning furnace in the primary copper smelting process, can be used as a raw material in a zinc recovery plant. Dust or sludge from the treatment of gases are used as raw materials for the production of other metals, such as Ge, Ga, In.

The sludge from a scrubber that contains metals is usually drained, for example in a chamber filter press, and sent back to the smelter. The released water is passed for treatment. If a wet scrubber is used for cleaning the process off-gas, wastewater is generated. This wastewater should be treated to reduce the amount of metal compounds. From this treatment, sludge is produced that may be rich in metals and at that case it is recycled to the process.

Any non-recyclable or non-reusable amounts of water have to be treated in order to minimize the concentration of pollutants before being discharged to the water bodies. Different techniques are applied to remove heavy metals from wastewater: chemical precipitation - by adding a reagent such as lime, sodium hydroxide, sodium sulfide or a combination of reagents, to adjust the pH value and promote the precipitation of soluble metals; biological treatment - a biological reduction of sulfate to sulfide and heavy metals precipitation as sulfides; electrolysis; ion exchange; membrane processes (reverse osmosis, electrodialysis). Usually several of the above mentioned techniques are used together.

Techniques for management of solid residues and wastes include minimization of residues from the metallurgical process, recycling of valuable components, treatment (to immobilize pollutants, eventually mobile under environmental conditions) before disposal and utilization of environmentally non-harmful solid waste. In addition, different techniques can be used to reduce heavy metal transport or bioavailability in soil, already contaminated by smelting activities. Examples for the both approaches are briefly presented below:

Three chemical immobilization materials - agricultural limestone (AL), mineral rock phosphate (RP), and diammonium phosphate (DAP), were evaluated using solute transport experiments to determine their ability to reduce subsurface heavy metal transport in a smelter contaminated soil. DAP applied at 10 g/kg was the most effective treatment for immobilizing the combination of Cd, Pb, and Zn, with mobility reductions of 94.6, 98.9, and 95.8%, respectively [102]. Different amendments were studied with respect of their ability to reduce metal uptake in crops grown on metal contaminated soils near a former metal smelter in Austria. In the pot experiment a reduction of uptake of heavy metals into barley grain was correspondingly Cd < 62%, Pb < 68%. In the field tests extractable fractions of Cd, Pb, and Zn were reduced by up to 96%, 99%, and 99%, respectively in amended soils [103].

4.2. Pollution Caused by Hydrometallurgy

4.2.1. Leaching in Place and Heap/Dump Leaching

Leaching in place (in situ leach mining - ISL mining) is a technique, where the ore body is dissolved under ground and then production solution pumped out, leaving minimal surface disturbance and no tailings or waste rock on the surface. There is no ore dust or direct ore exposure to the environment (consequently, no facilitated formation of ARD). Dilute sulfuric acid is used for leaching copper or nickel ores. Dilute sulfuric acid and sodium chlorate, or ammonium carbonate and hydrogen peroxide are used for leaching uranium deposits. Acids applied to dissolve the ore body usually dissolve metals in the host rock as well. The fluids, remaining after the leaching process in the pores of the rock leached, commonly contain elevated concentrations of heavy metals (such as Cd, Se, V, Pb, and others) and radioactive isotopes, posing a significant risk to nearby ground and surface water sources. In addition, the low pH of ISL mining wastewater can result in acidification of the surrounding environment and additional leaching of heavy metals. The most critical part of the ISL process is to control the movement of the chemical solutions within the aquifer. Any escape of these solutions outside the ore zone can lead to contamination of surrounding groundwater with heavy metals and other harmful pollutants. Uranium in-situ leaching produces certain amounts of waste slurries and wastewater during recovery of the uranium from the production solution. They are often with high salinity and contain toxic levels of heavy metals, process chemicals, and radionuclides.

In heap or dump leaching most often dilute sulfuric acid is used for leaching ores of copper, nickel, uranium. Alkaline cyanide solution is applied for precious metals. Environmental problems related to heap or dump leaching are mainly due to the failure to keep process solutions, charged with the corresponding heavy metal, within the leaching circuit. Potential problems include: a) Infiltration of leaching solutions through pile or collection pond liners; b) Release of solutions due to failures in the pile or pond liners and associated solution transfer equipment. (The liners beneath the ore heaps - usually made of high density polyethylene - have a tendency to tear and rip from a variety of factors including carelessness in depositing the ore, and from natural factors such as ice build-up which is a particular problem in northern regions). In the both cases the result would be contaminated groundwater; c) Leakage from solutions' transfer equipment or spills from broken pipes; d) Overflow of solutions containing toxic concentrations of heavy metals and/or cyanide after a heavy rainfall or rapid snowmelt.

During gold leaching operation, most of the barren cyanide solution is recycled to leaching activities. However, the technology requires a portion of the solution volume to be bled off and disposed of. Also, barren cyanide must be disposed of following operations closure. These solutions have to be treated and discharged to tailings impoundments. Improper treatment may lead to mobilization by leaching of metallo-cyanide complexes of Cu, Fe, Ni, and Zn, as well as other impurities, such as As and Sb.

Heap or dump leaching often generates piles of spent ore which has been in contact with the leaching solution. The spent ore typically contains residual leaching solution, unleached metals and other minerals characteristic for the ore body. The spent ore in most heaps is left in place for detoxification and disposal. Uncollected leachate from disposed material is a potential source of contamination of groundwater, surface water and soil. Over time, in the leachate may be mobilized As, Se, Hg and other heavy metals, which are present in the spent ore. Oxidation of sulfide ores in the spent material can generate acidic water, which additionally dissolves available heavy metals. Wind and/or storm can lead to physical contamination outside the site. In cases where the waste materials from the leaching are transported to a treatment facility, they stay and are precipitated, and during this process there is time and possibility to potentially further leaching of chemicals to the soil. This could lead to contamination of surrounding water bodies with heavy metals.

After the operation has been closed or reclaimed, runoff from the spent ore may occur if proper design and construction measures are not taken. This runoff may contain constituents associated with the ore, such as heavy metals, and total suspended solids. If sulfide ores are present, they may generate acidic leachate which may additionally mobilize the metals that are present in the spent material. The constituents associated with the leachate (metals and As) can cause degradation of ground and surface water quality.

4.2.2. Leaching in Vessels

Treatment of ores by wet methods, carried out in different vessels (vats with false bottom, Pachuca tanks, Dorr agitators, tanks, autoclaves, digesters) beside the production solution, produces residues and waste solutions. Table 14 presents heavy metals in wastewater from hydrometallurgical production of base metals. Liquid effluents, containing toxic reagents, must be treated before being discharged. Residues are disposed of in a tailings impoundment. Some problems may be related to leakage from solutions' preparation and transfer equipment or spills from broken pipes.

Table 14. Heavy metals in wastewater from hydrometallurgical production of base metals

Hydrometallurgical treatment	Major pollutants
Copper production - workshop for contact with acids	Cu, Cd, Pb, Zn, Hg, Ni, Se, As
Copper production - As extraction	As, Cu, Pb, Zn, Cd, Hg, Ni, Se
Copper production - electrolytic refining	Cu, As, Sb, Ni
Lead production - workshop for contact with acids	Cu, Cd, Pb, Zn, Hg, Ni, Se, As
Zinc production	Zn, Cd, Cu, Pb

The disposal of spent ore (tailings) from tank leaching operations requires a secure permanent tailings impoundment. Residues filtered off from aqueous solutions are not stable because they usually contain soluble ingredients. Seepage of contaminated water through the bottom or sides of the pond can lead to contamination of groundwater in the area. When the tailings pond is not planned with adequate capacity, or not properly constructed, an overflow is possible of solutions containing toxic concentrations of heavy metals after a heavy rainfall or rapid snowmelt. This causes heavy pollution of surroundings and of downstream water. The problems appear in heavier manner at a break of the dike of a tailings pond.

Leach residues of primary ore, containing $ZnCO_3$, bear 12 - 13 wt % Zn, 15 -16 wt % Pb and 6-7 wt % Fe [104]. If they are not properly stabilized, heavy metals can be leached out to the environment. The concentrations of the metals leached exceeded the US EPA's toxicity characteristic threshold for Pb (5 mg/L) and Cd (1 mg/L). Also, Pb, Zn, Cd can be leached above the drinking water standards thus possibly posing a potential risk to groundwater [105].

The toxicity characteristic leaching procedure (TCLP) was used to study the environmental activity of Zn leaching residue from leaching zinc sulfide concentrates (in China). The results indicated that the amounts of Zn and Cd in the leaching solution exceed 40 and 90 times the limit, respectively, which demonstrated that this residue is unstable in weak acidic environment at a short contact time [106].

Hakanson Potential Ecological Risk Index Method was used to access the environmental activity and potential ecological risks of heavy metals in the zinc leaching residue. The following order was obtained: Cd > Zn > Cu > As > Pb [107].

In cyanide gold leaching the spent ore exits the leach circuit as a slurry composed of gangue and process water bearing cyanide and cyanide complexes of different heavy metals. The tailings have to be treated to detoxify cyanide prior to disposal. Then, the slurry is typically disposed of in a tailings impoundment. Technology requires an amount of the leaching solution to be bled off and disposed of. The solution contains free cyanide and metallo-cyanide complexes of Cu, Fe, Ni, and Zn, as well as other impurities, such as As and Sb, that are mobilized during leaching. The solution has to be treated to detoxify the cyanide prior discharging to tailings impoundment. Activated charcoal is applied mainly for extraction of Au and Ag from cyanide solution. It is regenerated numerous times and generally an acid wash is involved. The acid wash residues may contain cyanide and heavy metals. They are usually neutralized prior to be released to tailings pond. Often gold is precipitated from the production solution by addition of metallic Zn. The wastes from zinc precipitation include a filter cake, and spent leaching solution, containing $[Zn(CN)_4]^{2-}$. That solution is deposed after detoxification in tailings pond.

The amalgamation waste from gold extraction is a slurry composed of mercury- and other heavy metals bearing solution and gangue. This waste was pumped to a tailings ponds.

In all above described cases improper treatment and disposal of wastewater and different technogenic solutions, slurry, and gangue material will cause environment pollution with different pollutants, including heavy metals.

"Red mud" is the final waste from the Bayer process for producing aluminium. Red mud (which is a suspension) is collected in special ponds. Solids from the slurry precipitate on the bottom of the tailings pond and water on the top is removed, treated, and discharged or reused. Dry red mud is solid, with very fine particles. The red mud may contain a small quantity of Ba, B, Cd, Cr, Co, Ga, Sc, V, Pb, and radionuclides. During the operating phase of the ponds the mud is typically submerged in the liquid, which can serve as the leaching medium conveying potential contaminants to groundwater. When tailings pond is closed, the liquid phase is evaporated or removed and the possibility of leaching and migration of contaminants depends strongly on the degree of infiltration of rainfall through the dry red mud. Data from tests made by the Environmental Protection Agency of the United States [87], showed that As and Se were the main contaminants that were leached from red mud by precipitation, in concentrations exceeding the screening criteria. Arsenic and Se are able to migrate easily after leaching. The contaminating potential of red mud leachate with respect to groundwater varies depending on the specific conditions available in the tailings pond. The concentrations of heavy metal ions under field conditions are expected to be low due to the alkaline nature of the waste. Hazardous components of red mud can enter the surface waters through discharge of groundwater, contaminated by leachate, or through direct discharge of dissolved or suspended ingredients at heavy rain. Arsenic and Se are expected to migrate in surface water. The pollutants' potential to migrate into surface waters and their impact depends on the specific site. The problems are severe at a break of the dike of a tailings pond.

4.3. Pollution Problems Related to Electrometallurgy

In general, the electrolysis plant is less polluting compared to mining, mineral beneficiation and pyrometallurgy activities. There are two eventually directly polluting streams in the electrolysis plant: a) evaporation of the electrolyte and gaseous electrode products and b) wastewaters. Usually different gaseous products evolved and fine electrolyte drops formed during the electrolysis do not bear heavy metals. Wastewater originates by the change of the used electrolyte and by the rinsing of the produced metal after electrolysis. This wastewater usually is acidic and contains heavy metals. If not properly treated when discharged, it can contaminate the receiving water body.

Aluminium, magnesium, copper, zinc and manganese are the most widely used metals in whose manufacturing at a given stage electrometallurgy is used.

Aluminium is produced from bauxite in two generalized steps: a) hydrometallurgical treatment of bauxite to obtain pure Al_2O_3, and b) electrolytic reduction of Al_2O_3 by carbon in fused fluoride salt to obtain metal. Graphite cathodes applied in the electrolytic cell for aluminium production deteriorate after certain period of use. They are removed, the material is known as spent potlining (SPL). SPL was listed by the EPA as a hazardous waste, because of its corrosive properties, reactivity with water - producing inflammable, toxic and explosive gases, content of toxic fluoride and cyanide compounds that are leachable in water. However, it does not bear increased concentrations of heavy metals.

Nearly 80% of the magnesium (Mg) produced annually is obtained by the hydrometallurgy–electrometallurgy route. The process of electrolytic Mg production from Mg acid-soluble minerals consists of the following stages: Raw material leaching by hydrochloric acid and purification of the solution produced; Separation of obtained magnesium chloride in the form of synthetic carnallite or mixture of chlorides from the solution; Dehydration of the obtained product in a fluidized bed by the stream of hot gases, containing hydrogen chloride; Feeding of the dehydrated product into electrolyzers, its electrolysis, with production of Mg and chlorine. The main water pollutants in the Mg production process are metal compounds under the form of suspended solids and chlorinated hydrocarbons. As the acidic wastewater is neutralized, metals further may precipitate as hydroxides. A water treatment system that includes settling of these residues in a lagoon would constitute a land pollution and eventual source of heavy metals pollution.

Nearly 80% of the zinc produced annually is obtained by the hydrometallurgy–electrometallurgy route. The first stage in most widely used electrolytic Zn production is to roast the zinc sulfide (ZnS) concentrate to zinc oxide (ZnO). The so obtained impure ZnO is then leached with sulfuric acid to produce impure zinc sulfate ($ZnSO_4$). The $ZnSO_4$ is separated from the remaining solids and passes to purification aimed at removing the impurity metals that have dissolved with the zinc. Usually, the principal ones are Cd, Cu, Co and Ni, more rarely Mn and Sb. Powdered Zn is added to the solution to reduce the dissolved impurity metals back to their metallic form. If not used for production of minor elements such as Cd and/or Cu, the solid material, bearing reduced impurity metals is dumped in tailing ponds and may become source of heavy metals pollution of the environment. The purified $ZnSO_4$ flows continuously through tanks containing a series of alternate Al cathodes and Pb anodes. Current is passed between opposite electrodes and metallic Zn is plated onto the cathodes by a reduction reaction. The solid residues separated from the leaching process contain substantial amounts of Zn, together with other components - Ag, Pb, other heavy metals. Most commonly the solids are leached with more concentrated hot sulfuric acid to dissolve most of the remaining Zn together with Fe and other impurities, and leaving a lead-silver residue that may be further treated for recovery of these metals. The dissolved Fe is removed from solution most often by precipitation and the obtained solid Fe residue is dumped. If it is not properly stabilized it may be source of heavy metals pollution.

Copper is often electrowon from sulfate solutions. Electrolytic Cu refining generates as waste streams slimes containing impurities such as Au, Ag, Sb, As, Bi, Fe, Pb, Ni, Se, S and Zn and spent acidic solutions bearing heavy metals. Spent acidic solutions should be treated prior to discharge or disposal. The most commonly used treatment method is addition of lime to neutralize the wastewater and to precipitate the dissolved metals. The solid residue from these operations is calcium sulfate, bearing heavy metals and requiring further handling. Analysis of calcium sulfate sludge, obtained by treating wastewater from electro-refining of Cu, shows that usually As, Pb, Cd, Sb, Ag and Cu are in dangerous concentrations. Analyzes of EPA [87] and data from industrial tests indicate that the following pollutants may be leached from the obtained calcium sulfate and may enter into groundwater: As, Se, Pb, Cd, Cu, Cu, Hg, Ni, Ag, Zn, Sb, Al, Mn. Having in mind the elements mobility in groundwater, it could be concluded that As, Se, Cd and Hg are the pollutants representing a major hazard to human health through groundwater. Surface waters can be polluted by discharge of groundwater, contaminated by leachate from the calcium sulfate sludge or by direct runoff of suspended calcium sulfate at heavy rain/storms. Arsenic, Hg, Se and Cd are the main possible pollutants of surface water. The other pollutants from calcium sulfate waste could potentially migrate to surface water through runoff.

Analyses of the pollution materials generated from electrolytic manganese industries in China gave the following results: (1) for Mn, 71.9% enters the product (i.e., electrolytic Mn), 12.6% enters anode mud, 13.7% enters residues and 1.8% enters wastewater; (2) for Cr, 2.4% enters the product and 97.6% enters wastewater; (3) for Se, 60.7% enters the product, 22.3% enters anode mud and 17% enters residues. The residues and wastewater are potentially dangerous to the environment [108].

4.4. Pollution Problems Related to Extraction of Minor High-Tech Metals

Further processing of residues of base metals' production, aimed at extraction of high-tech metals may be an additional source of heavy metals mobilization to the environment. Recovery of gallium (Ga), germanium (Ge) and indium (In) represents an example.

Around 90% of Ga is obtained as a by-product of aluminium extraction from bauxite [109]. In the Bayer process Al is extracted from bauxite ores by leaching with hot concentrated sodium hydroxide. At that around 70% of the Ga, available in bauxite, is leached into the liquid [110]. The solution is cooled to precipitate $Al(OH)_3$. Gallium remains in the solution that is recycled in the Bayer-process. After multiple cycles, the Bayer liquor becomes enriched in Ga. The solution can then be treated with different methods (fractional precipitation, electrolysis, cementation, solvent extraction, ion exchange) to extract 99.9% pure Ga metal [109]. By using several leaching and separation stages and electrolysis, Ga with purity of 99.9% is also extracted from solid residues from hydrometallurgical production of Zn.

Nowadays, germanium is extracted mainly as a by-product of Zn production from sphalerite (75% of worldwide production). Germanium production is carried out in two steps. Initially, germanium is recovered from the leachate of zinc residues by precipitation of a germanium concentrate. The concentrate, consisting mainly of germanium dioxide, is then refined. It is converted into germanium tetrachloride, hydrolyzed to the germanium oxide and then reduced to the pure metal by means of carbon or hydrogen [111].

Generally speaking, indium is produced mainly by leaching with hydrochloric acid or sulfuric acid of dusts, fumes, residues, and slag from Zn and Pb-Zn smelting. The solutions are concentrated with respect to In through a precipitation-leaching process and In is recovered - mainly through electrolysis and/or solvent extraction [112].

Wastewater from Ga and Ge production bear Pb, Zn, As in addition to produced metals and the wastewater from In production - Cd, Pb, Zn and In. If not properly treated, it could be source of heavy metals [113].

4.5. Technologies for Mitigating the Environment Pollution Caused by Hydro- and Electrometallurgy

Different technologies are applied for preventing the environment pollution caused by hydro- and electrometallurgy. Practically, techniques applied to remove heavy metals from wastewater are the same like for the wastewater from pyrometallurgy. Stabilization/ solidification (S/S) and thermal treatment of sludges are also applied. S/S processes reduce the mobilization of hazardous components, including heavy metals, through converting the contaminants into less soluble, mobile or toxic forms by using stabilization additives and binding materials such as cement, clay, zeolite, red mud, fly ash. By thermal treatment heavy metals are immobilized due to formation of a glass matrix in which the metals may be stabilized - the process known as vitrification. The vitrification has the potential to reduce leachability of hazardous constituents from waste. Other possibility to reduce the negative environmental impact is to process residues - thus extracting valuable metals and decreasing the content of heavy metals, able to be leached.

Zinc leaching residues were subjected to acid leaching and at the optimum conditions, 71.9% of initial Zn content was extracted. For Pb extraction, brine leaching was applied and Pb extraction up to 98.9% was attained [104].

A combination of reduction roasting, acid leaching and magnetic separation was proposed for treating zinc leaching residue. Under the optimum operation, 61.38% of zinc was recovered and 80.9% of iron recovery was achieved [114].

When the soil is already polluted by heavy metals, due to application of different technologies for metals extraction (but also due to mining and mineral beneficiation) different techniques for soil remediation may be applied [115].

The physical remediation includes soil replacement and thermal desorption. Clean soil may be used to replace or partly replace the contaminated soil with aim to dilute the pollutant concentration.

Chemical remediation includes chemical leaching, chemical fixation, and electrokinetic remediation. Chemical leaching is washing the contaminated soil using fresh water, reagents, and others fluids that can leach the pollutant from the soil. Trough the ions exchange, precipitation, adsorption and chelation, the heavy metals in soil are transferred from soil to liquid phase, and then recovered from the leachate. Chemical fixation represents adding reagents or materials into the contaminated soil in order to form insoluble compounds of heavy metals, thus decreasing the migration of heavy metals to environmental media. The soil conditioning materials include clays, metallic oxides, biomaterials. In electrokinetic remediation a voltage is applied at the two sides of soil thus forming electric field gradient. The pollutant is carried to space near to the two poles of the treatment room via electromigration, electroosmotic flow or electrophoresis, then collected and further treated.

The biological remediation includes phytoremediation, bioremediation and the combining remediation. Phytoremediation can be defined as the use of plants to remove pollutants from the environment. The phytostabilization, phytovolatilization and phytoextraction are the main three types of phytoremediation. Phytostabilization is fixing heavy metals by plants through the adsorption, precipitation and reduction of root, and thus reducing their migration and bioavailability and preventing them migrating into the groundwater and food chain. Phytovolatilization is transferring the heavy metals into volatile state or adsorbing the metals (such as mercury) and transferring into gaseous phase, using special polymers secreted by root. Phytoextraction is adsorbing the heavy metals from plants and then transferring and storing them at the over-ground parts. The applied plants have high accumulating efficiency under the low contaminants concentration, accumulate high concentrations of the contaminants, accumulate different kinds of heavy metals, grow fast and have pest and disease resistance ability. Then plants are collected and processed to extract metals.

Bioremediation uses the microorganisms' activity. The microorganisms can not degrade and destroy the heavy metals, but can affect their migration and transformation through changing their physical and chemical characteristics. The remediation mechanisms include extracellular complexation, precipitation, oxidation-reduction reaction and intracellular accumulation.

REFERENCES

[1] Zhang, X. W.; Yang L.S.; Li Y. H., Li H.R.; Wang W.Y.; Ge Q.S. Estimation of lead and zinc emissions from mineral exploitation based on characteristics of lead/zinc deposits in China *T Nonferr Metal Soc* 2011, *21*, 2513–2519.

[2] Rey, J.; Martinez, J.; Hidalgo, M. C.; Rojas, D. Heavy metal pollution in the Quaternary Garza basin: A multidisciplinary study of the environmental risks posed by mining (Linares, southern Spain) *CATENA* 2013, *110*, 234–242.

[3] Lee, C. G.; Chon, H. T.; Jung, M. C. Heavy metal contamination in the vicinity of the Daduk Au–Ag–Pb–Zn mine in Korea *Appl Geochem* 2001, *16*, 1377–1386.

[4] Larios, R.; Fernández-Martínez, R.; Álvarez, R.; Rucandio, I. Arsenic pollution and fractionation in sediments and mine waste samples from different mine sites *Sci Total Environ* 2012, *431*, 426–435.

[5] Otones, V.; Álvarez-Ayuso, E.; García-Sánchez, A.; Santa-Regina, I.; Murciego, A. Mobility and phytoavailability of arsenic in an abandoned mining area *Geoderma* 2011, *166*, 153–161.

[6] Chen, T.; Yan, B.; Lei, C.; Xiao, N.; Pollution control and metal resource recovery for acid mine drainage *Hydrometallurgy* 2014, *147–148*, 112–119.

[7] UNEP Chemicals Branch; Global Mercury Assessment 2013: Sources, Emissions, Releases and Environmental Transport; UNEP, Geneva, Switzerland, 2013.

[8] Taylor, M. P.; Davies, P. J.; Kristensen, L. J.; Csavina, J. L. Licenced to pollute but not to poison: The ineffectiveness of regulatory authorities at protecting public health from atmospheric arsenic, lead and other contaminants resulting from mining and smelting operations *Aeolian Research* 2014, *14*, 35–52.

[9] Castillo, S.; D. de la Rosa, J.; Sánchez de la Campa, A. M.; González-Castanedo, Y.; Fernández-Caliani, J. C.; Gonzalez, I.; Romero, A. Contribution of mine wastes to

atmospheric metal deposition in the surrounding area of an abandoned heavily polluted mining district (Rio Tinto mines, Spain) *Sci Total Environ* 2013, *449*, 363–372.

[10] Fontúrbel, F. E.; Barbieri, E.; Herbas, C.; Barbieri, F. L.; Gardon, J. Indoor metallic pollution related to mining activity in the Bolivian Altiplano *Environ Pollut* 2011, *159*, 2870–2875.

[11] Barbieri, E.; Fontúrbel, F. E.; Herbas, C.; Barbieri, F. L.; Gardon, J. Indoor metallic pollution and children exposure in a mining city, *Sci Total Environ* 2014, *487* 13–19.

[12] The International Network for Acid Prevention *Global Acid Rock Drainage Guide (GARD Guide);* INAP, http://www.gardguide.com, 2009.

[13] Smith, K. S. Strategies to predict metal mobility in surficial mining environment, *The Geological Society of America Reviews in Engineering Geology,* XVII, 2007.

[14] Schaider, L. A.; Senn, D. B.; Estes, E. R.; Brabander, D. J.; Shine, J. P. Sources and fates of heavy metals in a mining-impacted stream: Temporal variability and the role of iron oxides, *Sci Total Environ* 2014, *490*, 456–466.

[15] Khorasanipour, M.; Tangestani M. H.; Naseh, R.; Hajmohammadi, H. Hydrochemistry, mineralogy and chemical fractionation of mine and processing wastes associated with porphyry copper mines: A case study from the Sarcheshmeh mine, SE Iran *Appl Geochem* 2011, *26*, 714–730.

[16] Navarro, M. C.; Pérez-Sirvent, C.; Martínez-Sánchez, M. J.; Vidal, J.; Tovar, P. J.; Bech, J. Abandoned mine sites as a source of contamination by heavy metals: A case study in a semi-arid zone *J Geochem Explor* 2008, *96*, 183–193.

[17] Santos, S.; Ungureanu, G.; Boaventura, R.; Botelho, C. Selenium contaminated waters: An overview of analytical methods, treatment options and recent advances in sorption methods, *Sci Total Environ* 2015, *521–522*, 246–260.

[18] Concas, A.; Ardau, C.; Cristini, A.; Zuddas, P.; Cao, G. Mobility of heavy metals from tailings to stream waters in a mining activity contaminated site *Chemosphere* 2006, *63*, 244–253.

[19] Ninga, L.; Liyuana, Y.; Jiruib, D.; Xugui, P. Heavy Metal Pollution in Surface Water of Linglong Gold, Mining Area, China *Procedia Environmental Sciences* 2011, *10*, 914 – 917.

[20] Luís, A. T.; Teixeira, P.; Almeida, S. F. P.; Matos, J. X.; da Silva, E. F. Environmental impact of mining activities in the Lousal area (Portugal): Chemical and diatom characterization of metal-contaminated stream sediments and surface water of Corona stream *Sci Total Environ* 2011, *409*, 4312–4325.

[21] Rahman, S.; Kim, K-H.; Saha, S. K.; Swaraz, A.M.; Paul, D. K. Review of remediation techniques for arsenic (As) contamination: A novel approach utilizing bio-organisms *J Environ Manage* 2014, *134,* 175-185.

[22] Wurl, J.; Mendez-Rodriguez, L.; Acosta-Vargas, B. Arsenic content in groundwater from the southern part of the San Antonio-El Triunfo mining district, Baja California Sur, Mexico *J Hydrol* 2014 *518*, 447–459.

[23] Gemici, U.; Tarcan, G.; Helvac, C.; Somay A. M. High arsenic and boron concentrations in groundwaters related to mining activity in the Bigadic borate deposits (Western Turkey) *Appl Geochem* 2008, *23*, 2462–2476.

[24] Nieto J. M.; Sarmiento A. M.; Olías, M.; Canovas, C. R.; Riba, I.; Kalman, J., Delvalls T. A. Acid mine drainage pollution in the Tinto and Odiel rivers (Iberian Pyrite Belt,

SW Spain) and bioavailability of the transported metals to the Huelva Estuary, *Environ Int* 2007, *33*, 445–455.

[25] Maftei, A. E.; Iancu, O. G.; Buzgar, N. Assessment of minor elements contamination in Bistriţa River sediments (upstream of Izvorul Muntelui Lake, Romania) with the implication of mining activity *J Geochem Explor* 2014, *145*, 25–34.

[26] Zhao, Y.; Marriott, S. B. Dispersion and remobilisation of heavy metals in the River Severn system, UK, *Procedia Environmental Sciences* 2013, *18*, 167 – 173.

[27] Lusilao- Makiese, J.G.; Cukrowska, E. M.; Tessier, E.; Amouroux, D.; Weiersbye, I. The impact of post gold mining on mercury pollution in the West Rand region, Gauteng, South Africa *J Geochem Explor* 2013, *134*, 111–119.

[28] Stefanowicz, A. M.; Woch, M. W.; Kapusta, P. Inconspicuous waste heaps left by historical Zn–Pb mining are hot spots of soil contamination, *Geoderma* 2014, *235–236*, 1–8.

[29] Zhou, J. M.; Dang, Z.; Cai, M. F.; Liu, C. Q. Soil Heavy Metal Pollution Around the Dabaoshan Mine, Guangdong Province, China *Pedosphere* 2007, *17*, 588–594.

[30] Liu, G.; Tao, L.; Liu, X.; Hou, J.; Wang, A.; Li, R. Heavy metal speciation and pollution of agricultural soils along Jishui River in non-ferrous metal mine area in Jiangxi Province, China *J Geochem Explor* 2013, *132*, 156–163.

[31] Monterroso, C.; Rodríguez, F.; Chaves, R.; Diez, J.; Becerra-Castro, C.; Kidd, P. S.; Macías, F. Heavy metal distribution in mine-soils and plants growing in a Pb/Zn-mining area in NW Spain *Appl Geochem* 2014, *44*, 3–11.

[32] Boussen, S.; Soubrand, M.; Bril, H.; Ouerfelli, K.; Abdeljaouad, S. Transfer of lead, zinc and cadmium from mine tailings to wheat (Triticum aestivum) in carbonated Mediterranean (Northern Tunisia) soils *Geoderma* 2013, *192*, 227–236.

[33] García-Sánchez, A.; Alonso-Rojo, P.; Santos-Francés, F. Distribution and mobility of arsenic in soils of a mining area (Western Spain) *Sci Total Environ* 2010, *408*, 4194–4201.

[34] Thrinidad, S. M.; Silva, G. H.; Islas, M. E. R.; Reyes, J. M.; Munguia, G. S.; Valdez, S. S.; Martinez, R. G. Total mercury in terrestrial systems (air-soil-plant-water) at the mining region of San Joaquin region, Queretaro, Mexico *Geofisica International* 2013, *52-1*, 43-56.

[35] Li, Z.; Ma, Z.; Jan van der Kuijp, T.; Yuan, Z.; Huang, L. A review of soil heavy metal pollution from mines in China: Pollution and health risk assessment *Sci Total Environ* 2014, *468–469*, 843–853.

[36] Favas, P. J. C.; Pratas, J.; Elisa, M.; Gomes, P.; Cala, V. Selective chemical extraction of heavy metals in tailings and soils contaminated by mining activity: Environmental implications *J Geochem Explor* 2011, *111*, 160–171.

[37] Hozhina, E. I.; Khramov, A.A.; Gerasimov, P.A.; Kumarkov, A. A.; Uptake of heavy metals, arsenic, and antimony by aquatic plants in the vicinity of ore mining and processing industries *J Geochem Explor* 2001, *74*, 153–162.

[38] Muscatello, J. R.; Belknap, A. M.; Janz, D. M.; Accumulation of selenium in aquatic systems downstream of a uranium mining operation in northern Saskatchewan, Canada, *Environ Pollut* 2008, *156,* 387-393.

[39] Schmitt, C. J.; Brumbaugh, W. G.; May, T. W. Accumulation of metals in fish from lead–zinc mining areas of southeastern Missouri, USA, *Ecotox Environ Safe* 2007, *67*, 14–30.

[40] Liu, C.; Yu, H. Y.; Liu, C.; Li, F.; Xu, X.; Wang, Q. Arsenic availability in rice from a mining area: Is amorphous iron oxide-bound arsenic a source or sink? *Environ Pollut* 2015, *199*, 95-101.

[41] Hu, X. F.; Jiang, Y.; Shu, Y.; Hu, X.; Liu, L.; Luo, F. Effects of mining wastewater discharges on heavy metal pollution and soil enzyme activity of the paddy fields *J Geochem Explor* 2014, *147*, 139–150.

[42] Sun, H. F.; Li, Y. H.; Ji, Y. F.; Yang, L. S.; Wang, W. Y.; Li, H. R. Environmental contamination and health hazard of lead and cadmium around Chatian mercury mining deposit in western Hunan Province, China *T Nonferr Metal Soc* 2010, *20*, 308–314.

[43] Li, Y.; Wang, Y. B.; Gou, X.; Su, Y. B.; Wang, G. Risk assessment of heavy metals in soils and vegetables around non-ferrous metals mining and smelting sites, Baiyin, China *J Environ Sci* 2006, *18*, 1124–1134.

[44] Estival, J. R.; Barasona, J. A.; Mateo, R.; Blood Pb and δ-ALAD inhibition in cattle and sheep from a Pb-polluted mining area *Environ Pollut* 2012, *160*, 118–124.

[45] Carrera, J. P.; Mateo, R.; Estival, J. R. Lead (Pb) in sheep exposed to mining pollution: Implications for animal and human health *Ecotox Environ Safe* 2014, *108*, 210–216.

[46] Drouhot, S.; Raoul, F.; Crini, N.; Tougard, C.; Prudent, A. S.; Druart, C.; Rieffel, D.; Lambert, J. C.; Tête, N.; Giraudoux, P.; Scheifler, R. Responses of wild small mammals to arsenic pollution at a partially remediated mining site in Southern France, *Sci Total Environ* 2014, *470–471,* 1012–1022.

[47] Cho, Y. M.; Seo, S. C.; Choi, S. H.; Lee, S. K.; Kim, K. H.; Kim, H. J.; Choi, J. W. Association of arsenic levels in soil and water with urinary arsenic concentration of residents in the vicinity of closed metal mines, *Int J Hyg Envir Heal* 2013, *216*, 255–262.

[48] Mangalur greenstone belt of Northeastern Karnataka, India, Chakraborti, D.; Rahman, M. M.; Murrill, M.; Das, R.; Siddayy, S. G. P., Sarkar, A.; Dadapeer H. J.; Yendigeri, S.; Ahmed, R.; Das, K. K. Environmental arsenic contamination and its health effects in a historic goldmining area of the *J Hazard Mater* 2013, *262*, 1048– 1055.

[49] Misra, M.; Yang, K.; Mehta, R. K. Application of fly ash in the agglomeration of reactive mine tailings *J Hazard Mater* 1996, *51*, 181–192.

[50] Gomez-Ros, J. M.; Garcia, G.; Peñas, J. M. Assessment of restoration success of former metal mining areas after 30 years in a highly polluted Mediterranean mining area: Cartagena-La Unión *Ecol Eng* 2013, *57*, 393–402.

[51] Ghaderian, M.; Ravandi, A. A. G. Accumulation of copper and other heavy metals by plants growing on Sarcheshmeh copper mining area, Iran *J Geochem Explor* 2012, *123*, 25–32.

[52] Schwab, P.; Zhu, D.; Banks, M. K. Heavy metal leaching from mine tailings as affected by organic amendments *Bioresource Technol* 2007, *98*, 2935–2941.

[53] Overview of analytical methods, treatment options and recent advances in sorption methods, *Sci Total Environ* 2015, *521–522*, 246–260.

[54] United States Environmental Protection Agency; *Reference Guide to Treatment Technologies for Mining-Influenced Water;* EPA 542-R-14-001; EPA-USA, 2014.

[55] North American Metals Council; *Review of available technologies for the removal of selenium from water - final report*; CH2MHILL, 2010.

[56] Song, S.; Lopez-Valdivieso, A.; Hernandez-Campos, D. J.; Peng, C.; Monroy-Fernandez, M. G.; Razo-Soto, I. Arsenic removal from high-arsenic water by enhanced coagulation with ferric ions and coarse calcite *Water Res* 2006, *40*, 364 – 372.

[57] Lizama, K. A.; Fletcher, T. D.; Guangzhi Sun, G. Removal processes for arsenic in constructed wetlands *Chemosphere* 2011, *84*, 1032–1043.

[58] Taylor, J.; Pape, S.; Murphy, N. A summary of passive and active treatment technologies for acid and metalliferous drainage (AMD), In *Earth Systems*, Proceedings of the 5th Australian Workshop on Acid Mine Drainage, August 29-31, 2005, ACMER; Fremantle, Australia, 2005.

[59] Pinto, P. X.; Al-Abed, S. R.; Reisman, D. J. Biosorption of heavy metals from mining influenced water onto chitin products *Chem Eng J* 2011, *166*, 1002–1009.

[60] Name, T.; Sheridan, C. Remediation of acid mine drainage using metallurgical slags *Miner Eng* 2014, *64*, 15–22.

[61] Alcolea, A.; Vazquez, M.; Caparros, A.; Ibarra, I.; Garcia, C.; Linares, R.; Rodriguez, R.; Heavy metal removal of intermittent acid mine drainage with an open limestone channel *Miner Eng* 2012, *26*, 86–98.

[62] Sheoran, A. S.; Sheoran, V. Heavy metal removal mechanism of acid mine drainage in wetlands: A critical review, *Miner Eng* 2006, *19,* 105-116.

[63] Santos-Jallath, J.; Castro-Rodríguez, A.; Huezo-Casillas, J.; Torres-Bustillos, L.; Arsenic and heavy metals in native plants at tailings impoundments in Queretaro, Mexico *Phys Chem Earth* 2012, *37–39*, 10–17.

[64] Andrers, N. F.; Francisco, M. S. Effects of sewage sludge application on heavy metal leaching from mine tailings impoundments *Bioresource Technol* 2008, *99*, 7521–7530.

[65] Martínez, S. M.; Acosta, J. A.; Cano, A. F.; Carmona, D. M.; Zornoza, R.; Cerda, C. Assessment of the lead and zinc contents in natural soils and tailing ponds from the Cartagena-La Unión mining district, SE Spain *J Geochem Explor* 2013, *124*, 166–175.

[66] Liu, X. Q.; Cai, Z. L.; Kang, M. X. Study on the Distribution Regularity of Accompanying Harmful Elements in the Copper-Bearing Skarn Mine at the Fengshan in the Mineral Processing System, *Procedia Earth and Planetary Science* 2011, *2*, 139 – 144.

[67] Fawcett, S. E.; Jamieson, H. E.; Nordstrom, D. K.; McCleskey, R. B. Arsenic and antimony geochemistry of mine wastes, associated waters and sediments at the Giant Mine, Yellowknife, Northwest Territories, Canada *Appl Geochem* 2015, http://dx.doi.org/10.1016/j.apgeochem. 2014.12.012.

[68] Guo, Y. G.; Huang, P.; Zhang, W. G.; Yuan, X. W.; Fan, F. X.; Wang, H. L.; Liu, J. S.; Wang, Z. H. Leaching of heavy metals from Dexing copper mine tailings pond, *T Nonferr Metal Soc* 2013, *23*, 3068–3075.

[69] Santiago, E. C.; Africa, C. R. Trace metal concentrations in the aquatic environment of Albay Gulf in the Philippines after a reported mine tailings spill *Mar Pollut Bull* 2008, *56*, 1650–1667.

[70] Roychoudhury, A. N.; Petersen, J. Geochemical evaluation of soils and groundwater affected by infiltrating effluent from evaporation ponds of a heavy mineral processing facility, West Coast, South Africa *J Geochem Explor* 2014, *144*, 478–491.

[71] Palapa, T. M.; Maramis, A. A. Heavy Metals in Water of Stream Near an Amalgamation Tailing Ponds in Talawaan –Tatelu Gold Mining, North Sulawesi, Indonesia *Procedia Chemistry* 2015, *14,* 428 – 436.

[72] Karczewska, A.; Bogda, D.; Krysiak, A. Arsenic in soils in the areas of former mining and mineral processing in Lower Silesia, southwestern Poland *Trace Metals and other Contaminants in the Environment* 2007, *9*, 411–440.

[73] Kříbek, L.; Majer, V.; Pašava, J.; Kamona, F.; Mapani, B.; Keder, J.; Ettler, V. Contamination of soils with dust fallout from the tailings dam at the Rosh Pinah area, Namibia: Regional assessment, dust dispersion modeling and environmental consequences *J Geochem Explor* 2014, *144*, 391–408.

[74] Taylor, M. P.; Atmospherically deposited trace metals from bulk mineral concentrate port operations, *Sci Total Environ* 2015, *515–516*, 143–152.

[75] Habashi, F. Extractive Metallurgy Today. Progress and Problems, Métallurgie Extractive Québec, Quebec City, Canada, 2000. Distributed by Laval University Bookstore www.zone.ul.ca.

[76] Ye, X.; Hu, D.; Wang, H.; Chen, L.; Xie, H.; Zhang, W.; Deng, C.; Wang, X. Atmospheric mercury emissions from China's primary nonferrous metal (Zn, Pb and Cu) smelting during 1949–2010 *Atmos Environ* 2015, *103*, 331–338.

[77] Kukurugya, F.; Havlik, T.; Parilak, L. Characterization of fine-grained wastes generated during carbon steel and stainless steel production, Proceedings of the XX International Student's Day of Metallurgy, Cracow, March 14-16 2013; 1-10, 2013.

[78] Helsen, L. Sampling technologies and air pollution control devices for gaseous and particulate arsenic: A review, *Environ Pollut* 2005, *137*, 305-315.

[79] JRC, Institute for Prospective Technological Studies, Sustainable Production and Consumption Unit, *Best Available Techniques (BAT) Reference Document for the Non-Ferrous Metal Industries, Industrial Emissions Directive 2010/75/EU (Integrated Pollution Prevention and Control)* MR/GC/EIPPCB/NFM_Draft_3 July 2014 European IPPC Bureau, http://www.eurometaux.eu/.

[80] Fernández-Camacho, R.; de la Rosa, J.; Sánchez de la Campa, A. M.; González-Castanedo, Y.; Alastuey, A.; Querol, X.; Rodríguez, S. Geochemical characterization of Cu-smelter emission plumes with impact in an urban area of SW Spain 2010, *96*, 590–601.

[81] Soto-Jimenez, M. F.; Flegal, A. R. Metal-contaminated indoor and outdoor housedust from a neighborhood Smelter area in Torreon, Mexico, *Procedia Environmental Sciences* 2011, *4*, 134–137.

[82] Shukurov, N.; Kodirov, O.; Peitzsch, M.; Kersten, M.; Mouratov, S.; Steinberger, Y. Coupling geochemical, mineralogical and microbiological approaches to assess the health of contaminated soil around the Almalyk mining and smelter complex, Uzbekistan *Sci Total Environ* 2014, *476–477*, 447–459.

[83] Ettler, V.; Mihaljevič, M.; Kříbek, B.; Majer, V.; Šebek, O. Tracing the spatial distribution and mobility of metal/metalloid contaminants in Oxisols in the vicinity of the Nkana copper smelter, Copperbelt province, Zambia *Geoderma* 2011, *164*, 73–84.

[84] Li, Z.; Feng, X.; Li, G.; Bi, X.; Sun, G.; Qin, H.; Wang, J. Mercury and other metal and metalloid soil contamination near a Pb/Zn smelter in east Hunan province, China *Appl Geochem* 2011, *26*, 160–166.

[85] Martínez-Villegas, N.; Briones-Gallardo, R.; Ramos-Leal, J. A.; Avalos-Borja, M.; Castañón-Sandoval, A. D.; Razo-Flores, E.; Villalobos, M. Arsenic mobility controlled by solid calcium arsenates: A case study in Mexico showcasing a potentially widespread environmental problem *Environ Pollut* 2013, *176,* 114-122.

[86] Gomez-Gonzalez, M. A.; Serrano, S.; Laborda, F.; Garrido, F. Spread and partitioning of arsenic in soils from a mine waste site in Madrid province (Spain), *Sci Total Environ* 2014, *500–501*, 23–33.

[87] Office of Solid Waste, United States Environmental Protection Agency R*eport to Congress on specials wastes from mineral processing, Volume II: Methods and Analyses*, US EPA, 1990.

[88] Navarro, A.; Cardellach, E.; Mendoza, J. L.; Corbella, M., Metal mobilization from base-metal smelting slag dumps in Sierra Almagrera (Almería, Spain), *Appl Geochem* 2008, *23,* 895–913.

[89] Yuan, G. L.; Sun, T. H.; Han, P.; Li, J. Environmental geochemical mapping and multivariate geostatistical analysis of heavy metals in topsoils of a closed steel smelter: Capital Iron & Steel Factory, Beijing, China *J Geochem Explor* 2013, *130*, 15–21.

[90] Šajn, R.; Aliu, M.; Stafilov, T.; Alijagić, J. Heavy metal contamination of topsoil around a lead and zinc smelter in Kosovska Mitrovica/Mitrovicë, Kosovo/Kosovë *J Geochem Explor* 2013, *134,* 1–16.

[91] Lima L.R.P.; Bernardez, L. A.; Characterization of the lead smelter slag in Santo Amaro, Bahia, Brazil *J Hazard Mater* 2011, *189*, 692–699.

[92] Haffert, L.; Craw, D. Processes of attenuation of dissolved arsenic downstream from historic gold mine sites, New Zealand *Sci Total Environ* 2008, *405,* 286-300.

[93] Moiseenko, T. L.; Kudryavtseva, L. P. Trace metal accumulation and fish pathologies in areas affected by mining and metallurgical enterprises in the Kola Region, Russia *Environ Pollut* 2001, *114*, 285-297.

[94] Perkins, W. T. Extreme selenium and tellurium contamination in soils - An eighty year-old industrial legacy surrounding a Ni refinery in the Swansea Valley, *Sci Total Environ* 2011, *412-413*, 162–169.

[95] Bi, X.; Feng, X.; Yang, Y.; Qui, G.; Li, G.; Li, F.; Liu, T., Fu, Z.; Jin, Z. Environmental contamination of heavy metals from zinc smelting areas in Hezhang County, western Guizhou, China *Environ Int* 2006, *32*, 883–890.

[96] Zhan, H. Y.; Jiang, Y. F.; Yuan, J.; Hu, X. F.; Nartey, O. D.; Wang, B. L. Trace metal pollution in soil and wild plants from lead–zinc smelting areas in Huixian County, Northwest China *J Geochem Explor* 2014, *147*, 182–188.

[97] Svoboda, L.; Zimmermannová, K.; Kalač, P. Concentrations of mercury, cadmium, lead and copper in fruiting bodies of edible mushrooms in an emission area of a copper smelter and a mercury smelter *Sci Total Environ* 2000, *246*, 61–67.

[98] Eeva, T.; Belskii, E.; Gilyazov A. S.; Kozlov, M. V. Pollution impacts on bird population density and species diversity at four non-ferrous smelter sites *Biol Conserv* 2012, *150*, 33–41.

[99] Liu, Z. P. Lead poisoning combined with cadmium in sheep and horses in the vicinity of non-ferrous metal smelters *Sci Total Environ* 2003, *309*, 117–126.

[100] Cai, Q.; Long, M.L.; Zhu, M.; Zhou, Q. Z.; Zhang, L.; Liu, J. Food chain transfer of cadmium and lead to cattle in a lead–zinc smelter in Guizhou, China *Environ Pollut* 2009, *157*, 3078–3082.

[101] Goix, S.; Point, D.; Oliva, P.; Polve, M., Duprey, J. L.; Mazurek, H.; Guislain, L.; Huayta, C.; Barbieri, F. L.; Gardon, J. Influence of source distribution and geochemical composition of aerosols on children exposure in the large polymetallic mining region of the Bolivian Altiplano, *Sci Total Environ* 2011, *412–413*, 170–184.

[102] Basta, N. T.; McGowen, S. L. Evaluation of chemical immobilization treatments for reducing heavy metal transport in a smelter-contaminated soil *Environ Pollut* 2004, *127*, 73–82.

[103] Friesl, W.; Friedl, J.; Platzer, K.; Horak, O.; Gerzabek, M. H. Remediation of contaminated agricultural soils near a former Pb/Zn smelter in Austria: Batch, pot and field experiments *Environ Pollut* 2006, *144*, 40–50.

[104] Ruşen, A.; Sunkar, A. S.; Topkaya, Y. A. Zinc and lead extraction from Çinkur leach residues by using hydrometallurgical method *Hydrometallurgy* 2008, *93*, 45–50.

[105] Özverdi, A.; Erdem, M. Environmental risk assessment and stabilization/solidification of zinc extraction residue: I. Environmental risk assessment *Hydrometallurgy* 2010, *100*, 103-109.

[106] Li, M.; Peng, B.; Chai, L. Y.; Peng, N.; Xie, X. D.; Yan, H. Technological mineralogy and environmental activity of zinc leaching residue from zinc hydrometallurgical process *T Nonferr Metal Soc* 2013, *23*, 1480–1488.

[107] Min, X. B.; Xie, X. D.; Chai, L. Y.; Liang, Y. J., Li, M.; Ke, Y. Environmental availability and ecological risk assessment of heavy metals in zinc leaching residue *T Nonferr Metal Soc* 2013, *23*, 208–218.

[108] Duan, N.; Fan, W.; Changbo, Z.; Chunlei, Z.; Hongbing, Y. Analysis of pollution materials generated from electrolytic manganese industries in China *Resour Conserv Recy* 2010, *54*, 506–511.

[109] Zhao Z.; Yang Y.; Xiao Y.; Fan Y. Recovery of gallium from Bayer liquor: A review. *Hydrometallurgy* 2012, *125–126*, 115–124.

[110] Habashi F. *Aluminium – history and metallurgy*; Laval University: Quebec City, 2008.

[111] Moskalyk, R. R. Review of germanium processing worldwide *Miner Eng* 2004, 17, 393–402.

[112] Alfantazi, A. M.; Moskalyk R. R., Processing of Indium: A Review *Miner Eng* 2003, 16, 687–694.

[113] Office of the Federal Register National Archives and Records Administration *Protection of Environment*; Code of federal regulations, 40, Parts 400 to 424, A Special edition of Federal Register revised as of July 1, 2008.

[114] Yan, H.; Chai, L.Y.; Peng, B.; Li, M.; Peng, N.; Hou, D. K. A novel method to recover zinc and iron from zinc leaching residue *Miner Eng* 2014, *55,* 103–110.

[115] Yao, Z.; Li, J.; Xie, H.; Yu, C. Review on remediation technologies of soil contaminated by heavy metals *Procedia Environmental Sciences* 2012, *16*, 722 – 729.

In: Heavy Metals
Editor: Deepak Pathania
ISBN: 978-1-63484-740-7

Chapter 4

HEAVY METALS IN THE ENVIRONMENT: FATE, TRANSPORT, TOXICITY AND REMEDIATION TECHNOLOGIES

Pavan Kumar Gautam**, Ravindra Kumar Gautam, Sushmita Banerjee, M. C. Chattopadhyaya and J. D. Pandey

Department of Chemistry, University of Allahabad,
Allahabad, India

ABSTRACT

Mining, manufacturing, industrialization and the use of synthetic products have resulted in heavy metal contamination of all the segments of environment. Due to their non-biodegradability, it is very difficult to eliminate metals from the environment. Their sources can be categorized as natural and anthropogenic activities. Natural sources are seepage and weathering from rocks, volcanic activity and forest fires. Some of the heavy metals like Cr, Cd, Se, Pb, As, Hg, Cu, Co, Ni and Zn are of severe concern for the researchers in terms of their environmental load and health effects. Several scientific groups, chemical engineers, and environmentalists are making several attempts for minimizing the effects caused by excessive release of heavy metals in the environment and on human being.

Therefore, different remediation technologies such as phytoremediation, adsorption, ion-exchange, coagulation, precipitation, flocculation, ultra-filtration and electrochemical methods have been developed and applied for the scavenging of the toxic metals. In the present chapter, a thorough discussion about the sources, adverse impacts on living beings as well as on environment and the remediation technologies of heavy metals have been systematically compiled.

Keywords: heavy metals, toxicity, environmental fate, adsorption, remediation

---

* E-mail: pavanchemau@gmail.com, Tel.: +91-9198137294.

1. Introduction

The fast pace of industrialization, galloping demand for energy and reckless exploitation of natural resources during the last century are the key reasons for aggravating the problem of environmental pollution, which is now set to pose serious threat to biodiversity and ecosystem processes. A variety of toxic organic and inorganic pollutants are discharged into the water bodies which further pollutes the soil and sediments. One of the most hazardous pollutants that are affecting the environment includes heavy metals [1]. Metals, a major category of globally-distributed pollutants, are natural elements that have been extracted from the earth and harnessed for human industry and products for millennia. The tremendous increase in the use of heavy metal ions over the last few decades has inevitably resulted in an increased flux of metallic substances in the aquatic and terrestrial environment [2]. Acid mine drainage, industrial and domestic effluents, agricultural run-off, acid rain etc. have all contributed to the metal loads in all the segments of environment [3]. Heavy metals are of specific concern because they are non-biodegradable and therefore persistent. The metals are abounded on earth for long time and about 75% of the elements are classified as metals in the periodic table. Most of the contaminants listed by USEPA as trace inorganic contaminants are heavy metals [4]. Metals fall into the following categories: alkali metals, alkaline earth metals, transition metals, noble metals, platinum metals, rare metals, rare earth metals, actinide metals and light metals [5]. The heavy metal pollution in the environment originates from geological weathering, industrial processing of ores and metals, use of metal and its compounds, leaching of metals from garbage and solid waste dumps, landfill waste, livestock and poultry manure, animal and human excretion, urban runoff and reuse of drainage water and sewage effluent [6]. The principal source is the industrial wastes from manufacturing or metal finishing operations. The use of metal containing compounds as fertilizers, pesticides, etc., in agriculture is the second major source of heavy metal pollution. Due to leaching, diffusion and infiltration, these metals are transported to surface and ground waters [7]. In recent years, public awareness has increased because of the long term toxic effects of water containing dissolved metal ions. The heavy metals are not usually eliminated from the aquatic ecosystem by natural process, in contrast to most organic pollutants. Consequently, the concentration of heavy metals increases and creates biological and physiological problems. They have a unique property of accumulation along a food chain for a long period and can cause acute or chronic toxicity to aquatic and human lives [8]. The permissible limits of some toxic in-organics set by United States Environmental Protection Agency (USEPA) and Indian Standards Institution (ISI) for their level in drinking water is given in the Table 1. Some of the common industrial units releasing toxic heavy metals into environment are listed in Table 1.

2. Sources of Heavy Metals in Environment

Metals differ from other toxic substances in that they are neither created nor destroyed by humans. Environmental pollution from hazardous metals and minerals can arise from natural as well as anthropogenic sources [9]. Natural sources are: seepage from rocks into water, volcanic activity, forest fires etc., [10].

Table 1. Permissible limits of different Heavy metals in drinking water

Trace inorganic Contaminants	USEPA (mg/L)	Indian Standards (mg/L)
Arsenic, as As	0.05	0.05
Cadmium, as Cd	0.005	0.01
Chromium, as Cr(VI)	0.05	0.05
Iron, as Fe	0.3	0.3
Lead, as Pb	0.05	0.1
Manganese, as Mn	0.05	0.1
Mercury, as Hg	0.002	0.001
Nickel, as Ni	0.1	0.05

Heavy metal pollution mainly arises from the effluents of industrial units [11]. Irrigation by effluents released from paper mills and fertilizer factories are adding various alkalies, ammonia, cyanides and heavy metals into the water resources [12, 13]. The waste water from the dyes and pigment industries, film and photography, galvanometry, metal cleaning, electroplating, leather and mining industries contains considerable amounts of heavy metal ions [14].

In contrast to herbicides, pesticides and other potential toxicants which undergo break down, albeit extremely slowly, heavy metals cannot be eliminated from a water body and thus persist in sediments where these are slowly released into the water [15]. Burning of fossil fuel, mining and metallurgy, industries and transport sectors redistribute toxic heavy metals into the environment [16, 17]. The distribution of contaminants released to soils by human activities is related to how and where they are added [18]. Apart from industries, roadways and automobiles contribute substantially to the environmental burden of heavy metals since particulate matter in traffic emissions include heavy metals like lead, cadmium and arsenic [19].

Generally, metals enter the aquatic environment through atmospheric deposition, erosion of geological milieu or due to anthropogenic activities caused by industrial effluents, domestic sewage and mining waste [20-22].

Then also through urban storm, water runoff, landfill, mining of coal and ore. But naturally metals get to waters by chemical weathering of minerals and soil leaching [23]. Heavy metals are always present at background levels of non-anthropic origin where their source in soils could be related to weathering of parent rocks and pedogenesis. The metals dissolve and move downstream to lower reaches of the water bodies while others settle into the sediments [24].

Mention must be made that the specific type of metal contamination found in a contaminated soil is directly related to the operation that occurred at the site. The range of contaminant concentrations and the physical and chemical forms of contaminants will also depend on activities and disposal patterns for contaminated waste on the site [25]. Certain factors that may have effect on the form, concentration, and distribution of contaminants include soil and ground-water chemistry and local transport mechanisms.

3. Toxicity of Heavy Metals

Due to their mobility in aquatic ecosystems and their toxicity to higher life forms, heavy metals in surface and groundwater supplies have been prioritised as major inorganic contaminants in the environment [26]. Table 2 depicts the common sources of some toxic metals in environment.

Heavy metals are dangerous because these tend to bioaccumulate. Bioaccumulation means an increase in the concentration of a chemical in an organism over time, compared to its concentration in the environment. At low concentrations some of the heavy metals stimulate some biological processes, but at threshold concentration these become toxic. Being nonbiodegradable, these metals accumulate at various trophic levels through food chain and can cause human health problems [27]. Next section presents the harmful effects to the some heavy metals that are prevalent in the environment.

3.1. Chromium

The name Chromium is derived from the Greek word "Chroma" meaning color because of its nature of giving color to its compounds. Humans are exposed to chromium through breathing, eating or drinking and through skin contact with chromium or chromium compounds [28].

Table 2. Anthropogenic sources of some toxic heavy metals

Heavy metals	Sources
Chromium	Chrome plating, petroleum refining, electroplating industry, leather, tanning, textile manufacturing and pulp processing units. It exists in both hexavalent and trivalent forms.
Nickel	Galvanization, paint and powder, batteries processing units, metal refining and super phosphate fertilizers
Lead	Petrol based materials, pesticides, leaded gasoline, and mobile batteries
Copper	Electroplating industry, plastic industry, metal refining and industrial emissions.
Zinc	Rubber industries, paints, dyes, wood preservatives and ointments, Batteries, electroplating industries, phosphate fertilizers, detergents,
Cadmium	refined petroleum products, paint pigments, pesticides, galvanized pipes, plastics, polyvinyl and copper refineries.
Iron	From metal refining, engine parts.
Aluminium	Industries preparing insulated wiring, ceramics, automotive parts, aluminum phosphate and pesticides.
Arsenic	Automobile exhaust/industrial dust, wood preservatives and dyes
Mercury	Electric/light bulb, wood preservatives, leather tanning, ointments, thermometers, adhesives and paints

For most people eating food that contains chromium (III), it is the main route of chromium uptake, as chromium (III) occurs naturally in many vegetables, fruits, meats, yeasts and grains [29]. Chromium (VI) is a danger to human health, mainly for people who work in the steel and textile industry. Chromium (VI) is known to cause various health effects [30]. Different Chromium species such as Cr(III) and Cr(VI) have differing effects on health. Cr(III) has been found to be an essential micronutrient for human and animal systems and has toxicity less than Cr(VI) [31]. Cr(III) has low solubility and hence relatively immobile in the aquatic systems.

The low solubility retains Cr(III) in the solid phase as colloids or precipitates. Cr(III) in low doses is an essential dietary mineral and found in drinking water as well as in most fresh foods including breads, meats and vegetables [32].

Excessive uptake of Cr(VI) could lead to negative health effects. It has been reported that chromium based diseases such as chromium poisoning, ulcer, allergic dermatitis, lung cancer, liver necrosis, brain damage, premature death and kidney problems were prevalent among mine workers and others surrounding the mine areas [33]. Several investigations have revealed that chromium is mutagenic in bacteria, mutagenic and carcinogenic in higher organisms including humans [34]. Chromium mediated carcinogenicity is the reduction of Cr(VI) to Cr(V) and Cr(IV) by cellular reductants [35]. Cr(VI) also causes birth defects and decrease the reproductive health. Ingestion of chromium in large doses may cause death in animals and humans. Inhalation of hexavalent chromium compounds can result in ulceration and perforation of the mucous membranes of the nasal septum, irritation of the pharynx and larynx, asthmatic bronchitis, bronchospasms and edema. Respiratory symptoms may include coughing and wheezing, shortness of breath, and nasal itch.

3.2. Mercury

No other metal better illustrate, the diversity of effects caused by different biochemical forms than does mercury. Mercury is generally considered to be one of the most toxic metals found in the environment [36]. The major sourse of mercury is the natural degassing on earth crust including land areas, rivers and the ocean and is estimated to be in the order of 25000 to 15000 tons per year.

Chlor-alkali, paints, pulp and paper, oil refining, rubber processing and fertilizer, batteries, thermometers, fluorescent light tubes and high intensity street lamps, pesticides, cosmetics and pharmaceuticals comprise anthropogenic sources of mercury. {toxicity to various forms or salts of Hg is related to cationic mercury whereas solubility, biotransformation and tissue distribution are influenced by valence state and anionic component [37].

With chronic exposure to mercury vapour major effects are on the central nervous system. Methyl mercury is the most important form of mercury in terms of toxicity and health effects from environmental exposure. Methyl mercury causes deformities in the offspring, mainly affecting the nervous system (teratogenic effects) [38].

Children suffer from mental retardation, cerebral palsy and convulsions. Mercury also brings about genetic defects causing chromosome breaking and interference in cell division, resulting in abnormal distribution of chromosome [39]. Mercury causes impairment of pulmonary function and kidney, chest pain and dyspnoea.

3.3. Nickel

Nickel is present in crude oil in varying concentrations and burning of petroleum products. Nickel contaminations in the soil are metal plating industries, combustion of fossil fuels, nickel mining and electroplating [40]. Humans may also be exposed to nickel (Ni) by inhalation, drinking water, and eating contaminated food.

It is released into the air by power plants and trash incinerators and settles to the ground after undergoing precipitation reactions. Nickel can also end up in surface water when it is a part of wastewater streams [41]. This can cause various kinds of cancer on different sites within the bodies of animals, mainly of those that live near refineries. Higher concentrations of nickel cause cancer of lungs, nose and bone.

Dermatitis (Ni itch) is the most frequent effect of exposure to Ni. Acute poisoning of Ni (II) causes headache, dizziness, nausea and vomiting, chest pain, tightness of the chest, dry cough and shortness of breath, rapid respiration, cyanosis and extreme weakness [42].

3.4. Lead

Lead is a naturally occurring and found as a mineral combined with other elements such as sulphur (PbS, $PbSO_4$) and oxygen ($PbCO_3$). Also waste incineration contributes to a greater amount of lead available in urban areas. Its uses include storage batteries, solders, bearings, cable covers, ammunition, plumbing, pigments, caulking, sound and vibration absorbers [43]. Lead can exist in several valence states- Pb(0)-metal, Pb(I), Pb(II) and Pb(1V) all of which with the possible exception of Pb(I) are of environmental importance. Lead has a tendency to concentrate in the water surface microlayer, especially when surface organic materials were present in their films. Lead is a serious cumulative body poison [44]. Lead enters water bodies from industrial, mine and smelter discharges or from dissolution of old leads plumbing. The presence of lead (Pb) may also affect the gastrointestinal tract, kidneys, and the central nervous system [45]. For instance, children exposed to lead (Pb) suffer from impaired development, lower IQ, shortened attention span, hyperactivity and mental deterioration. Those at substantial risk are the children under the age of six [46]. Then in the case of adults" decreased reaction time, loss of memory, nausea, insomnia, anorexia, weakness of the joints, failures of reproduction, inhibition of heam synthesis, irritation, and producing tumor are all caused by exposure to lead [47]. Encephalopathy is one of the most severe acute clinical disorders of lead poisoning. The symptoms of acute encephalopathy syndrome include vomiting, drowsiness, apathy, etc. in acute cases. Convulsion and coma may also appear as associated symptoms. The milder symptoms of acute encephalopathy include mental deterioration, aggressive behaviours, headache, vertigo and tremors.

3.5. Arsenic

Arsenic is commonly found throughout the environment in a wide array of chemical species that vary in toxicity and mobility [48]. Arsenic is found in a wide variety of chemical forms throughout the environment and can be readily transformed by microbes, changes in geochemical conditions, and other environmental processes. Arsenic is a well-known toxic chemical that the Environmental Protection Agency (EPA) and the World Health Organization (WHO) list as a known carcinogen [49]. Arsenic toxicity depends upon oxidation state, the chemical form and solubility of the chemical [50]. According to WHO, maximum permissible limit of arsenic is O.05 mg/L of water [51]. The lethal dose of arsenic is 1-4mg/kg of body weights and thus it belongs to super-toxic category. Arsenic affects the mucus membrane in the respiratory system. There is also reported history of broncho pulmonary disease among people who have ingested arsenic affected water [52]. Cough, breathing difficulty respiratory sign of rales and/or rhonchi are common. Arsenic is found damaging the cells of pancreas islets of Langerhans, which produces insulin in body. It has also been observed that chromosomal abnormalities including chromatic breaks and gaps, chromosome breaks, translocation, rings and dicentrics were common among prolonged arsenic exposed people. They opined that arsenic in man is both mutagenic and carcinogenic [53]. Various types of neoplastic changes of the skin, including Bowen's diseases and basal and squamous cell carcinomas have been associated with arsenic exposure [54]. Normally the daily intake of inorganic arsenic should not exceed 50 μg (0.05 mg). According to the Bureau of Indian Standards (BIS), the permissible upper limit of arsenic in drinking water should not exceed 0.05 mg/L [55].

3.6. Cadmium

Cadmium is distributed in the earth's crust at an average concentration of about 0.1 mg/kg-0.2 mg/kg and is commonly found in association with zinc [56]. The major route of exposure to cadmium for the non-smoking segment of population is via food. Even in the absence of environmental contamination cadmium as a trace constituent is present in most food commodities. High inhalation exposure to cadmium oxide fume may result in acute pneumonitis with pulmonary oedema, which may be cause of death in extreme cases [57]. High ingestion exposure of soluble cadmium salts causes acute gastroenteritis. Long-term occupational exposure to cadmium has caused severe chronic effects, predominantly in the lungs and kidneys [58]. Long-term inhalation exposure and intratracheal administration causes chronic inflammatory changes in the lungs fibrosis. Long-term parenteral or oral administration produces effects primarily on the kidneys, but also on the liver and the hematopoietic, immunity, skeletal, and cardiovascular systems [59]. The most severe from Cd (II) toxicity in humans is "itai- itai", a disease characterized by excruciating pain in the bone [60]. Cadmium may interfere with the metallothionein's ability to regulate zinc and copper concentrations in the body. Epidemiological studies have revealed that Cd^{2+} may contribute to some forms of cancer in humans and low exposures may result in kidney damage [61].

3.7. Zinc

Zinc is ubiquitous in the environment and occurs in the earth's crust at an average concentration of about 70 mg/kg [62]. Both, anthropogenic and natural sources are responsible for zinc contamination in environment. The sources include combustion exhaust, galvanized metal parts and railings, tire dust, motor oil and hydraulic fluid, brake linings and cement production [63]. Zinc is a metallic element that is found in the body as a divalent cation. Accordingly, it does not undergo metabolism. It interacts electrostatically with anions (i.e., carbonate, hydroxide, oxalate, phytate) and negatively charged moieties on macromolecules such as proteins [64]. It can also form soluble chelation complexes with amino acids and multidentate organic acids such as ethylenediaminetetraacetic acid. Zinc is essential to maintain a good health but increased concentrations may cause skin irritations, vomiting, nausea and stomach cramps [65]. Even higher concentrations cause liver damage, respiratory disorders and disturb protein metabolism [66]. Some of the industrial activities such as coal combustion, mining and galvanizing steel cause water pollution with zinc which increases the acidity of water. Aquatic organisms may accumulate zinc in their bodies. Zinc polluted soil may affect plants, cattle and can contaminate the ground water. Breakdown of the organic matter is also affected by this as zinc has a negative influence on the activities of earthworms and other microorganisms [67]. The activity of the soil is greatly influenced.

3.8. Cobalt

Natural sources of cobalt to the environment include volcanic eruptions, seawater spray and forest fires. Anthropogenic sources of cobalt to the atmosphere include coal-fired power plants and incinerators, and exhaust from vehicles [68]. Cobalt mining and processing activities, the production of alloys and chemicals containing cobalt, sewage effluents, urban run-off, and agricultural run-off are major anthropogenic contributors of cobalt to the aquatic environment [69]. Cobalt has both beneficial and harmful effects on human health. Cobalt is beneficial for humans because it is part of vitamin B_{12}, which is essential to maintain human health. Cobalt (0.16–1.0 mg cobalt/kg of body weight) has also been used as a treatment for anemia. When too much cobalt is taken into your body, however, harmful health effects can occur [70]. Serious effects on the lungs, including asthma, pneumonia, and wheezing, have been found in people exposed to 0.005 mg cobalt/m3 while working with hard metal, a cobalt-tungsten carbide alloy [71]. People exposed to 0.007 mg cobalt/m^3 at work have also developed allergies to cobalt that resulted in asthma and skin rashes. The general public, however, is not likely to be exposed to the same type or amount of cobalt dust that caused these effects in workers [72]. Being exposed to radioactive cobalt may be very dangerous to your health by damaging cells depending on the amount of radiation to which you are exposed [73]. Exposure to lower levels of radiation might cause nausea, and higher levels can cause vomiting, diarrhea, bleeding, coma, and even death.

3.9. Iron

Iron is one of the major constituents of the lithosphere and comprises approximately 5%. It is routinely detected in municipal waste effluent, particularly in cities where iron and steel are manufactured [74]. Iron readily complexes with sulphates in the sediments of many surface waters. The primary concern about the presence of iron in drinking water is its objectionable taste. The taste of iron in drinking water can be easily detected even at low concentrations of about 1.8 mg/L [75]. Iron has broad industrial applications.

Anthropogenically, entry of iron into the environment is mainly through effluents of industries like ink, paint and pharmaceutical products, mining industries, ore processing, electroplating industries, zinc and steel processing plants. The presence of Fe(II) in water leads to the growth of iron bacteria which in turn leads to undesirable colour, taste and odour to water [76]. Abundant growth of iron bacteria in water leads to a slimy condition. So the water becomes unfit for domestic and industrial purposes. The recommended iron content in stream water is 0.3 mg/L, at higher concentrations it causes vomiting [77]. Presence of iron in water makes the water turbid, besides causing staining of plumbing fixtures and laundry. Moreover, iron precipitates in the gills of fishes, which is deleterious to them [78].

Iron is essential for human health and necessary for proper functioning of the biological system. Deficiency of iron causes anemia. However, excess of iron is toxic to human beings. Excessive iron intake leads to haemochromatosis resulting in tissue damage due to iron accumulation [79]. Initial symptoms of iron toxicity on human are vomiting, diarrhea and damage to the intestine.

4. Remediation Technologies for Metal Decontamination

Over the last few decades, several methods have been devised for the treatment and removal of heavy metals. Numerous industries (e.g., electroplating, metal finishing operations, electronic –circuit production, steel and non-ferrous processes and fine-chemical and pharmaceutical production) discharge a variety of toxic metals into the environment. Pollution by metal ions has become a major issue throughout many countries [80] because the contents of metal ions in potable waters and wastewaters often exceed the permissible sanitary standards. Therefore, there is a requirement for newer and effective methods which are also cost–effective and there is a continuing interest in the development of more economic and environmentally-benign processes of selective removal of heavy metals from dilute wastewater streams. The methods include precipitation, evaporative and electrolytic recovery, solvent extraction, cementation, reverse osmosis, electrodialysis, ion exchange and adsorption.

4.1. Chemical Precipitation

It is one of the oldest and most widely used methods for the removal of metals from wastewater. Chemical precipitation involves the addition of chemicals to modify the physical state of dissolved and suspended solids and facilitate their removal by sedimentation [81]. Precipitation of metals is achieved by the addition of coagulants such as alum, lime, iron salts and other organic polymers [82]. Chemical precipitation is employed for most of the metals. Common precipitants include hydroxide $(OH)^{-}$, carbonate $(CO3)^{2-}$ and sulfide $(S)^{2-}$. On the basis of precipitation agent used, the technique may be categorized into following types:

4.1.1. Hydroxide Precipitation

Chemical precipitation of heavy metals as their hydroxides using lime or sodium hydroxide is commonly used [83]. Lime is generally preferred for precipitation purposes due to the low cost of precipitant, ease of pH control in the range of 8.0 – 10.0 and the excess of lime also serves as an adsorbent for the removal of metal ions [84]. The efficiency of the process depends on a number of factors, which include the ease of hydrolysis of the metal ion, nature of the oxidation state, pH, presence of complex forming ions, standing time, degree of agitation and settling and filtering and characteristics of the precipitate. The removal of metals can be understood by following reaction:

$$M(Metal)^{2+} + 2OH^{-} + M(OH)_2$$

The limitations of this method include difference between metals in the optimum pH for hydroxide formation may lead to the problems in the treatment of effluents containing combined metal ions. Variability in metal hydroxide solubility at a fixed pH and generation of large amount of sludge are other drawbacks.

4.1.2. Carbonate Precipitation

Carbonate precipitation of metals using calcium or sodium carbonate is very limited. Different researchers reported improved results using carbonate precipitate for Cd (II) and Pb (II) from electroplating effluents. When the pH was brought to 7.5, residual concentration of Pb (II) and Cd (II) were 0.60 and 0.25 mg/L respectively.

4.1.3. Sulphide Precipitation

Since most of the heavy metals form stable sulphides, excellent metal removal can be obtained by sulphide precipitation [85]. Treatment with sulphides is most advantageous when used as a polishing step after conventional hydroxide precipitation or when very high metal removals are required [86].

4.2. Chemical Reduction

Reduction of hexavalent chromium can also be accomplished with electro-chemical units [87]. The electrochemical chromium reduction process uses consumable iron electrodes and an electric current to generate ferrous ions that react with hexavalent chromium to give trivalent chromium as follows:

$$3Fe^{2+} + CrO_4^{2-} + 4H_2O \rightarrow 3Fe^{3+} + Cr^{3+} + 8OH^{-}$$

Another application of reduction process is the use of sodium borohydride, which has been considered effective for the removal of mercury, cadmium, lead, silver and gold.

4.3. Xanthate Process

Insoluble starch xanthate (ISX) is made from commercial cross linked starch by reacting it with sodium hydroxide and carbon disulphide. To give the product stability and to improve the sludge settling rate, magnesium sulphate is also added. ISX works like an ion exchanger, removing the heavy metals from the wastewater and replacing them with sodium and magnesium. Average capacity is 1.1-1.5 meq of metal ion per gram of ISX [88]. ISX is most commonly used by adding to it the wastewater as slurry for continuous flow operations or in the solid form for batch treatments. It should be added to the effluent at pH ≥ 3. Then the pH should be allowed to rise above 7 for optimum metal removal. Residual metal ion level below 50 μg/L has been reported. The effectiveness of soluble starch xanthate (SSX) for removal of Cd (II), Cr (VI) and Cu (II) and insoluble starch xanthate (ISX) for Cr (VI) and Cu (II) have been evaluated under different aqueous phase conditions. Insoluble starch xanthate had better binding capacity for metals. The binding capacity of SSX and ISX respectively for different metal ions follows the sequence of Cr (VI)> Cu (II)> Cd(II) and Cr (VI)> Cu (II) [89].

4.4. Solvent Extraction

Liquid-liquid extraction (also frequently referred as solvent extraction) of metals from solutions on a large scale has experienced a phenomenal growth in recent years due to the introduction of selective complexing agents [90]. In addition to hydrometallurgical applications, solvent extraction has gained widespread usage for waste reprocessing and effluent treatment. Solvent extraction involves an organic and an aqueous phase. The aqueous solution containing the metal or metals of interest is mixed with the appropriate organic solvent and the metal passes into the organic phase [91]. In order to recover the extracted metal, the organic solvent is contacted with an aqueous solution whose composition is such that the metal is stripped from the organic phase and is reextracted into the stripping solution. The concentration of the metal in the strip liquor may be increased, often 110 to 100 times over that of the original feed solution. Once the metal of interest has been removed, the organic solvent is recycled either directly or after a fraction of it has been treated to remove the impurities.

4.5. Membrane Process

Important examples of membrane process applicable to inorganic wastewater treatment include reverse osmosis and eletrodialysis [92]. These processes involve ionic concentration by the use of selective membrane with a specific driving force. For reverse osmosis, pressure difference is employed to initiate the transport of solvent across a semipermeable membrane and electro dialysis relies on ion migration through selective permeable membranes in response to a current applied to electrodes [93]. The application of the membrane process described is limited due to pretreatment requirements, primarily, for the removal of suspended solids. The methods are expensive and sophisticated, requiring a higher level of technical expertise to operate [94]. liquid membrane is a thin film that selectively permits the passage of a specific constituent from a mixture. Unlike solid membranes, however liquid membranes separate by chemistry rather than size, and thus in many ways liquid membrane technology is similar to solvent extraction.

Since liquid membrane technology is a fairly recent development, a number of problems remain to be solved. A major issue with the use of supported membranes is the long term stability of the membranes, whereas the efficient breakup of microspheres for product recovery is one of the difficulties encountered frequently with emulsion membranes [95].

4.6. Cementation

Cementation is the displacement of a metal from solution by a metal higher in the electromotive series [96]. It offers an attractive possibility for treating any wastewater containing reducible metallic ions. In practice, a considerable spread in the electromotive force between metals is necessary to ensure adequate cementation capability [97]. Due to its low cost and ready availability, scrap iron is the metal used often. Cementation is especially suitable for small wastewater flow because a long contact time is required. Some common examples of cementation in wastewater treatment include the precipitation of copper from printed etching solutions and the reduction of Cr (VI) in chromium plating and chromate-inhibited cooling water discharges [98]. Removal and recovery of lead ion by cementation on iron sphere packed bed has been reported [99]. Lead was replaced by a less toxic metal in a harmless and reusable form.

4.7. Ion Exchange

Ion exchange resins are available selectively for certain metal ions. The cations are exchanged for H^+ or Na^+. The cation exchange resins are mostly synthetic polymers containing an active ion group such as SO_3H. The natural materials such as zeolites can be used as ion exchange media [100]. The modified zeolites like zeocarb and chalcarb have greater affinity for metals like Ni and Pb [101]. The limitations on the use of ion exchange for inorganic effluent treatment are primarily high cost and the requirements for appropriate pretreatment systems. Ion exchange is capable of providing metal ion concentrations to parts per million levels [102]. However, in the presence of large quantities of competing mono-and divalent ions such as Na and Ca, ion exchange is almost totally ineffective.

4.8. Electrodeposition/Electrochemical Recovery

Electrolytic metal recovery is one of a number of technologies capable of removing metals from process wastewaters. The technology has been used for many years in the mining industry for electrorefining of ores, and has been used to recover copper from pickle liquors [103]. During the last 25-30 years, electrolytic metal recovery has been investigated for recovery of metals plating tanks. In electrolytic recovery, a direct current is passed through an aqueous solution containing metal ions between cathode plates and insoluble anodes. The positively charged metallic ions adhere to the negatively charged cathodes leaving a metal deposit that can be stripped off and recovered [104]. Some metals found in waste solution can be recovered by electrodeposition using insoluble anodes. For example, spent solutions resulting from sulphuric acid cleaning of Cu may be saturated with copper sulphate in the presence of residual acid. These are ideal for electro-winning where the high quality cathode copper can be electrolytically deposited while free sulphuric acid is regenerated.

4.9. Floatation

Floatation is employed to separate solids or dispersed liquids from a liquid phase using bubble attachment. The attached particles are separated from the suspension of heavy metal by the bubble rise. Floatation can be classified as: (i) dispersed-air flotation, (ii) dissolved-air flotation (DAF), (iii) vacuum air flotation, (iv) electro-floatation, and (v) biological floatation [105]. Among the different types of flotation, dissolved-air flotation is the most commonly used for the purification of metal-contaminated wastewater. Adsorptive bubble separation employs foaming to separate the metal impurities from water [106]. The target floated substances are separated from bulk water in a foaming phase. Although it is only a kind of physical separation process, heavy metal removal by floatation has the potential for industrial application. Low-cost materials such as zeolite and chabazite have been found to be effective collectors with removal efficiency of higher than 95% for an initial metal concentration ranging from 60 to 500 mg/L [107]. Floatation can be employed to treat inorganic effluent with a metal concentration of less than 50 mg/L or higher than 150 mg/L. Advantages like better removal of small particles, shorter hydraulic retention times and low cost make flotation one of the promising alternatives for the treatment of metal contaminated wastewater [108].

4.10. Biological Approaches to Metal Removal

Biological approach of remediation includes: (i) the use of microorganisms to detoxify the metals by valence transformation, extra-cellular chemical precipitation or volatilization, and (ii) use of specific plants to decontaminate soil or water, by inactivating metals in rhizosphere or translocating them into aerial parts. The second approach is called phytoremediation, which is considered as a new and highly promising technique for the reclamation of polluted sites, and is cheaper than physicochemical techniques (Garbisu and Alkorta, 2001).

4.10.1. Phytoremediation

Phytoremediation, also referred to as the "botanical bioremediation", is becoming an area of current research, which has tremendous prospects towards decontamination of polluted soils and water [109]. The term phytoremediation actually refers to diverse collection of plant based technologies that use either naturally occurring or genetically engineered plants, for cleaning contaminated environment [110]. Phytoremediation is generally viewed as more efficient and cost effective, compared with other methods of decontamination such as engineering and/or microbial approaches [111]. The history of metal hyperaccumulation in plants started at the end of 19th century when it was observed that Thlaspi caerulescens and Viola calaminaria growing on soils naturally enriched with Zn, contained extraordinary high levels of Zn [112]. This prompted research on the identification of metal hyperaccumulating plants. The term "hyperaccumulation" was interdicted by Jaffre [113] describing abnormal levels of Ni accumulation in the plant and this term was later extended to other metals. At present, the criteria used for hyperaccumulation vary with metal, and range from 100 mg kg-1 dw for Cd, to 1000 mg kg^{-1} dw for Ni, Cu, Co, Cr and Pb, to 10,000 mg kg^{-1} dw for Zn and

Mn [114]. It may be mentioned that these values have to be found in any of the above ground parts of plants growing in their natural habitat, but not under artificial conditions. As usual, these plants exhibit shoot–to–soil metal concentration ratio, the so-called bioaccumulation factor (BCF) higher than 1. Hyper-accumulator plants are usually found on metalliferrous soils; calamine soils rich in Zn and Pb; serpentine soils derived from Fe and Mg rich ultramafic rocks, enriched also in Ni, Cr and Co [115]. According to Ernst [116] natural exposure of plants to a surplus of various metals has driven the evolution of metal hyperaccumulation, as well as plant resistance to heavy metals. To date, more than 400 metal accumulating taxa belonging to at least 45 plant families have been identified [117]. Some families and genra are known as sources of specific metal hyperaccumulator: Ni (Brassicaceae–Thlaspi), and Cu and Co (Lamiaceae, Scrophulariaceae). Ipomoea alpina, T. rodundifolium T. caerulescens etc. have been extensively studied for their metal accumulation potential [118]. The ability to accumulate a heavy metal varies significantly between species and among cultivars within a species, as different mechanisms of ion uptake are operative in each species, based on their genetic, morphological, physiological and anatomical characteristics. There are different techniques of phytoremediation including phytoextraction, phytostabilization, phytofilteration and phytovolatilization depending upon the mechanism of remediation.

4.10.1.1. Phytoextraction

Phytoextraction is a technique which comprises use of plants to absorb, translocate and store toxic contaminants from a soil matrix into their root and shoot tissue [119]. After the plants have been allowed to grow for some time, they are harvested and either incinerated or composted to recycle the metals. If plants are incinerated, the ash must be disposed of in a hazardous waste landfill, but the volume of ash will be less than 10% of the volume that would be created if the contaminated soil itself were dug up for treatment [120]. Metals such as nickel, zinc, and copper are the best candidates for removal by phytoextraction.

4.10.1.2. Rhizofiltration

Rhizofiltration ('rhizo' means 'root') is the adsorption or precipitation onto plant roots (or absorption into the roots) of conta-minants that are in solution surrounding the root zone [121]. Rhizofiltration is similar to phytoextraction, but the plants are used to clean up contaminated groundwater rather than soil. The plants to be used for cleanup are raised in greenhouses with their roots in water. Contaminated water is either collec-ted from a waste site and brought to the plants or the plants are planted in the contaminated area, where the roots then take up the water and the contaminants dissolved in it [122]. As the roots become saturated with contaminants, they are harvested.

4.10.1.3. Phytostabilisation

Phytostabilisation is the use of certain plant species to immobilise contaminants in the soil and groundwater through absorption and accumulation by roots, adsorption onto roots, or precipitation within the root zone of plants (rhizosphere) [123]. This process reduces the mobility of the contaminant and prevents migration to the groundwater or air, and also reduces bioavailability for entry into the food chain. This technique can be used to re-establish a vegetative cover at sites where natural vegetation is lacking due to high metal concentrations in surface soils or physical disturbances to surficial materials. Metal-tolerant species can be used to restore vegetation to the sites, thereby decreasing the potential migration of contamination through wind erosion and transport of exposed surface soils and leaching of soil contamination to groundwater [124].

4.10.2. Microbial Remediation of Metals

Microorganisms able to survive well in high concentrations of heavy metals are of great interest as bioremediation agents because they can achieve different transformation and immobilization processes [125]. To survive under metal-stressed conditions, bacteria have evolved several types of mechanisms to tolerate the uptake of heavy metal ions. These mechanisms include the efflux of metal ions outside the cell, accumulation and complexation of the metal ions inside the cell, and reduction of the heavy metal ions to a less toxic state [126]. Most of the organisms have detoxification abilities (i.e., mineralization, transformation and/or immobilization of pollutants), and thereby maintain sustenance of the biosphere [127]. The mechanism by which microorganisms remove heavy metals can be divided into three categories; the first mechanism is the biosorption of metal ions on the cell surface, second intracellular uptake of metal ions and third, chemical transformation of metal ions by microorganism [128]. Among the different technique employed for removal of metals from multi elemental system, biosorption has been found to be highly selective [129]. Furthermore, metal accumulating bacteria can be used to remove metals from industrial effluents as they utilize the same for their metabolism to an extent possible. Heavy metal pollution exerts a selective pressure on microbial community leading to the emergence of resistant strains with apparent reduction in the extracellular enzyme activity of that particular ecosystem. Microorganisms, including bacteria, algae, fungi and yeast, serve as very constructive models for studying the harmful effects of metals at the cellular level [130]. These microorganisms can also be used to remove toxic metals from contaminated sites as they can efficiently accumulate the heavy metals from the environment. There are several researches carried out on the removal of heavy metals from industrial wastes [131]. Soltan, [132] isolated 240 Bacillus strains which were resistant to Pb, Cd, Hg, Zn, Ag and Cu, and reported some strains could accumulate high concentration of these metals. Workentine et al. [133] reported bacterial isolates of Bacillus and Pseudomonas were resistant to mercury and the genes conferring the resistance to mercury. Martinez et al. [134] and Silver and Phung [135] reported Multiple metal-resistant bacterias isolated from metal contaminated soil in Indonesia, among the 43 isolates 93% were gram negative and among them 26% were pigmented. They belonged to nine genera-Moraxella, Pseudomonas, Bacillus Flavobacterium, Vibrio, Xanthomonas, Alcaligens, Micrococcus, Aeromonas and Acinetobacter.

4.11. Adsorption

Among other processes for removal of metal ions from industrial/mining wastewater, adsorption process is an economically feasible alternative [136]. Adsorption is a process in which a single or a group of ions/compounds get accumulated on the surface of another solid or liquid [137]. The substance on which the adsorption takes place is known as adsorbent and the substance, which is adsorbed, is called adsorbate. Adsorption is a treatment which can economically meet today's higher effluent standards and water reuse requirements [138].

Adsorption of metals on efficient solid support represents a simple and economically viable cleaning technology for metal removal especially, where the sorbents are inexpensive and easily available [139]. More importantly, it imparts no side effect or toxicity to the water and this account for the superior removal of organic contaminants as compared to conventional treatments [140].

Therefore, it is preferred over other techniques in terms of cost, ease of operation, flexibility, negligible disturbances with diurnal variations, reusability and recyclability of the sorbent [141]. Different materials have been tried for the minimisation of heavy metals from aqueous matrices and soils viz. activated carbon, agricultural residues, industrial by products, natural materials like clays, zeolites, silicaceous material etc., [142].

Activated Carbon

Activated carbon has been undoubtedly the most popular and widely used adsorbent in waste water treatment throughout the world [143]. Charcoal, the front runner of modern activated carbon has been recognized as the oldest adsorbent known in waste water treatment. Activated carbon is produced by a process of consisting raw material dehydration and carbonization followed by activation [144]. Carbonization converts this organic material to primary carbon, which is a mixture of ash, tars, amorphous carbon and crystalline carbon.

However due to the higher cost of activated carbon, attempts are being made to prepare the activated carbon from abundantly available and low cost materials [145].

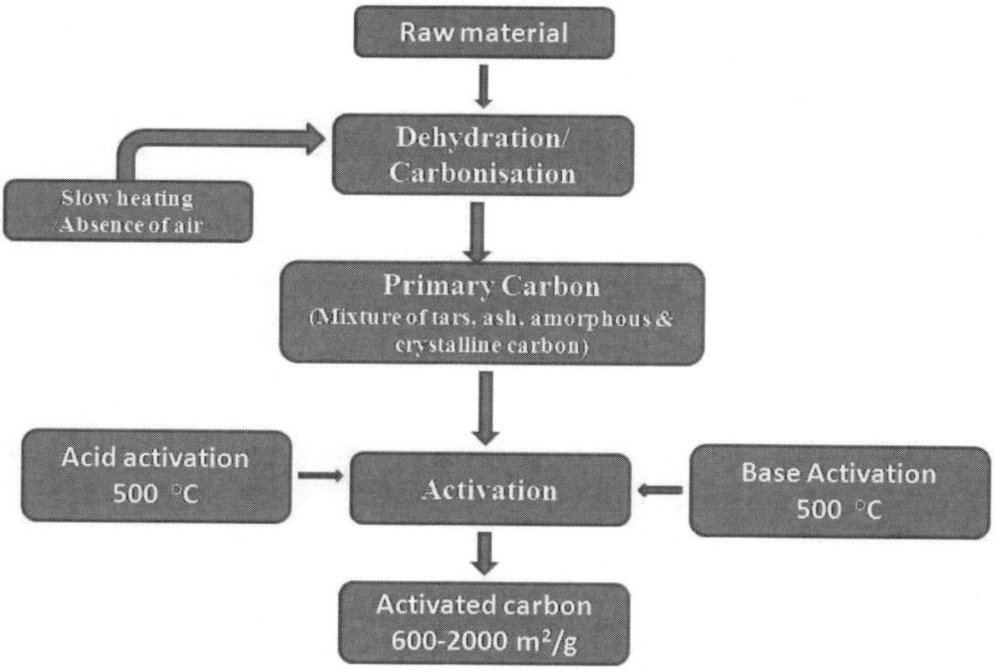

Figure 1. Block diagram of preparation of activated carbon.

Activated carbons can be prepared from a variety of carbons containing materials such as coke, olive stones, pinewood, rice hulls, palm shell, sawdust, anthracite, plum kernels, peat, bituminous coal, coconut shell, palm fruit bunch particles etc., [146].

A schematic representation of the preparation of activated carbon is shown in Figure (1).

Recently, Bohli et al. [147] chemically synthesized microporous activated carbon and evaluates its efficiency in removal of Cu(II), Cd(II) and Pb (II) ions from single and binary aqueous solutions. The study on process variables suggest that the adsorbent reached saturation phase within 50 minutes for all metal ions and establishment of complete equilibrium takes place in about 200 minutes moreover solution pH also influences the removal process maximum sorption of metal ion takes place at pH 5.0. It has been also investigated that for single aqueous solution the adsorbent shows maximum adsorption capacity of 0.654 mmol/g for Pb(II) ions followed by Cd(II) ion, 0.559 mmol/g and Cu(II) ion, 0.268 mmol/g.

This trend however found to be reversed in case of binary aqueous solutions where similar adsorbent shows relatively high affinity for Cu(II) ions.

4.11.1. Other Materials for Metal Decontamination

Agricultural Residues

Use of agricultural wastes in wastewater treatment found to be an attractive alternative treatment technique that grabs the attention of several researchers from past decades. Agricultural by-products usually are composed of lignin and cellulose as major constituents and may also include other polar functional groups of lignin, which includes alcohols, aldehydes, ketones, carboxylic, phenolic and ether groups. These groups are capable to bind with heavy metals usually by donation of an electron pair from these groups to form complexes with the metal ions in solution [148]. Among agricultural wastes, rice and wheat husk were one of the most explored waste material. Sobhanardakani et al. [149] reported removal of Cr(III) and Cu(II) ions from contaminated wastewater. The removal efficiency found to be strongly depends on pH, initial metal ion concentration, contact time and sorbent dose. The maximum percent removal of 58% and 67% have been observed for Cr(III) and Cu(II) ions respectively at given experimental condition such as sorbent dose of 2.5g/L, initial metal concentration of 100 mg/L and pH value of 5.0-6.0. The percent removal of Cr(III) and Cu(II) reported to reached maximum 100% when adsorbent dose of 4g/L and 5g/L was employed. Farajzadeh and Monji [150] have utilized wheat bran as a natural metal adsorbent for removal of toxic metal ions such as Hg(II), Pb(II), Cu(II), Cd(II) and Ni(II) from aqueous solution. Adsorption equilibrium was achieved rapidly in about 10 min for all cations. The removal efficiency remain more than 82% for all studied cations, except Ni(II) ion. The probable mechanism behind adsorption presumed as ion exchange, complexation or size exclusion. According to Yin et al. [151] naturally available adsorbents exhibits low metal removal capacity and slow process kinetics. Therefore surface functionalization proves to be an effective solution. Yin et al. reported hydroxylethylidenediphosphonic acid functionalized agricultural waste, buckwheat hull for adsorption of Au(III) ions from aqueous media. The experimental results revealed that the adsorption property of functionalized buckwheat hulls with 120 mesh for Au(III) sorption was very excellent, and the maximum adsorption capacity for Au(III) calculated as 450.45mg/g at 35°C which is much higher than that of value obtained for raw adsorbent.

Industrial By-Products

The use of industrial by-products in wastewater treatment considered to be the most suitable practice of waste management. Fly ash is one of the most popular industrial by product that has been extensively investigated as an adsorbent material for the removal of various organic and inorganic contaminants from aqueous solution. Beside fly ash various other industrial by product such as metal hydroxide sludge, red mud, biosolids and waste slurry were also used as adsorbents for metal removal. Rozada et al. [152] has examined the adsorption efficiency of heavy metals onto sewage sludge derived materials. The metal adsorption found to be highly pH dependent and preferential order of metal adsorption followed as Hg(II) > Pb(II) > Cu(II) > Cr(III) with respective maximum sorption capacity of 64.9 mg/g > 30.1 mg/g > 3.6 mg/g > 1.5 mg/g. Further Soco and Kalembkiewicz [153] investigate the efficiency of coal fly ash sample for removal of Ni(II) and Cu(II) ions from aqueous solutions. The maximum degree of metal removal was found to be in the ranges of 98.0– 98.4% at pH 8.0–10.0 in the case of Cu(II) and 97.7% at pH 8.0 in the case of Ni(II).

Natural Inorganic Materials

Clay is one of the potential adsorbent and its removal efficacy is comparable to that of activated carbon as well. Their sorption capabilities come from their high surface area and ion exchange capacities. The negative charge on the structure of clay minerals gives clay the capability to attract metal ions. The clay minerals such as kaolinite and montmorillonite have been comprehensively tested as adsorbents in the removal of toxic heavy metals from aqueous solutions. Bhattacharya and Sen Gupta [154] found that kaolinite and montmorillonite clay minerals can efficiently remove Fe(III), Co(II) and Ni(II) from aqueous medium. Metal adsorption increased with pH till precipitation of insoluble hydroxides. Montmorillonite had a much higher adsorption capacity for the metal ions having maximum sorption capacity of 28.4 to 28.9 mg/g compared to that of 10.4 to 11.2 mg/g for kaolinite. The large difference in sorption capacities of the clay minerals attributed to charge density as montmorillonite had a high surface charge density resulting from isomorphous substitution in tetrahedral and octahedral sheets, whereas kaolinite had little isomorphous substitution accounting for its low cation adsorbing capacity.

Chitosan has received great interest for heavy metals removal due to its excellent metal-binding capacities. These wastes could be easily procured from local fishery industries. As these wastes are abundantly available, chitosan may be produced at low cost. Thus, chitosan offers a lot of promising benefits in wastewater treatment. An effective approach towards remediation of Cu(II) and Pb(II) ions researched by Haider and Park [155] by employing chitosan nanofibers. The adsorbent shows excellent metal sorption capacity of 485 mg/g and 263 mg/g for Cu(II) and Pb(II) respectively. The high sorption capacity of the adsorbent possibly due to the high surface area per unit mass, spongy nature and wet absorption that cause fast and increase adsorption due to more exposure of the chitosan nanofiber's functional groups to the metal ions.

Natural zeolites also gained a significant interest among scientist, mainly due to their valuable properties. Natural zeolites are highly porous hydrated alumina silicate materials having three dimensional crystal structures. These materials are intensively studied for their applicability in the removing of various pollutants, such as heavy metal ions, inorganic anions, phenols, pesticides and dyes in water. High ion-exchange capacity, relatively high specific surface areas, and more importantly, their low cost make zeolites attractive adsorbents.

Erdem et al. [156] make detailed investigation related to using of zeolites for removal of Co(II), Cu(II), Zn(II), and Mn(II) ions from metal finishing wastewaters. The adsorption phenomena found to be depends on charge density and hydrated ion diameter. Therefore the selectivity sequence follows the order Co(II) > Cu(II) > Zn(II) > Mn(II).

Weeds and Dried Algal Biomass

Application of dead plant biomass could be efficiently used as low cost and ecofriendly adsorbent for the removal and recovery of toxic metals from contaminated wastewater even at low concentrations.

These plant based sorbent material possess diverse functional groups such as carboxyl, hydroxyl, sulfate, and amino groups which are responsible for possible binding with metal cations. Till date various aquatic weeds namely *Pistia*, *Salvinia*, *Spirodela*, *Lemna*, *Eichhornia*, algae such as *Caulerpa*, *Ulva*, *Sargassum*, *Nostoc*, *Spirogyra*, *Gelidium* etc., have been successfully employed in scavenging of metal ions from liquid phase.

FUTURE OUTLOOK

Several advantages in using low cost biosorbents for removal of metallic species from aqueous solutions are discussed below together with the main drawbacks.

- The metal sorption rates from aqueous solution by biosorption are generally faster than those by other metabolically mediated processes. This property of biosorption would make it the most effective in treatment of voluminous water bodies containing low concentrations of metal ions, such as the final wastewater treatment process to meet regulation standards of metal ions before being discharged to the environment [157].
- The costs for formulating biosorbents and operating the treatment systems are generally low, especially waste bacteria and fungi from various industrial and agricultural byproducts, and the biomass that exists abundantly such as exotic weeds can be utilized as cheap raw materials of biosorbents.
- It is not necessary to feed any substrates or energy for propagation of the raw biomass of biosorbents if these microbial wastes and abundant biomass are to be utilized. They are, however, often required to receive some modifications before introduction into the real wastewater treatment systems, and they will take additional costs according to the reagents and equipments used for the modifications.
- Performances of biosorbents can be improved through various kinds of physical and chemical pretreatments. The main purposes of modification of raw biosorbent materials are to (a) shape and impart mechanical strength and resistance to chemical and microbial degradation, and (b) improve biosorption selectivity for targeted metal ions. Pretreatments of raw biomass by physical or chemical methods have been also found to improve the adsorption performances of the biosorbents, although this is not always the case.

Conclusion

In the development of new separation and purification technologies, such as removal of toxic heavy metals through biosorption mechanism can play a key enabling role. A breakthrough in sorbents development is needed to solve the critical problem of removal and recovery of metal ions. Inexpensive, locally available and effective materials could be used in place of commercial activated carbon for the removal of metal ions from industrial effluents. Undoubtedly biosorbents offer a lot of promising benefits for commercial purposes in the future. Although a lot has been done in the areas of low cost sorbents and much work is necessary to better understand the adsorption mechanism.

References

[1] D. C. Adriano, Trace Elements in Terrestrial Environments: Biogeochemistry, Bioavailability and Risks of Metals, second ed., Springer, New York, 2001.

[2] B. Knopf, H. Konig, Biomethylation of Heavy Metals in Soil and Terrestrial Invertebrates: Soil Heavy Metals, Springer, Berlin/Heidelberg, 2010, pp. 315 – 328.

[3] R. J. Haynes, G. Murtaza, R. Naidu, Inorganic and organic constituents and contaminants in biosolids: implications for land application, *Adv. Agron.*, 104 (2009) 165 – 267.

[4] R. Wuana, F. Okieimen, J. Imborvungu, Removal of heavy metals from a contaminated soil using organic chelating acids, *Int. J. Environ. Sci. Technol.*, 7 (2010) 485 – 496.

[5] S. Tandy, K. Bossart, R. Mueller, J. Ritschel, L. Hauser, R. Schulin, B. Nowack, Extraction of heavy metals from soils using biodegradable chelating agents, *Environ. Sci. Technol.,* 38 (2004) 937 – 944.

[6] C. M. Davidson, A. L. Duncan, D. Littlejohn, A. M. Ure, L. M. Garden, A critical leval- uation of the three stage BCR sequential extraction procedure to assess the potential mobility and toxicity of heavy metals in industrially-contaminated land, *Anal. Chim. Acta*, 363(1998) 45 – 55.

[7] M. Pueyo, G. Rauret, D. Lück, M. Yli-Halla, H. Muntau, P. Quevauviller, J. López-Sánchez, Certification of the extractable contents of Cd, Cr, Cu, Ni, Pb and Zn in a freshwater sediment following a collaboratively tested and optimized three-step sequential extraction procedure, *J. Environ. Monit.*, 3 (2001) 243 – 250.

[8] S. K. Papageorgiou, F. K. Katsaros, E. P. Kouvelos, J. W. Nolan, H. LeDeit, N. K. Kanellopoulos, Heavy metal sorption by calcium algi-nate beads from Laminaria digitata, *J. Hazard. Mater.*, B137 (2006)1765 – 1772.

[9] S. L. McGowen, N. T. Basta, G. O. Brown, Use of diammonium phosphate to reduce heavy metal solubility and transport in smelter-contaminated soil, *J. Environ. Qual.*, 30 (2001) 493 – 500.

[10] E. A. Alvarez, M. C. Mochón, J. Sanchez, M. T. Rodrıguez, Heavy metal extractable forms in sludge from wastewater treatment plants, *Chemosphere,* 47(2002) 765 – 775.

[11] N. K. Srivastava, C. B. Majumder, Novel biofiltration methods for the treatment of heavy metals from industrial wastewater, *J. Hazard. Mater.,* 151 (2008) 1 – 8.

[12] D. E. Guven, G. Akinci, Heavy metals partitioning in the sediments of Izmir Inner Bay, *J. Environ. Sci.*, 20 (2008) 413 – 418.

[13] N. T. Basta, R. Gradwohl, Estimation of Cd, Pb, and Zn bioavailability in smelter contaminated soils by a sequential extraction procedure, *J. Soil Contam.*, 9 (2000) 149 – 164.

[14] R. Naidu, N. S. Bolan, Contaminant chemistry in soils: key concepts and bioavailability, in: R. Naidu (Ed.), Chemical Bioavailability in Terrestrial Environment, Elsevier, Amsterdam, The Netherlands, 2008, pp. 9 – 38.

[15] D. C. Adriano, W. W. Wenzel, J. Vangronsveld, N. S. Bolan, Role of assisted natural remediation in environmental cleanup, *Geoderma,* 122 (2004) 121 – 142.

[16] K. Anastasiadou, K. Christopoulos, E. Mousios, E. Gidarakos, Solidification/stabilization of fly and bottom ash from medical waste incineration facility, *J. Hazard. Mater.*, 207–208 (2012) 165 – 170.

[17] Y. J. Park, J. Heo, Vitrification of fly ash from municipal solid waste incinerator, *J. Hazard. Mater.*, 91 (2002) 83 – 93.

[18] A. M. Chaudri, C. M. Allain, V. L. Barbosa-Jefferson, F. A. Nicholson, B. J. Chambers, S. P. McGrath, A study of the impacts of Zn and Cu on two rhizobial species in soils of a long-term field experiment, *Plant Soil,* 221 (2000) 167 – 179.

[19] B. Sun, F. Zhao, E. Lombi, S. McGrath, Leaching of heavy metals from contami- nated soils using EDTA, *Environ. Pollut.,* 113 (2001) 111 – 120.

[20] K.-Y. Lee, K.-W. Kim, Heavy metal removal from shooting range soil by hybrid electrokinetics with bacteria and enhancing agents, *Environ. Sci. Technol.,* 44 (2010) 9482 – 9487.

[21] R.-S. Juang, L.-D. Shiau, Ion exchange equilibria of metal chelates of ethylenediaminetetraacetic acid (EDTA) with Amberlite IRA-68, Ind. *Eng. Chem. Res.*, 37 (1998) 555 – 560.

[22] B. Nowack, Environmental chemistry of aminopolycarboxylate chelating agents, *Environ. Sci. Technol.*, 36 (2002) 4009 – 4016.

[23] D. Kołody′ nska, The chelating agents of a new generation as an alternative to conventional chelators for heavy metal ions removal from different waste waters, *Desalination*, (2011)339 – 371.

[24] S. M. Ross, Retention transformation and mobility of toxic metals in soils, in: S.M. Ross (Ed.), Toxic Metals in Soil–Plant Systems, Wiley, New York, 1994, pp. 63 – 152.

[25] R. D. R. Harter, R. Naidu, Role of metal–organic complexation in metal sorption by soils, *Adv. Agron.,* 55 (1995) 219 – 264.

[26] J. S. Thayer, F. E. Brinckman, The biological methylation of metals and metalloids, *Adv. Organomet. Chem.*, 20 (1982) 313 – 356.

[27] P. Alvarenga, A. P. Goncalves, R. M. Fernandes, A. Varennes, G. Vallini, E. Duarte, Organic residues as immobilizing agents in aided phytostabilization: (I) effects on soil chemical characteristics, *Chemosphere*, 74 (2009) 1292 – 1300.

[28] S. L. McGowen, N. T. Basta, G. O. Brown, Use of diammonium phosphate to reduce heavy metal solubility and transport in smelter-contaminated soil, *J. Environ. Qual.*, 30 (2001) 493 – 500.

[29] R. J. Bartlett, J. M. Kimble, Behavior of chromium in soils: II. Hexavalent forms, *J. Environ. Qual.,* 5 (1976) 383 – 386.

[30] M. T. Aide, M. F. Cummings, The influence of pH and phosphorus on the adsorption of chromium (VI) on boehmite, *Soil Sci.,* 162 (1997) 599–603.

[31] M. G. M. Alam, S. Tokunaga, F. Stagnitti, Removal of arsenic from contaminated soils using different salt extractants, *J. Environ. Sci. Health A,* 42 (2007) 447 – 451.

[32] R. J. Bartlett, Chromium cycling in soils and water: Links, gaps, and methods, Environmental Health Perspectives. 92 ((1991) pp 17 - 24.

[33] A. D. Covington, Modern tanning chemistry. *Chem. Soc. Rev*., 26 (1997) 111 - 126.

[34] M. E. Losi, C. Amrhein and W. T. Frankenberger Jr. Factors affecting chemical and biological reduction of hexavalent chromium in soil. *Environ. Toxicol. Chem*., 13 (1994) 1727 - 1735.

[35] R. K. Kanojia, M. Junaid and R. C. Murthy. Embryo and fetotoxicity of hexavalent chromium: A long-term study. *Toxicol. Lett.*, 95 (1998) 165-172.

[36] K. P. Nickens, S. R. Patierno and S. Ceryak. Chromium genotoxicity: A double-edged sword. *Chemico-Biological Interactions*, 188 (2010) 276-288.

[37] M. Megharaj, S. Avudainayagam and R. Naidu, Toxicity of hexavalent chromium and its reduction by bacteria isolated from soil contaminated with tannery waste, *Curr. Microbio*., 47 (2003) 51 - 54.

[38] F. N. Moreno, C. W. N. Anderson, R. B. Stewart, B. H. Robinson, M. Ghomshei, J. A. Meech, Induced plant uptake and transport of mercury in the presence of sulphur-containing ligands and humic acids, *New Phytol.*, 166 (2005) 445–454.

[39] F. N. Moreno, C. W. N. Anderson, R. B. Stewart, B. H. Robinson, Mercury volatilization and phytoextraction from base metal mine tailings, *Environ. Pollut*., 136 (2005) 341 – 352.

[40] S. P. Bizily, C. L. Rugh, A. O. Summers, R.B. Meagher, Phytoremediation of methyl mercury pollution: MerB expression in Arabdopsis thaliana confers resistance to organomercurials, *Proc. Natl. Acad. Sci. U. S. A.*, 96 (1999) 6808 – 6813.

[41] S. Lyyra, R. B. Meagher, T. Kim, A. Heaton, P. Montello, R. S. Balish, S. A. Merkle, Coupling two mercury resistance genes in Eastern cottonwood enhances the processing of organomercury, *Plant Biotechnol. J.,* 5 (2007) 254 – 262.

[42] L. Jean-Soro, F. Bordas, J. C. Bollinger, Column leaching of chromium and nickel from a contaminated soil using EDTA and citric acid, *Environ. Pollut*., 165 (2012) 175 – 181.

[43] J. C. Seaman, J. S. Arey, P. M. Bertsch, Immobilization of nickel and other metals in contaminated sediments by hydroxyapatite addition, *J. Environ. Qual.*, 30 (2001) 460 – 469.

[44] L. Jean-Soro, F. Bordas, J. C. Bollinger, Column leaching of chromium and nickel from a contaminated soil using EDTA and citric acid, *Environ. Pollut.,* 165 (2012) 175 – 181.

[45] M. R. Dudzinska, D. A. Clifford, Anion exchange studies of lead-EDTA complexes, *React. Polym*., 16 (1991) 71 – 80.

[46] C. E. Halim, R. Amal, D. Beydoun, J. A. Scott, G. Low, Evaluating the applicabilityof a modified toxicity characteristic leaching procedure (TCLP) for the classifi-cation of cementitious wastes containing lead and cadmium, *J. Hazard. Mater.,* 103 (2003) 125 – 140.

[47] J. W. Drexler, W. J. Brattin, An in vitro procedure for estimation of lead relative bioavailability: with validation, *Hum. Ecol. Risk Assess.,* 13 (2007) 383 – 401.

[48] F. J. Peryea, Phosphate-induced release of arsenic from soils contaminated with lead arsenate, *Soil Sci. Soc. Am. J.,* 55 (1991) 1301 – 1306.

[49] J. A. Ryan, K. G. Scheckel, W. R. Berti, S. L. Brown, S. W. Casteel, R. L. Chaney, J. Hallfrisch, M. Doolan, P. Grevatt, M. Maddaloni, D. Mosby, Reducing children's risk from lead in soil, *Environ. Sci. Technol.,* 38 (2004) 18 – 24.

[50] J. Kumpiene, P. Desogus, S. Schulenburg, M. Arenella, G. Renella, E. Brannvall, A. Lagerkvist, L. Andreas, R. Sjoblom, Utilisation of chemically stabilized arsenic-contaminated soil in a landfill cover, *Environ. Sci. Pollut. Res.,* (2013), http://dx.doi.org/10.1007/s11356-013-1818-3.

[51] A. Kouras, I. Katsoyiannis, D. Voutsa, Distribution of arsenic in groundwater inthe area of Chalkidiki, Northern Greece, *J. Hazard. Mater.*, 147 (2007) 890–899.

[52] D. M. Sahabi, M. Takeda, I. Suzuki, J. Koizumi, Adsorption and abiotic oxidationof arsenic by aged biofilter media: equilibrium and kinetics, *J. Hazard. Mater.*, 168 (2009) 1310 – 1318.

[53] M. J. Kosnett. Clinical treatment of chronic arsenic intoxication. This paper was presented in International Seminar on Arsenic Poisoning held in Bangladesh, Feb. 1998.

[54] A. M. Carcía, C. Montero-Ocampo, Improvement of arsenic electro-removal from underground water by lowering the interference of other ions, *Water Air Soil Pollut.*, 205 (2010) 237 – 244.

[55] V. K. Sharma, M. Sohn, Aquatic arsenic: toxicity, speciation, transformations, and remediation, *Environ. Int.,* 35 (2009) 743 – 759.

[56] D. Masih, Y. Seida, Y. Izumi, Arsenic removal from dilute solutions by highsurface area mesoporous iron oxyhydroxide, *Water Air Soil Pollut.,* 9 (2009)203 – 211.

[57] E. M. Leonard, L. M. Pierce, P. L. Gillis, C. M. Wood, M. J. O'Donnell, Cadmium trans-port by the gut and Malpighian tubules of Chironomus riparius, *Aquat. Toxicol.*, 92 (2009) 179 – 186.

[58] Z. L. Zhu, L. H. Zhang, H. Zhang, Y. L. Qiu, R. H. Zhang, J. F. Zhao, Extraction of cadmium from sewage sludge using polyepoxysuccinic acid, *Pedosphere,* 19 (2009) 137 – 142.

[59] N. Basta, R. Gradwohl, K. Snethen, J. Schroder, Chemical immobilization of lead, zinc, and cadmium in smelter-contaminated soils using biosolids and rock phosphate, *J. Environ. Qual.*, 30 (2001) 1222 – 1230.

[60] P. Loganathan, M. J. Hedley, N. D. Grace, Pasture soils contaminated with fertilizer derived cadmium and fluoride: livestock effects, *Rev. Environ. Contam. Toxicol.*, 192 (2008) 29 – 66.

[61] J. Klimczak, J. M. Wisniewska-Knypl, and J. Kolakowski, Stimulation of lipid peroxidation and heme oxygenase activity with inhibition of cytochrome P-450 monooxygenase in the liver of rats repeatedly exposed to cadmium, *Toxicology,* 32 (1984) 267 – 276.

[62] C. O. Ikediobi, V. L. Badisa, L. T. Ayuk-Takem, L. M. Latinwo, and J. West, Response of antioxidant enzymes and redox metabolites to cadmium-induced oxidative stress in CRL- 1439 normal rat liver cells, *Int. J. Mol. Med.*, 14 (2004) 87 – 92.

[63] D. Hanahan and R. A. Weinberg, The hallmarks of cancer, *Cell,* 100 (2000) 57 – 70.

[64] R. S. Stahl, B. R. James, Zinc sorption by B horizon soils as a function of pH, *Soil Sci. Soc. Am. J.,* 55 (1991) 1592 – 1597.

[65] M. B. McBride, M. Pitiranggon, B. Kim, A comparison of tests for extractable copper and zinc in metal-spiked and field-contaminated soil, *Soil Sci.,* 174 (2009) 439 – 444.
[66] L. Beesley, M. Marmiroli, The immobilisation and retention of soluble arsenic, cadmium and zinc by biochar, *Environ. Pollut.*, 159 (2011) 474–480.
[67] J. M. Llobet, J. L. Domingo, M. T. Colomina, Subchronic oral toxicity of zinc in rats, *Bull. Environ. Contam. Toxicology*, 41(1988) 36 - 43.
[68] B. Lönnerdal, Dietary factors influencing zinc absorption, *J. Nutr.,* 130 (2000) 1378 - 1383.
[69] J. L. Malo, J. Malo, A. Cartier, Acute lung reaction due to zinc inhalation. *Eur. Respir. J.,* 3 (1990) 111 - 114.
[70] S. A. Abbasi, P. C. Nipaney, R. Soni. Environmental status of cobalt and its micro determination with 7-nitroso-8-hydroxyquinoline-5-sulfonic acid in waters, aquatic weeds and animal tissues, *Anal. Lett.,* (1989) 22 (1):225 - 235.
[71] L. Alessio, A. DellOrto, Biological monitoring of cobalt. In: Biological monitoring of toxic metals. New York, NY: Plenum Press, (1988) 407-417.
[72] J. L. Abraham, A. Hunt, Environmental contamination by cobalt in the vicinity of a cemented tungsten carbide tool grinding plant, *Environ. Res.*, 69 (1995) 67 - 74.
[73] Z. Adamis, E. Tatrai, K. Honma,. A study on lung toxicity of respirable hard metal dusts in rats, *Ann. Occup. Hyg.,* 41(1997) 515 - 526.
[74] C. S. Alexander. Cobalt and the heart. *Ann. Intern. Med.*, 70 (1969) 411-413.
[75] C. Adam, J. P. Baudin, J. Garnier-Laplace Kinetics of 110mAg, 60Co, 137Cs and 54Mn bioaccumulation from water and depuration by the crustacean Daphnia magna, *Water Air Soil Pollut.*, 125 (2001) 171 - 188.
[76] D. Mohan and S. Chander, Single, binary and multicomponent sorption of iron and manganese on lignite, *J. Colloid Interface Sci.,* 299 (2006) 76 - 87.
[77] A. Edwin Vasu, Adsorption of Ni(II), Cu(II) and Fe(III) from aqueous solutions using activated carbon, *E.-J. Chem.*, 5 (2008) 1 - 9.
[78] FSTRAC. 1999. Summary of state and federal drinking water standards and guidelines 1998-1999. Federal-State Toxicology and Risk Analysis Committee. U.S. Environmental Protection Agency.
[79] A. K. Singh, D. P. Singh, K. K. Panday, and V. N. Singh, Wollastonite as adsorbent for removal of Fe(II) from water", *J. Chem. Technol. Biot.*, 42 (1988) 39 - 49.
[80] M. K. Mukhopadhyay, and S. K. Konar, Lethal effects of iron on fish and aquatic ecosystem, *Ind. Assoc. Wat. Pollut. Contr. Technol.,* Annual 9 (1982) 49 - 162.
[81] H. Barman, and K. G. Bhattacharya, Metal contamination of groundwater in a predominantly rural area in Assam, Proceedings of the 10th national symposium on environment, IIT, Mumbai, (2001) 161-166.
[82] R. Wuana, F. Okieimen, J. Imborvungu, Removal of heavy metals from a contaminated soil using organic chelating acids, *Int. J. Environ. Sci. Technol.*, 7(2010) 485 – 496.
[83] Matthew M. Matlock, Brock S. Howerton, David A. Atwood, Chemical precipitation of heavy metals from acid mine drainage, *Water Res.,* 36 (2002) 4757 – 4764.
[84] M. M. Matlock, B. S. Howerton, D. A. Atwood. Chemical precipitation of lead battery recycling plant wastewater, *Ind. Eng. Chem. Res.,* 41 (2002)1579 – 82.
[85] G. Chaudhuri, P. Dey, D. Dalal, P. Venu-Babu, W. R. Thilagaraj, A novel approachto precipitation of heavy metals from industrial effluents and single-ion solu-tions using bacterial alkaline phosphatase, *Water Air Soil Pollut.,* 224 (2013)1625.

[86] D. Feng, C. Aldrich, H. Tan, Treatment of acid mine water by use of heavy metal precipitation and ion exchange. *Miner. Eng.,* 13 (2000) 623-642.
[87] P. Kousi, E. Remoudaki, A. Hatzikioseyian, M. Tsezos, A study of the operating parameters of a sulphate-reducing fixed-bed reactor for the treatment of metal-bearing wastewater. In: 17th International Biohydrometallurgy Symposium, Germany, Frankfurt am Main (2007).
[88] M. T. Alvarez, C. Crespo, B. Mattiasson, Precipitation of Zn(II), Cu(II) and Pb(II) at bench-scale using biogenic hydrogen sulfide from the utilization of volatile fatty acids. *Chemosphere,* 66 (2007) 1677 - 1683.
[89] Y. Ku, Jung, I. L., 2001. Photocatalytic reduction of Cr(VI) in aqueous solutions by UV irradiation with the presence of titanium dioxide. *Water Res.*, 35, 135 - 142.
[90] Y. K. Chang, J. E. Chang, T. T. Lin, Y. M. Hsu, Integrated copper-containing wastewater treatment using xanthate process, *J. Hazard. Mater.*, 94 (2002) 89 - 99.
[91] J. C. Duan, Q. Lu, R. W. Chen, Y. Q. Duan, L. F. Wang, L. Gao, Pan, S. Y., Synthesis of a novel flocculant on the basis of crosslinked Konjac glucomannan-graftpolyacrylamide-co-sodium xanthate and its application in removal of Cu2+ ion. *Carbohydr. Polym.,* 80 (2010) 436-441.
[92] N. T. Basta, R. Gradwohl, Estimation of Cd, Pb, and Zn bioavailability in smelter contaminated soils by a sequential extraction procedure, *J. Soil Contam.*, 9 (2000) 149 – 164.
[93] A. Khanmirzaei, K. Bazargan, A. A. Moezzi, B. K. Richards, K. Shahbazi, Single and sequential extraction of cadmium in some highly calcareous soils of Southwestern Iran, *J. Soil Sci. Plant Nutr.*, 13 (2013) 153 – 164.
[94] E. Sampera, M. Rodrígueza, M. A. De la Rubia, D. Prats, Removal of metal ions at low concentration by micellar-enhanced ultrafiltration (MEUF) using sodium dodecyl sulfate (SDS) and linear alkylbenzene sulfonate (LAS). *Sep. Purif. Technol.*, 65 (2009) 337 - 342.
[95] A. Nemeh, A. P. Van Peteghem, Extraction of multicomponent system metals from simulated and industrial effluents by liquid surfactant membranes, *Hydrometallurgy,* 31 (1992) 149 – 159.
[96] M. Ruppert, J. Draxler, R. Marr, Liquid-membrane permeation and its experiences in pilot-plant and industrial scale, *Sep. Sci. Technol.*, 23 (1988) 1659 – 1666.
[97] T. Gallego-Lizon, E. S. Pérez de Ortiz, Drop sizes in liquid membrane dispersions, *Ind. Eng. Chem. Res.*, 39 (2000) 5020 – 5028.
[98] E.-S. Z. El-Ashtoukhy, M. H. Abdel-Aziz, Removal of copper from aqueous solutions by cementation in a bubble column reactor fitted with horizontal screens, *Int. J. Min. Eng.,* 121 (2013) 65 – 69.
[99] M. Karavasteva, Kinetics and deposit morphology of copper cementation onto zinc, iron and aluminum. *Hydrometallurgy,* 76 (2005) 149 – 152.
[100] O. P. Case, Metallic recovery from wastewaters utilising cementation, EPA-270/2-74-008, U.S. Washington DC (1974).
[101] T. Angelidis, K. Fytianos, and G. Vasilikiotics, Lead removal from wastewater by cementation utilising a fixed bed of iron spheres. *Environ. Pollut.*, 50 (1988) 243 - 251.
[102] B. Pakzadeh, J. R. Batista, Chromium removal from ion-exchange waste brines with calcium polysulfide, *Water Res.,* 45 (2011) 3055 – 3064.

[103] M. R. Dudzinska, D. A. Clifford, Anion exchange studies of lead-EDTA complexes, *React. Polym.,* 16 (1991) 71 – 80.

[104] R.-S. Juang, L.-D. Shiau, Ion exchange equilibria of metal chelates of ethylene-diaminetetraacetic acid (EDTA) with Amberlite IRA-68, *Ind. Eng. Chem. Res.*, 37 (1998) 555–560.

[105] A. M. Carcía, C. Montero-Ocampo, Improvement of arsenic electro-removal fromunderground water by lowering the interference of other ions, *Water Air Soil Pollut.*, 205 (2010) 237 – 244.

[106] K. Juttner, U. Galla and H. Schmieder, Electrochemical approaches to environmental problems in the process industry, *Electrochim. Acta.,* 45 (2000) 2575 – 2594.

[107] P. Jokela and P. Keskitalo, Plywood mill water system closure by dissolved air flotation treatment, *Water Sci. Technol.*, 40 (1999) 33 – 42.

[108] K. A., Matis, A. I., Zouboulis, N. K. Lazaridis and I. C. Hancock, Sorptive flotation for metal ions recovery, *Int. J. Miner. Process*, 70 (2003) 99 – 108.

[109] T. Zabel, Floatation in Water Treatment in the scientific basis of flotation. Martinus Nijhoff Publishers, *The Hague*, (1984) 349 – 378.

[110] L. K Wang, E. M. Fahey and Z. C. Wu, Dissolved air flotation in Physicochemical Treatment Processes, Humana Press, New Jersey, (2004) 431 – 500.

[111] M. B. Gabos, C. A. Abreu, A. R. Coscione, EDTA assisted phytoremediation of a Pb contaminated soil: metal leaching and uptake by jack beans, *Sci. Agric.*, 66 (2009) 506–514.

[112] T. Vesely′, P. Tlusto˘S., J. Szakova, Organic acid enhanced soil risk element (Cd, Pb and Zn) leaching and secondary bioconcentration in water lettuce (Pistia stratiotes L.) in the rhizofiltration process, *Int. J. Phytoremediation,* 14 (2012) 335 – 349.

[113] C. Neunhauserer, M. Berreck, H. Insam, Remediation of soils contaminated with molybdenum using soil amendments and phytoremediation, *Water Air Soil Pollut.*, 128 (2001) 85 – 96.

[114] S. P. Bizily, C. L. Rugh, A. O. Summers, R. B. Meagher, Phytoremediation of methyl mercury pollution: MerB expression in Arabdopsis thaliana confers resistance to organomercurials, *Proc. Natl. Acad. Sci. U. S. A.,* 96 (1999) 6808 – 6813.

[115] T. Jaffre, R. R. Brooks, J. Lee, and R. D. Reeves, (1976). Sebertia acuminate: A nickel accumulating plant from New Caledonia. *Science.,* 193 (1976) 579 - 580.

[116] M. B. Kirkham, EDTA-facilitated phytoremediation of soil with heavy metals from sewage sludge, *Int. J. Phytoremediation*, 2 (2000) 159 – 172.

[117] X. Cao, L. Q. Ma, A. Shiralipour, Effects of compost and phosphate amendments on arsenic mobility in soils and arsenic uptake by the hyperaccumulator, Pteris vittata L., *Environ. Pollut.,* 126 (2003) 157–167.

[118] W. H. O. Ernst, (1996). Bioavailability of heavy metals and decontamination of soils by plants. *Appl. Geochem.*, 11 (1996) 163 - 167.

[119] N. Caille, S. Swanwick, F. J. Zhao, S. P. McGrath, Arsenic hyperaccumulation by Pteris vittata from arsenic contaminated soils and the effect of liming and phosphate fertilization, *Environ. Pollut.*, 132 (2004) 113 – 120.

[120] F. Bartoli, D. Coinchelin, C. Robin, G. Echevarria, Impact of active transport and transpiration on nickel and cadmium accumulation in the leaves of the Ni hyperaccumulator Leptoplax emarginata: a biophysical approach, *Plant Soil,* 350 (2012) 99 – 115.

[121] W. W. Wenzel, R. Unterbrunner, P. Sommer, P. Sacco, Chelate-assisted phytoextraction using canola (Brassica napus L.) in outdoors pot and lysimeter experiments, *Plant Soil,* 249 (2003) 83 – 96.

[122] H. Grcman, D. Velikonja-Bolta, D. Vodnik, B. Kos, D. Leˇstan, EDTA enhanced heavy metal phytoextraction: metal accumulation, leaching, and toxicity, *Plant Soil,* 235 (2001) 105 – 114.

[123] T. Vesely´, P. Tlustos, J. Szakova, Organic acid enhanced soil risk element (Cd, Pb and Zn) leaching and secondary bioconcentration in water lettuce (Pistia stratiotes L.) in the rhizofiltration process, *Int. J. Phytoremediation,* 14 (2012) 335 – 349.

[124] Y. X. Chen, Y. P. Wang, W. X. Wu, Q. Lin, S. G. Xue, Impacts of chelate-assisted phytoremediation on microbial community composition in the rhizosphere of a copper accumulator and non-accumulator, *Sci. Total Environ.*, 356 (2006) 247 – 255.

[125] A. Ruttens, J. Colpaert, M. Mench, J. Boisson, R. Carleer, J. Vangronsveld, Phytostabilization of a metal contaminated sandy soil. II. Influence of compost and/or inorganic metal immobilizing soil amendments on metal leaching, *Environ. Pollut.,* 144 (2006) 533 – 539.

[126] A. Chlopecka, D. C. Adriano, Mimicked in-situ stabilization of metals in a cropped soil: bioavailability and chemical form of zinc, *Environ. Sci. Technol.,* 30 (1996) 3294–3303.

[127] N. Miyata, Y. Tani, M. Sakata, K. Iwahori, Microbial manganese oxide formation and interaction with toxic metal ions, *J. Biosci. Bioeng*., 104 (2007) 1 – 8.

[128] R. Zimmerman, Abiotic and microbial oxidation of laboratory-produced black carbon (Biochar), *Environ. Sci. Technol*., 44 (2010) 1295 – 1301.

[129] R. S. Oremland, J. T. Hollibaugh, A. S. Maest, T. S. Presser, L. G. Miller, C. W. Culbertson, Selenate reduction to elemental selenium by anaerobic bacteria in sediments and culture: biogeochemical significance of a novel sulfate-independent respiration, *Appl. Environ. Microbiol*., 55 (1989) 2333 – 2343.

[130] H. A. Wiatrowski, P. M. Ward, T. Barkay, Novel reduction of mercury (II) by mercury-sensitive dissimilatory metal reducing bacteria, *Environ. Sci. Technol.*, 40 (2006) 6690–6696.

[131] G. Renella, L. Landi, J. Ascher, M. T. Ceccherini, G. Pietramellara, M. Mench, P. Nannipieri, Long-term effects of aided phytostabilisation of trace elements on microbial biomass and activity enzyme activities, and composition of microbial community in the Jales contaminated mine spoils, *Environ. Pollut.,* 152 (2008) 702 – 712.

[132] M. M. Urrutia, T. J. Beveridge, Remobilization of heavy metals retained as oxyhydroxides or silicates by Bacillus subtilis cells, *Appl. Environ. Microbiol.,* 59 (1993) 4323 – 4329.

[133] L. R. Baker, P. M. White, G. M. Pierzynski, Changes in microbial properties after manure, lime, and bentonite application to a heavy metal-contaminated mine waste, *Appl. Soil Ecol*., 48 (2011) 1 – 10.

[134] E. S. Soltan, Isolation and Charaterization of antibiotic and heave metal resistance P. Stutzeri. *Biometals.*, 7 (2001) 30 - 40.

[135] M. L. Workentine, J. J. Harrison,. P. U. Stenroos, H. Ceri, and R. J. Turner, Pseudomonas fluorescence view of the periodic table. *Environ. Microbiol.*, 10 (2008) 238 - 250.

[136] J. L. Martinez, M. B. Sanchez, L. Martinez-Solano, A. Hernandez, L. Germendia, A. Fajardo, C. Alvarez-Ortega, Functional role of bacterial multidrug efflux pumps in microbial natural ecosystems, *FEMS Microb. Rev.*, 33 (2009) 430 - 449.

[137] S. Silver, L. T. Phung, Bacterial heavy metal resistance: new surprises. *Ann. Rev. Microb.*, 50 (1996) 753 - 789.

[138] V. K. Gupta, B. Gupta, A. Rastogi, S. Agarwal, A. Nayak, A comparative investigation on adsorption performances of mesoporous activated carbon prepared from waste rubber tire and activated carbon for a hazardous azo dye-Acid Blue 113, *J. Hazard. Mater.*, 186 (2011) 891 – 901.

[139] S. Banerjee, M. C. Chattopadhyaya, V. Srivastava, Y. C. Sharma, Adsorption studies of methylene blue onto activated saw dust: kinetics, equilibrium, and thermodynamic studies, *Environ. Prog. Sust. Ener.*, 33 (2014) 790 - 799.

[140] R. K. Gautam, A. Mudhoo, M. C. Chattopadhyaya, Kinetic, equilibrium, thermodynamic studies and spectroscopic analysis of Alizarin Red S removal by adsorption onto mustard husk, *J. Environ. Chem. Eng.*, 1 (2013) 1283 - 1291.

[141] R. K. Gautam, P. K. Gautam, S. Banerjee, V. Rawat, S. Soni, S. K. Sharma, M. C. Chattopadhyaya, Removal of tartrazine by activated carbon biosorbents of Lantana camara: Kinetics, equilibrium modeling and spectroscopic analysis, *J. Environ. Chem. Eng.*, 3:79 (2015) 79 - 88.

[142] N. Barka, M. Abdennouri, M. E. Makhfouk, S. Qourzal, Biosorption characteristics of cadmium and lead onto eco-friendly dried cactus (Opuntiaficusindica) cladodes. *J. Environ. Chem. Eng.,* 1 (2013) 144–149.

[143] V. K. Gupta, C. K. Jain, I. Ali, M. Sharma, V. K. Saini, Removal of cadmium and nickel from wastewater using bagasse fly ash—a sugar industry waste, *Water Res.,* 37 (2003) 4038 – 4044.

[144] B. H. Hameed, Spent tea leaves: A new non-conventional and low-cost adsorbent for removal of basic dye from aqueous solutions, *J. Hazard. Mat.*, 161 (2009) 753 - 759.

[145] H. G. Park, T. W. Kim, M. Y. Chae, I. K. Yoo, Activated carbon-containing alginate adsorbent for the simultaneous removal of heavy metals and toxic organics, *Process Biochem.*, 42 (2007) 1371 – 1377.

[146] A. Erto, L. Giraldo, A. Lancia, J. C. Moreno-Piraján, A comparison between a low-cost sorbent and an activated carbon for the adsorption of heavy metals from water, *Water Air Soil Pollut.*, 224 (2013) 1531.

[147] P. C. C. Faria, J. J. M. Orfao, M. F. R. Pereira, Adsorption of anionic and cationic dyes on activated carbons with different surface chemistries, *Water Res.,* 38 (2004) 2043 – 2052.

[148] Thouraya Bohli, Abdelmottale Ouederni, Nuria Fiol, Isabel Villaescusa, Evaluation of an activated carbon from olive stones used as an adsorbent for heavy metal removal from aqueous phases, *C. R. Chimie*, 18 (2015) 88 – 99.

[149] Ayhan Demirbas, Heavy metal adsorption onto agro-based waste materials: A review, *J. Hazard. Mater.*, 157 (2008) 220 – 229.

[150] S. Sobhanardakani and H. Parvizimosaed and E. Olyaie, Heavy metals removal from wastewaters using organic solid waste—rice husk, *Environ. Sci. Pollut. Res.*, (2013) 20:5265 – 5271.

[151] Mir Ali Farajzadeh, Akbar Boviery Monji, Adsorption characteristics of wheat bran towards heavy metal cations, *Separation and Purification Technology*, 38 (2004) 197–207.

[152] Ping Yin, Zengdi Wang, Rongjun Qu, Xiguang Liu, Jiang Zhang, and Qiang Xu, Biosorption of Heavy Metal Ions onto Agricultural Residues Buckwheat Hulls Functionalized with 1-Hydroxylethylidenediphosphonic Acid, *J. Agric. Food Chem.*, 2012, 60, 11664 – 11674.

[153] F. Rozada, M. Otero, A. Morana, A. I. Garcia, Adsorption of heavy metals onto sewage sludge-derived materials, *Bioresource Technology,* 99 (2008) 6332 – 6338.

[154] E. Soco, J. Kalembkiewicz, Adsorption of nickel(II) and copper(II) ions from aqueous solution by coal fly ash, *Journal of Environmental Chemical Engineering,* 1 (2013) 581– 588.

[155] Krishna G. Bhattacharyya, Susmita Sen Gupta, Kaolinite and montmorillonite as adsorbents for Fe(III), Co(II) and Ni(II) in aqueous medium, *Appl. Clay Sci.,* 41 (2008) 1 – 9.

[156] Sajjad Haider, Soo-Young Park, Preparation of the electrospun chitosan nanofibers and their applications to the adsorption of Cu(II) and Pb(II) ions from an aqueous solution, *Journal of Memb. Sci.,* 328 (2009) 90 – 96.

[157] E. Erdem, N. Karapinar, R. Donat, The removal of heavy metal cations by natural zeolites, *J. Colloid Interface Sci.*, 280 (2004) 309 – 314.

[158] R. K. Gautam, A. Mudhoo, G. Lofrano, M. C. Chattopadhyaya, Biomass-derived biosorbents for metal ions sequestration: Adsorbent modification and activation methods and adsorbent regeneration. *J. Environ. Chem. Eng.*, 2 (2014) 239 – 259.

In: Heavy Metals
Editor: Deepak Pathania
ISBN: 978-1-63484-740-7

Chapter 5

BIOSORPTION OF HEAVY METALS

***Ibrahim Ismail*[1,*] *and Tarek Moustafa*[2]**
[1]Director of Renewable Energy Engineering Program,
University of Science and Technology, Zewail City, Giza, Egypt
[2]Chemical Engineering Department, Faculty of Engineering,
Cairo University, Giza, Egypt

ABSTRACT

Heavy metals are dangerous pollutants that may exist in marine, ground, industrial and even treated waste waters. Contamination of aqueous environments by heavy metals is a worldwide environmental problem due to their toxic effects and their accumulation through the food chain. Due to rapid development of industrial activities in recent years, the levels of heavy metals in water systems, which have substantially increased over time, will lead to excessive release of the heavy metals into the environment. Hence it's necessary to remove these metals from industrial effluents before discharging aqueous wastes into environment. Industrial wastewater is often contaminated with various types of heavy metals, it is imperative that it should be treated to an environmental acceptable limit. Eleven metals namely lead, chromium, mercury, uranium, selenium, zinc, arsenic, cadmium, cobalt, copper, nickel, out of other twenty classified toxic metals, are emitted into environment in quantities that pose risks, to human health. Even if they exist in low concentrations, they are often considered as toxic compounds and their removal from wastewater is crucial issue as the accumulation of these metals in human body can lead to severe disorders in the performance of kidneys, liver, sexual, brain and nerve system. Several technologies have been developed over the years to isolate and remove these heavy metals from wastewaters. Such methods include ion-exchange, electrochemical reduction, evaporation, solvent extraction, reverse osmosis, chemical precipitation, membrane filtration, cementation, coagulation and flocculation, complexation, adsorption, biosorption, etc. Some of these traditional metal removal methods have certain disadvantages like incomplete metal removal, high reagent and/or energy requirements, generation of toxic sludge or other waste products .Recently, Biosorption of heavy metals have attracted several researchers as one of the most convenient heavy

---

* Corresponding author: Ibrahim Ismail. Renewable Energy Program, Zewail City for Science and Technology, 6th of October, Giza, Egypt. E-mail: ibrahim.ismail@zewailcity.edu.eg.

metal removal techniques. Biosorption, which can be defined as the uptake of heavy metal ions and radionuclides from aqueous environments by biological materials, such as algae, bacteria, yeast, fungi, plant leaves and root tissues, has many advantages such as low capital and operating costs, selective removal of metals, possibility of biosorbent regeneration and metal recovery, rapid kinetics of adsorption and desorption and no sludge generation. In this chapter we shall explore biosorption methodologies, theoretical background, possibilities, limitations and different applications.

Keywords: heavy metals, adsorption, sorption, biosorption, hazardous waste, environment

1. Introduction

This chapter provides an overview of previous investigations dealing with the adsorption of heavy metals ions from solution on activated carbons and other adsorbents specially bio adsorbents. A comprehensive literature review related to the use of activated carbon and anaerobic biomass structure is included with a summary of processes where heavy metals are removed.

This chapter also includes a review on the different equilibrium isotherm models which are used to describe experimental adsorption data. The adsorption isotherm data are the corner stone in predictive modeling procedures used for analysis and design of adsorption process. A variety of different isotherm equations are presented in this chapter, some of which have a theoretical foundation and some being of more empirical nature [1].

A number of mathematical models have been presented in this chapter to explain the kinetic behavior, isotherms and to estimate the breakthrough curves.

1.1. Concept and Definition of Adsorption and Bio Sorption

1.1.1. Adsorption

There are several definitions reported for adsorption. Adsorption is a separation process based on the ability of a certain solid(s) to remove liquid or gaseous components from a flow stream [2]. It involves contact of solids with either liquid or gas, with mass transfer in the direction of fluid to solid; taking place on the interface [3]. Adsorption is sorption process, which exploits the ability of certain insoluble solid particles preferentially to concentrate specific substances, called solutes, onto their surfaces [4]. Adsorption is a mass transfer operation where a constituent in the gas/liquid phase is transferred to the solid phase at the interface [5]. In an adsorption process, molecules or atoms or ions in a gas or liquid diffuse to the surface of a solid, where these bond with the solid surface or are held there by weak intermolecular forces. The substance which is removed from the gas or liquid phase at the interface is termed adsorbate. A porous solid material of large surface area per unit mass where adsorption of liquids or gases occurs on its surface is known as adsorbent [6].

To achieve a very large surface area per unit volume or unit mass for adsorption, highly porous solid particles with small diameter interconnected pores are used [7]. Adsorption is always accompanied by the liberation of heat which is called heat of adsorption [8]. All adsorption processes involve a decrease in the free energy of the system [3]. Adsorption is one of the most commonly used processes in waste water treatment because it is fairly simple and convenient unit operation and its cost is relatively low compared to other treatment processes. Adsorption can be equally effective in removing trace component from the liquid phase and may be used either to recover the component or simply to get rid of an industrial effluent of a noxious substance [8]. It has wide applications including the purification of drinking water, removal of harmful pollutants from wastewater effluents. It also has application in air pollution control and many other processes in chemical engineering [9]. It is well known that liquid-phase adsorption is one of the most efficient methods for the removal of colors, odors, and organic pollutants from waste effluents [10]. The removal of such pollutants, particularly the synthetic variety, can be of the particular significance to the environment.

The relative advantages of adsorption over other conventional advanced treatment methods are:

1. It can remove both organic as well as inorganic constituents even at very low concentration.
2. It is relatively easy and safe to operate.
3. It can be operated in both batch and continuous mode.
4. It has no sludge formation.
5. The adsorbent can be regenerated and used again [11].

1.1.2. Biosorption

Biosorption can become a good weapon in the fight against toxic metals threatening the environment. Bio sorption, the process in which a microorganism and/or biological species are used to remove and recover heavy metals from aqueous solutions, has been known for a few decades but has emerged a low cost promising technology in the last decade. In this process, the uptake of heavy metals occurs as a result of physical-chemical interactions of metal ions with cellular compounds of biological species.

Bio sorption can be also described as the removal of materials (compounds, metal ions, etc.) by inactive, non-living biomass (materials of biological origin) due to "high attractive forces" present between the two [12].

The bio sorption process includes two phases. One phase is a solid phase (biomass/ sorbent/bio sorbent/biological material) and another is liquid phase (solvent, usually water) containing a dissolved species to be sorbed (sorbate/metal ion). Principally, bio sorption process, which is metabolism-independent accumulation of metals, is often rapid. On the contrary, bioaccumulation, which is metabolism-dependent intracellular uptake of metal ions by living microorganisms, is slower process compared to bio sorption [13].

Generally, Bio sorption is a property of certain types of inactive, dead, microbial biomaterials to bind and concentrate heavy metals from even very dilute aqueous solutions. Biomass exhibits this property, acting just as a chemical substance, as an ion exchanger of biological origin. It is particularly the cell wall structure of certain algae, fungi and bacteria, which was found responsible for this phenomenon.

2. Principle of Adsorption and Biosorption

2.1. Principle of Adsorption

The constituent that undergoes adsorption is referred to as the adsorbate and the solid into which the constituent is adsorbed referred to as the adsorbent. During the adsorption process, dissolved species are transported into the porous solid adsorbent granular by diffusion and are then adsorbed onto the extensive inner surface of the adsorbent. Dissolved species are concentrated on the surface by chemical reaction (chemisorption) or physical attraction (physical adsorption) to the surface. Physical adsorption is a rapid process caused by non-specific binding mechanisms such as van der Waals forces. Physical adsorption is reversible, that is, the adsorbate in response to a decrease in solution concentration. Physical adsorption is the most common mechanism by which adsorbates are removed in water treatment [14]. The physical adsorption process is exothermic with a heat of adsorption that is typically 4 to 40 KJ/mole [15].

Chemisorption is more specific because a chemical reaction occurs that entails the transfer of electrons between adsorbent and adsorbate and a chemical bond at the surface takes place. The heat of adsorption for chemisorption is typically above 200 KJ/mole [16]. Chemisorptions is usually not reversible and adsorption, if it occurs, is accompanied by a chemical change in the adsorbate [17].

While physical adsorption and chemisorption can be distinguished easily at their extremes, some cases fall between the two,as a highly unequal sharing of electrons may not be distinguishable from the high degree of distortion and electron cloud that occurs with physical adsorption [18-21].

Table 1 shows a comparison of physical and chemical adsorption [22].

In case the adsorption process is characterized by electrical attraction between the adsorbate and the adsorbent surface, ions with greater charge are attracted more strongly toward a site of opposite charge than are molecules with lesser charge. Also, the smaller size of the ion is the greater attraction with the adsorbent surface [15, 23]. This process is known as ion exchange.

Table 1. Comparison of physical adsorption and chemisorption mechanisms [22]

Parameter	Physical Adsorption	Chemisorption
Water treatment application	Most common type of adsorption mechanism	Rare in water treatment
Process speed	Limited by mass transfer	Variable
Type of bonding	Non-specific binding mechanisms such as van der Waals forces, vapor condensation	Specific exchange of electrons, chemical bond at surface
Type of reaction	Reversible, exothermic	Typical non reversible, exothermic
Heat of adsorption	4-40 KJ/mole	More than 200 KJ/mole

Table 2. Comparison of characteristic features of heavy metal biosorption and bioaccumulation processes

Feature	Biosorption process	Bioaccumulation process
Definition	The passive sorption of the dead microbial cell walls is called as biosorption	The active process of metal removal by living cells is referred to as bioaccumulation
Metal affinity	Higher under favorable conditions	Toxicity affects metal uptake by living cells, but in some instances, there is high metal accumulation
Rate of metal uptake	Generally rapid, a few seconds for outer cell wall accumulation	Normally slower than biosorption
Selectivity	Variety of ligands involved, hence poor	Better than biosorption, but less than some chemical technologies
Temperature	Within a modest range	Inhibited by low temperatures

2.2. Principles of Biosorption

Living as well as dead (metabolically inactive) biological materials have been sought to remove metal ions. It was found that various functional groups present on their cell wall offer certain forces of attractions for the metal ions and provide a high efficiency for their removal [24, 25]. Eccles suggested the following comparison of characteristic features of heavy metal biosorption and bioaccumulation processes [26].

3. Adsorption Process Design

The contact between solid and liquid can be made by at least five systems: batch contact, fixed bed, pulsed bed, steady state moving bed and fluidized bed [4]. Our study focused on the batch and fixed bed adsorption systems as given below.

3.1. Batch System

When a mass of adsorbent is contacted with a given volume of a liquid containing an adsorbate solute (heavy metals ions), for a sufficient long time, adsorption occurs until equilibrium is achieved between the amount of metal ions adsorbed and the amount of metal ions remaining in the solution. The uptake of metal ions in single and multi component systems was calculated by difference in their initial and final concentrations. For system under equilibrium conditions, a mass balance states the amount of solute adsorbed onto the solid must be equal to the amount of solute removed from the solution, or in mathematical terms:

Amount adsorbed = Initial amount of adsorbate present - Final amount of adsorbate

The symbolic representation at equilibrium can be written as follow:

$$q_e W = V C_o - V C_e \qquad (2.1)$$

Equation 2.2 can be written as:

$$q_e = \frac{V}{W}(C_o - C_e) \qquad (2.2)$$

where q_e is the amount of the material adsorbed at equilibrium/gm adsorbent, V is the volume of solution and W is the mass of the adsorbent. It is well known that q is sometimes expressed in units of mass or mole of solute per unit mass of the solid phase [27, 28], however in this chapter the mass of solute per unit mass of the solid phase is the selected definition. Accordingly, the initial metal ion concentration C_o and the metal ion concentration at equilibrium C_e are expressed in mg solute per liter and the adsorption graphs are presented on mg/g basis.

The percentage adsorption is calculated as follows:

$$\% \text{ Adsorption} = \frac{(C_o - C_e)}{C_o} * 100 \qquad (2.3)$$

The capacity mg/g of metal ions adsorbed from solution by adsorbent q_e is calculated from:

$$q_e = \frac{(\% \text{ Adsorption} * C_o)}{100} * \left[\frac{V}{W}\right] \qquad (2.4)$$

3.2. Fixed Bed Adsorption

In recent years there has been increasing concerns about the contamination of drinking water sources by organic and inorganic chemicals. The public awareness of the danger associated with water contamination has opened new horizons for developing effective unit process operation for industrial wastewater treatment [29]. Fixed bed adsorber is the most efficient arrangement for conducting adsorption operation for industrial applications in wastewater treatment [30]. Adsorption on granulated activated carbon has since been recognized as an effective treatment technology for removal of a wide spectrum of contaminates from industrial wastewater [31]. Adsorption is commonly performed in a packed bed where an adsorbent such as granulated activated carbon (GAC) is held stationary in a column and the contaminated water is contacted with the adsorbent by its continuous passage through the bed [32].

For long passage times, the adsorbent closer to the inlet of the fluid becomes saturated with the contaminants. This gives rise to a concentration front that divides the bed into a contaminated region and an intact region. This concentration front (also called the mass transfer zone) continues to travel along the bed as more influent is being fed into the bed. At the time when the leading edge of the concentration front has reached the exit end of the bed, breakthrough is said to have been attained and the first emergence of the contaminant (usually at very low concentration) has just taken place. This staged process is schematically depicted in Figure 1, which also shows the development of the S-shape breakthrough curve [33].

A breakthrough curve represents the effluent concentration history (i.e., effluent concentration versus time) along the course of the adsorption operation.

As shown in Figure 1, the point at which breakthrough occurs is called the breakthrough point. The location of the breakthrough point and the shape of the breakthrough curve (degree of concavity) are influenced by several factors pertaining to adsorbent-adsorbate interaction mechanisms and the hydrodynamic conditions [16]. Table 3 list most of these factors and their impact on the break point delay [33]. The breakthrough point is often taken as a relative concentration of 0.05 to 0.1 and since only the last portion of fluid processed has this high of concentration, the average fraction of solute removed from the start to the break point is often 0.99 or higher [34].

A number of factors determine the shape of breakthrough curve and the height of the adsorption zone [35]:

- Column dimensions
- Flow rate
- The type of the adsorbent, the shape and size of the adsorbent particle
- The feed concentration of impurities
- Temperature
- pH of solution

As a general rule, the time to break point is decreased by [36]

- Increase in flow rate
- Decrease in bed depth
- Increase in concentration of solute in the influent
- Increase in particle size
- Increase or decrease in pH of the solution

4. Competitive Adsorption

For systems containing more than one adsorbable species i.e., Multicomponent system, the concentrations of each species shifts relative to one another as the fluid travels. The shift, however, soon stabilize to form the concentration fronts [37]. In process having three components, four plateau zones are formed, which are separated by three transition zone. Roll-up phenomena is the widely known phenomenon in competitive adsorption, as humps on the weakly adsorbed components breakthrough curve where the concentration exceeds the feed concentration. This increase in concentration is caused by the displacement of a weaker adsorbate by a stronger one [38, 39].

As shown in Figure 1, at the onset of the adsorption process, there will be a plateau zone 1 or (feed zone) in which the adsorbent is saturated and the concentration of components is the same as in feed. Zone 2 is the first transfer zone (separation zone) in which the stronger component is adsorbed by solid causing an increase in concentration of the other components.

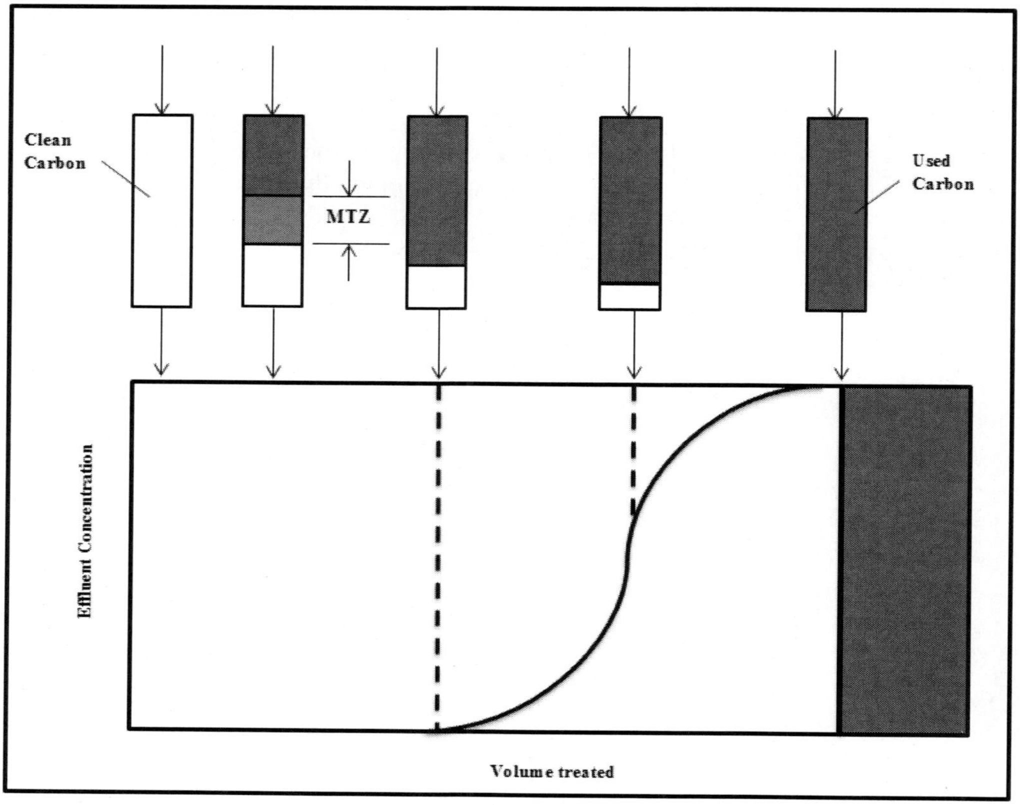

Figure 1. Adsorption column mass transfer zone and idealized breakthrough curve [33].

Table 3. Effect of increasing value of input parameters on break point [33]

Parameter	Breakthrough time
Influent concentration	Decrease
Influent flow rate	Decrease
Cross-sectional area	Increase
Weight of adsorbent	Increase
Bed depth	Increase
Solution temperature	Generally decrease
Solution pH	Variable
Bed voidage	Decrease
Particle diameter	Decrease
Intraparticle porosity	Increase
External mass transfer coefficient	Increase
Internal diffusion	Increase

Again, zone 3 is a plateau zone, where the weak components are in equilibrium with the adsorbed one. Zone 4 is the second transfer zone, in which the second stronger component is adsorbed, causing an increase in the concentration of the weak component in the fluid. Zone 5 is again a plateau zone, where the weak component is in equilibrium with the other adsorbed component. Zone 6 is the third transfer zone in which the last component (most weakly adsorbed), is adsorbed. Finally, zone 7 is a plateau zone, in which there is still a residual capacity for adsorption of further fluid. This system of zones travels along the bed until the feed zone 1 reaches the end of the bed i.e., saturation is reached. The shape of the breakthrough curve is determined by the nature or type of adsorption isotherm [38, 40].

5. Variables Affecting Adsorption Capacity

Adsorption capacity for specific single solute is thought to be direct function of:

1. The adsorbate properties, such as functionality, branching or geometry, polarity, hydrophobicity, dipole moment, molecular weight and size, and aqueous solubility.
2. The solution condition such as pH, temperature, pressure, adsorbate concentration and ionic strength.
3. The nature of the adsorbent, such as surface, pore size distribution and surface characteristics (i.e., number of $O\bar{H}$ bond /$A^{\circ 2}$) [41]

By using the same type of adsorbent under identical single solution conditions, different workers attempted to correlate single solute adsorption capacity with position of substitution [41], molecular weight [42, 43], solubility [43] and functionality [44]. The results indicated that capacity for adsorption will:

- Increase with increasing molecular weight as a homologous series is ascended.
- Decrease with increasing solute polarity and solubility.
- Decrease for the position of substitution for hydroxyl- and amino benzoic acid in the order ortho, para and meta.
- Decrease with increasing branching.

6. Adsorption Isotherm

The adsorption isotherm expresses the relation between the amount or mass of material adsorbed (q_e) and the material concentration in the solution (C_e) at equilibrium. Generally the amount of material adsorbed is determined as a function of the concentration at a constant temperature and the resulting function is called adsorption isotherm [45].

Adsorption isotherm can be defined as "the equilibrium relationship between the adsorbate concentration in the fluid phase and the adsorbate concentration in the adsorbent particles at a given temperature" [27]. Some typical isotherm shapes are shown as arithmetic graphs [46, 47]. The linear isotherm passes through the origin and the amount of adsorbed is proportional to the concentration in the fluid. The convex upward isotherm is considered favorable, because a relatively high loading can be obtained at low concentration in the fluid. The limiting case of a very favorable isotherm is irreversible adsorption, where the amount adsorbed is independent of concentration down to very low values. The concave upward isotherm is considered unfavorable because a relatively low solid loading is obtained and because it leads to quite long mass transfer zone in the bed [48]. From theses curves it will be noted that adsorption is a specific property depending upon the nature of the adsorbate-adsorbent system.

6.1. Adsorption Isotherm Models

Most of adsorption theories have been developed for gas-solid system because the gas state is better understood than liquid. Till now the statically theories developed for gas-solid system have been applied to liquid-solid systems with little confidence in designing of the equipment [4].

Adsorption isotherm equations are used to describe experimental sorption data. The equation parameters and the underlying thermodynamic assumptions of these equilibrium models often provide some insight into both the sorption mechanism and the surface properties and affinity of the sorbent [49].

6.1.1. Single Component Isotherm Models

6.1.1.1. Langmuir Isotherm

One practical mathematical form of an isotherm which is often found to fit experimental data are the so called Langmuir isotherm. The Langmuir model is probably the best known and most widely applied sorption isotherm. It has produced a good agreement with a wide variety of experimental data. The Langmuir isotherm is applied to homogeneous sorption [49].

Langmuir isotherm equation is represented as follows:

$$q_e = \frac{qmbC_e}{1+bC_e} \tag{2.5}$$

By arranging equation 2.5 to

$$\frac{C_e}{q_e} = \frac{1}{qmb} + \frac{1}{qm} C_e \tag{2.6}$$

The value of the Langmuir constants qm and b can be calculated by least square methods or graphically by plotting C_e/q_e against C_e [50].

The Langmuir isotherm is valid for description of single layer adsorption on a surface containing a finite number of binding sites. It is the simplest theoretical model for monolayer adsorption which was developed from either kinetic derivation or thermodynamic derivation [40]. The Langmuir isotherm is developed by assuming:

- Fixed number of accessible sites are available on the adsorbent surface.
- Each site can hold one adsorbate molecules.
- All sites are energetically equivalent.
- There is no interaction between molecules adsorbed on neighboring sites.
- Adsorption is reversible.

The adsorption isotherm of copper, nickel, cobalt and manganese from aqueous solution by kaolinite was studied [51]. The sorption of these metals on kaolinite conforms to linear form of Langmuir adsorption equation. The Langmuir adsorption model was applied to the adsorption of cadmium and zinc ions onto aquatic moss [52]. The experimental results obtained for each metal fit adequately with Langmuir isotherm model. The Langmuir isotherm can be extended for multicomponent system to give the following form [53].

$$q_{e,i} = \frac{q_{m,i} b_i C_{e,i}}{1+\sum_{i=1}^{N} b C_{e,k}} \qquad (2.7)$$

where C_{e}, i is the equilibrium concentration of component i in multicomponent system, q_{e}, i is the equilibrium uptake of the component i, and b_i is the single component Langmuir parameter for component i.

6.1.1.2. BET Isotherm

Brunauer, Emmet and Teller (BET) developed a simple model isotherm for multilayer adsorption [40]. The BET model assumes that a number of layers of adsorbate accumulate at the adsorbent surface and that the Langmuir isotherm model applies to each layer. BET equation is limited by the assumption of uniform energies of adsorption and it may be deduced either from kinetic consideration or the thermodynamics of the adsorption process [54]. The BET model is based on a number of rather serious idealization as follows [40]:

- Each molecule in the first adsorbate layer is considered to provide one site for the second and subsequent layers.
- The molecules in the second subsequent layers which are in contact with other adsorbate rather than with the surface of the adsorbent, are considered to behave as saturated liquid. The BET equilibrium isotherm equation takes the simplified form:

$$q_e = \frac{q m b C_e}{(Cs-C_e)\left[1+(b+1)^{C_e/Cs}\right]} \qquad (2.8)$$

where C_s is the saturation concentration of the adsorbent.

6.1.1.3. Freundlich Isotherm

Herbert Max Finley Freundlich, a German physical chemist, presented an empirical adsorption isotherm for non-ideal systems in 1906 [55]. The Freundlich isotherm is the earliest known relationship describing the sorption equation and has widely been used for many years. This fairly satisfactory empirical isotherm can be used for non-ideal sorption and involves heterogeneous sorption and is expressed by the following equation.

$$q_e = Kf\, C_e^{1/n} \tag{2.9}$$

And the equation may be linearized by taking logarithms as follows:

$$\log(q_e) = \frac{1}{n}\log(C_e) + \log(Kf) \tag{2.10}$$

where Kf and 1/n are empirical constants. These constants can be determined by plotting Log (q_e) vs. Log (C_e), and the linear line obtained gives a slope which is the value of 1/n and y-intercept is Log(Kf). The intercept is an indicator of adsorption capacity and the slope is adsorption intensity. A relatively slight slope (and hence a high value of n) indicates that adsorption is good over the entire range of concentration studied, while a steep slope (and hence small value of n) means that adsorption is effective at high value concentrations. A greater value of the intercept Kf indicates a higher capacity for adsorption.

Tito et al. showed that the equilibrium isotherms of zinc onto natural bentonite are well described by Langmuir and Freundlich model [50]. They found that the maximum adsorption capacity qm according to Langmuir model is 52.9 mg/g for an initial zinc concentration of 300 mg/L. Muhammad et al. found that the adsorption of Cu, Cr and Cd onto sand satisfied the Langmuir and Freundlich isotherms [55]. They determined the adsorption isotherm for heavy metals ions in aqueous solutions using activated carbon. These adsorption isotherms were seen to be consistent with Freundlich adsorption isotherm. The adsorption of cadmium and zinc in single and multi-component systems using activated carbon was studied [56]. Adsorption data were found to follow both the Freundlich and Langmuir models. The data are better fitted by the Freundlich isotherms as compared to Langmuir in both the single and multi-component systems. While, the adsorption data of cadmium and nickel in wastewater were found to follows the Langmuir model better than the Freundlich isotherm [54].

6.1.1.4. Linear Isotherm

The linear isotherm model describes the accumulation of solutes by sorbent as directly proportional to the solution concentration as follows:

$$q_e = K_D . C_e \tag{2.11}$$

The constant of proportionality or distribution coefficient *KD* is often referred to as partition coefficient. Linear, Langmuir and Freundlich isotherms are used for analysis of experimental data. All three models may give close results at low concentrations but the Freundlich isotherm gives the best fit for the data.

6.1.1.5. Radke-Prausnitz Isotherm

The Radke-Prausnitz isotherm model [57] expressed empirically by the following equation:

$$q_e = \frac{K_{RP}\, C_e}{1+\left(\frac{K_{RP}}{F_{RP}}\right) C_e^{\,1-N_{RP}}} \tag{2.12}$$

where KRP and NRP are the model parameters, normally obtained by non-linear statistical fit of the equation to the experimental data.

The Radke-Prausnitz equation has several important properties allowing it to be suitable for use in many adsorption systems. At low concentration it reduces to a linear isotherm. At high concentration it becomes the Freundlich isotherm and for the special cases of NRP = 0 it becomes the Langmuir isotherm. Thus Radke-Prausnitz model give a good fit over a wide range of concentrations.

6.1.1.6. Dubinin-Radushkevich Model

The Dubinin-Radushkevich equation [29] represents the isotherm of fractional filling (qe/qDR) of micropore volume in which the maximum adsorption capacity is qDR:

$$\frac{q_e}{qD_R} = e^{(-\beta \epsilon^2)} \tag{2.13}$$

where: qD_R is the maximum adsorption capacity in the micropore volume (mol/g), ϵ is the Polanyi adsorption potential, and β is the characteristic parameter related to adsorption energy for the given system (mol^2/kj^2).

The Polanyi adsorption potential ϵ is an adsorption affinity defined in the liquid phase by:

$$\epsilon = R.T \ln \frac{C_S}{C_e} \tag{2.14}$$

where T is the temperature in Kelvin and Cs is the solubility of solute (mol/l).

Compared with Langmuir theory, the Dubinin approach assumes that there is a surface area where the adsorption energy is homogeneous.

The characteristic adsorption energy E_o (kJ/mole) is given by:

$$E_o = 1 / \sqrt{\beta} \tag{2.15}$$

6.1.1.7. Redlich-Peterson Model

The Redlich-Peterson isotherm model [58] is expressed by the following equation:

$$q = \frac{A_R C_e}{1 + B_R C_e^{m_R}} \tag{2.16}$$

where: A_R, B_R and m_R are the model parameters.

6.1.1.8. Combination of Langmuir-Freundlich Model

The Sips model [59] for single component adsorption is given by:

$$q_e = \frac{bqmC_e^{1/n}}{1 + bC^{1/n}} \tag{2.17}$$

6.1.2. Multi-Component Isotherm Models

The experimental measurement of multi-component adsorption isotherm is time consuming because of large number of variables involved. Thus, the problem of predicting multicomponent adsorption isotherm from single component adsorption data had attracted a lot of investigators. Several isotherms have been proposed to describe the competitive adsorption. Most of these isotherms are based on single component isotherm parameters and correction factors extracted from the experimental competitive data [53]. In the following sections, some of these models will be presented.

6.1.2.1. Extended Langmuir Model

The Langmuir isotherm can be extended for multicomponent system to give the following form [27, 53, 60]:

$$q_{e,i} = \frac{qmb_iC_{e,i}}{1+\sum_{k=1}^{N} bkC_ek,} \quad (2.18)$$

where $C_{e,i}$ is the equilibrium concentration of the component i in the multicomponent solution, $q_{e,i}$ is the equilibrium uptake of the component i, qm, i and bi are the single component Langmuir parameters for component i.

6.1.2.2. Redlich-Peterson Model

The competitive Redlich-Peterson model [53] related to the individual isotherm parameters is given by the following equation:

$$q_{e,i} = \frac{K_{R_i}(b_{R_i})C_{e,i}}{1+\sum_{k=1}^{N} b_{R,k}C^{m_{R,i}}} \quad (2.19)$$

where $K_{R,i}$ and $m_{R,i}$ are the Redlich-Peterson single component parameter for component i.

6.1.2.3. Combination of Langmuir-Freundlich Model

The competitive Sips model [53] related to the individual isotherm parameters is expressed by the following equation:

$$q_{e,i} = \frac{qmb_iC_{e,i}{}^{1/n}}{1+\sum_{i=1}^{N} b_iC_{e,i}{}^{1/n}} \quad (2.20)$$

6.2. The Isotherm Shape

The effect of isotherm shape can be used to predict whether adsorption system is favorable or unfavorable. The essential feature of the Langmuir isotherm can be expressed in terms of a dimensionless constant separation factor or equilibrium parameter KR which is defined by the following relationship

$$KR = \frac{1}{1+bC_o} \quad (2.21)$$

This parameter (KR) indicates the shape of the isotherm [49] as shown in Table 4 below.

Table 4. Determination of the shape of the isotherm

Valuable of KR	Type of isotherm
KR > 1	Unfavorable
KR = 1	Linear
0 < KR < 1	Favorable
KR = 0	Irreversible

7. Adsorption Mechanisms

For the adsorption of a solute onto the porous surface of an adsorbent, the following steps are required [7].

- External (interphone) mass transfer of the solute from the bulk fluid by convection, through a thin film r boundary layer surrounding the adsorbent to the outer solid surface of the adsorbent (film diffusion).
- Inter (interphone) mass transfer of the solute by pore diffusion from the outer surface of the adsorbent to the inner surface of the internal porous structure.
- Surface diffusion along the porous surface.
- Adsorption of the solute onto the porous surface. This involves the attachment of solute (adsorbate) to the adsorbent at an available adsorption site.

Because adsorption occurs in series of steps, the slowest step in the series is defined as the rate limiting step.

In general if physical adsorption is the principal method of adsorption one of the diffusion transport steps will be the rate limiting because the rate of physical adsorption is rapid. Where chemical adsorption is the principle of the adsorption step has often been observed to be the rate limiting [5].

Adsorption process proceeds through various mechanisms such as external mass transfer of solute onto adsorbent followed by intraparticle diffusion.

7.1. Film Diffusion

External (interphone) mass transfer of the solute from the bulk of solution by convection took place through a thin film or boundary layer surrounding the adsorbent, to the outer solid surface of the adsorbent [7]. The driving force for mass transfer is the concentration gradient across the boundary layer that surrounds each adsorbent particle through which the mass transfer occurs by molecular diffusion. The boundary layer is affected by the hydrodynamic condition outside the particle [47]. It is convenient to correlate mass transfer rates in term of an effective coefficient (Kf), defined according to a linear driving force equation [40] as:

$$\rho p = \frac{\partial qi}{\partial t} = kfa(Cbi - Cpi, R = Rp) \tag{2.22}$$

where a is the external surface area per unit volume of the particle, defined as:

$$a = \frac{3}{Rp} \tag{2.23}$$

where Rp is the particle radius. Film diffusion generally describes the resistance to mass transfer at the surface of the particle. The film diffusion model is restricted to apply to low concentration and if mass transfer is governed only by external resistance (film diffusion), there is no concentration gradient within the particle. The appropriate dimensionless group characterizing film mass transfer is the Sherwood number, defined by [40], as:

$$Sh = \frac{2RpKf}{Dm} \tag{2.24}$$

This is an analogy of the Nuselt number for heat transfer. The limiting value of Sherwood number for low Reynolds number flows is 2.0. At higher Reynolds numbers, convection effect becomes significant and a correlation of the form Sh = (f (Sc, Re) is to be expected.

The correlation of Ranze and Marshall (1952) has been widely applied to packed beds as:

$$Sh = \frac{2RpKf}{Dm} = 2.0 + 0.6Sc^{1/3}Re^{1/2} \tag{2.25}$$

The above equation 2.25 is usually used in fixed bed catalytic applications as well and proved to fit well the experimental results. Equation 2.25 implies that the Sherwood number approaches a limiting value of 2.0 at low velocities. The recommended correlation of Wilson and Geankoplis [61], for external mass transfer coefficient in adsorption beds is given by:

$$Sh = \frac{1.09}{\varepsilon b} Re^{1/3} Sc^{1/3} \text{ for } 0.0015 < Re < 55 \tag{2.26}$$

For fixed bed fluid mass transfer coefficient, Williamson and Geanloplis (1963) used:

$$\frac{kf}{vs} Sc^{0.58} = 2.4Re^{-0.6}, \text{ for } 0.08 < Re < 125 \tag{2.27}$$

Molecular diffusion coefficient of the solute in aqueous systems can be obtained from this equation [27]:

$$Dm = 7.4 * 10^{-12} (\varphi.Mf)^{1/2} \frac{T}{\mu wVAi^{0.6}} \tag{2.28}$$

7.2. Intrapartical Diffusion

Mass transfer within the particle may be limited by [47];

- Pore diffusion (Macropore diffusion)
- Solid diffusion (Micropore diffusion)
- Combined pore and solid diffusion

7.2.1. Pore Diffusion (Macrospore Diffusion)

Diffusion of molecules in the liquid pores is called "Pore diffusion." The driving force for diffusion is the concentration gradient of diffusion species within the pore. This process is often referred to as macropore diffusion [62].

The equation of diffusion for macropore-controlled may be obtained from diffusional mass balance on a spherical shell element [40, 47]:

$$\rho p \frac{\partial qi}{\partial t} + \varepsilon p \frac{\partial Cpi}{\partial t} = Dpi\varepsilon p\left[\frac{\partial^2 Cpi}{r^2} + \frac{2}{r}\frac{\partial Cpi}{\partial r}\right] \tag{2.29}$$

where Dpi is the diffusivity of the liquid in pores of adsorbent and which is also called the effective pore diffusion coefficient, and is assumed to be independent of concentration.

7.2.2. Solid Diffusion (Micropore Diffusion)

Solid diffusion in the adsorbent is often called surface diffusion, micropore diffusion or intercrystaline diffusion. The driving force for adsorption is the concentration gradient of species in its adsorbed state [47].

For most particle shapes, representation as an equivalent sphere is an acceptable approximation and transport may therefore be described by a diffusion equation, written in spherical coordinates [40]:

$$\frac{\partial qi}{\partial t} = Dst\left[\frac{\partial^2 qi}{\partial r^2} + \frac{2}{r}\frac{\partial qi}{\partial r}\right] \tag{2.30}$$

7.2.3. Combined Pore and Solid Diffusion

In porous adsorbent and ion exchange resins, intrapartical diffusion can occur with pore and solid diffusion in parallel. The dominant transport process is the faster one, and this depends on the following [47]:

1. Relative diffusivity
2. Pore fluid concentration
3. Adsorbed phase concentration

It is assumed that the diffusion coefficients are constant; the diffusion within the particles is represented by the unsteady-state equation [63] as:

$$\rho p \frac{\partial qi}{\partial t} + \varepsilon p \frac{\partial Cpi}{\partial t} = Dpi\varepsilon p\left[\frac{\partial^{2Cpi}}{\partial r^2} + \frac{2}{r}\frac{\partial Cpi}{\partial r}\right] + \rho pDsi\left[\frac{\partial^2 qi}{\partial r^2} + \frac{2}{r}\frac{\partial qi}{\partial r}\right] \tag{2.31}$$

And for dilute solution [64]:

$$\rho p \frac{\partial qi}{\partial t} = Dpi\varepsilon p\left[\frac{\partial^{2Cpi}}{\partial r^2} + \frac{2}{r}\frac{\partial Cpi}{\partial r}\right] + \rho pDsi\left[\frac{\partial^2 qi}{\partial r^2} + \frac{2}{r}\frac{\partial qi}{\partial r}\right] \tag{2.32}$$

If the equilibrium isotherm is linear

$$qi = KiCpi \tag{2.33}$$

Equation 2.33 may be expressed in either C or q and the two right hand terms are added together. The coefficient of diffusion would then be:

$$De,si = \rho p Dsi + \frac{Dpi\varepsilon p}{Ki} \tag{2.34}$$

$$De,si = \rho pDsiKi + Dpi\varepsilon p \tag{2.35}$$

When using q_i and Cp_i as concentration variables respectively, the two effective coefficient of diffusion $D_{e,s}$ and $D_{e,p}$ will differ by a factor of Ki. For nonlinear equilibrium there is no such simple relationship.

8. Adsorbents

Most solids have some tendency to adsorbed species from gases or liquids onto their surface, however only some materials have a sufficient selectivity and capacity to make them serious candidates for commercial adsorbent [7]. The adsorbents should have the following properties:

- a high internal volume which is accessible to the components being removed from the fluids. Typically a high porous solid which may be carbonaceous or inorganic in nature, synthetic or naturally occurring, and in a certain circumstances may have true molecular sieving properties, can be used.
- good mechanical properties such as strength and resistance to attrition.
- good kinetic properties, that is, it must be capable of transferring adsorbing molecules rapidly to the adsorption sites.
- in most applications the adsorbent must be regenerated after use and therefore it is desirable that regeneration can be carried out efficiently and without damage to mechanical and adsorptive properties [52].

Two important characteristics of solid adsorbent:

1. It should have extremely large surface area to volume.
2. It should have preferential affinity for certain constituent in the liquid phase [5].

The high internal surface area of an adsorbent creates the capacity needed for a successful separation or purification process. Adsorbent can be made with internal surface area which range from about $100m^2/g$ to over $3000m^2/g$. For most adsorbents the internal surface area is created from pores of various sizes. For practical applications, however, the range is stricter from about 300 m^2/g to $1200m^2/g$.

Many adsorbent materials, such as carbons, silica gels and alumina are amorphous and contain complex network of interconnected micropores, mesopores and macropores. In contrast, in zeolite adsorbents the pores or channels have precise dimensions although a macroporous structure is created when pellets are manufactured from the zeolite crystals by the addition of the binder. Fluid molecules which are to be adsorbed on the internal surface must first pass through the fluid film which is external to the adsorbent particles, hence through the macroporous structure into the micropores where the bulk of the molecules are adsorbed [52]. Typically commercial adsorbent, may be granules, spheres, cylindrical pellets, flasks and/or powders of size ranging from 50 um to 1.2 cm [7]. Table 5 provides the summery of previous published work using different types of adsorbents. An overview of activated carbon, clay and other adsorbent is provided below:

Table 5. Some previous reported work of heavy metal sorption by different adsorbents

Adsorbate	Adsorbent	Variables studied	Note
Cu(II) Cd(II)	Bone char	Single component system; Flow rate Initial metal concentration Particle size	Fixed bed
Cu(II) Ni(II) Co(II) Mn(II)	Kaolinite	Adsorption isotherm	Batch
Cr(VI)	Activated carbon	Initial metal concentration pH Particle size	Batch
Pb(II)	Activated carbon	Initial ion conc.	Batch
Cd(II)	Kaolinite Montmorillonite	Mass of adsorbent pH Contact time Temperature Initial metal conc.	Batch
Pb(II) Cd(II) Zn(II)	Soil materials (Montmorillonite, Kaolinite and Calcite)	Contact time Initial metal conc. Type of natural soil	Batch
Pb(II)	Activated carbon Brick kiln ash River sand Egg shells Tea waste	pH Temperature Particle size	Batch

9. BIOSORBENTS

Microorganisms like, bacteria, fungi, yeast, and algae from their natural habitats are excellent biosorbent [65]. These biosorbents possess metal-sequestering property and can be used to decrease the concentration of heavy metal ions in solution from ppm to ppb level. It can effectively sequester dissolved metal ions out of dilute complex solutions with high efficiency and quickly, therefore it is an ideal candidate for the treatment of high volume and low concentration complex wastewaters [66]. The first major challenge for the biosorption field was to choose the most promising types of these biomass/biosorbent from enormously available and inexpensive biomaterials. Even though several materials of biological origin bind heavy metals, biomaterials with sufficiently high metal-binding capacity and selectivity for heavy metals are appropriate for full-scale biosorption process [65]. Biosorbents for the treatment of water and wastewater containing heavy metals are discussed by many reviews [67-74]. Biosorbents for the removal of metals mainly come under the following categories:

- Agricultural products, such as rice straw, coconut husks, waste coffee powder, dried plant leaves, wool, cotton seed hulls, waste tea, cork biomass [70, 72, 75]
- Sewage sludge and microbial cells, such as bacterial, fungi, algae and yeast.
- Peat moss [76, 77].
- Industrial wastes, such as S. cerevisiae waste biomass from fermentation and food industry), and other polysaccharide materials, etc. [70].

These biosorbents can be primarily fall into 5 categories as indicated in Table 6. The development and implementation of cost-effective process for removal/recovery of heavy metals from waste water is essential to improve the competitiveness of industrial operations. In recent years there has been a trend toward the implementation of passive treatment schemes. These take advantage of naturally occurring geochemical and biological processes to improve water quality with minimal operation and maintenance requirements. Biological removal includes the use of microorganisms (fungi, algae, and bacteria), plants (live or dead) and biopolymers may provide a suitable means for heavy metals removal from wastewater.

In particular, some waste mycelia are available in large quantities for the removal of heavy metals [66, 78]. When choosing biomass, for large-scale industrial uses, the main factor to be taken into account is its availability and cheapness [12, 79, and 80]. Considering these factors, native biomass can come from

- Industrial wastes, which should be available free of charge;
- Organisms easily obtainable in large amounts in nature; and
- Organisms that can be grown quickly or specially cultivated or propagated for biosorption purposes [12, 80]

Strong biosorbent behavior of certain micro-organisms towards metallic ions is a function of the chemical make-up of microbial cells. This type of biosorbent consists of dead and metabolically inactive cells.

Some types of biosorbents would be broad range, binding and collecting the majority of heavy metals with no specific activity, while others are specific for certain metals [81]. Feasibility studies for large-scale applications have demonstrated that biosorptive processes using non-living biomass are in fact more applicable than the bioaccumulative processes, which use living microorganisms, since the latter require a nutrient supply and complicated bioreactor system [70]. In addition, maintenance of a healthy microbial population is difficult due to toxicity of the pollutants being extracted, and other unsuitable environmental factors like temperature and pH of the solution being treated. Recovery of valuable metals is also limited in living cells since these may be bound intracellularly. For these reasons, attention has been focused on the use of non-living biomass as biosorbents [70, 82]. Recent bio sorption experiments have focused attention on waste materials, which are by–products or the waste materials from large-scale industrial operations e.g., the waste mycelia available from fermentation processes, olive solid residues, activated sludge from sewage treatment plants.

Table 6. Types of native biomass that have been used as biosorbents

Category	Examples
Bacteria	Gram-positive bacteria (Bacillus sp., Corynebacterium sp., etc.) Gram-negative bacteria (Escherichia sp., Pseudomonas sp., etc.) and Cyanobacteria (Anabaena sp., Synechocystis sp., etc.)
Algae	Micro-algae (Chlorella sp., Chlamydomonas sp., etc.) and Macro-algae (green seaweed (Enteromorpha sp., Codium sp., etc.), Brown seaweed (Sargassum sp., Ecklonia sp., etc., and red seaweed (Gelidium sp., Porphyra sp., etc.))
Fungi	Molds (Aspergillus sp., Rhizopus sp., etc.) Mushrooms (Agaricus sp., Trichaptumsp.,etc.) and Yeast (Saccharomyces sp., Candida sp., etc.
Yeast	
Other bio masses	Fermentation wastes, food/beverage wastes, activated sludge, anaerobic sludge, etc.; Fruit/Vegetable wastes, rice straws, wheat bran, soybean hulls, etc.; Plant residues, sawdust, tree barks, weeds, etc.; Chitosan-driven materials, cellulose-driven materials, etc.

9.1. Bacteria

Bacteria are the most abundant and versatile of microorganisms and constitute a significant portion of an entire living terrestrial biomass [83]. Bacteria have simple morphology. The most commonly bacteria present in three basic shapes: spherical or ovoid (coccus), rod (bacillus, with a cylindrical shape), and spiral (spirillum), although there are a great variety of shapes due to differences in genetics and ecology. Bacteria vary in size as much as in shape [65]. There is a great deal of heterogeneity among different bacterial species related to:

- Number of surface binding sites.
- Binding strength for different ions.
- Binding mechanisms.

Bacteria have polysaccharide slime layers and readily provide amino, carboxyl, phosphate and sulphate group for metals biosorption [83]. Bacterial biomass is generally produced as a waste by-product of industrial operations or can be specifically propagated in large scale. The uptake capacities of bacteria generally range between 0.23 to 0.90 mmol/g [84, 85]. Bacteria were used as biosorbents on account of their small size, ubiquity, and capability to grow under controlled conditions, and their resistance against a wide range of varying environmental conditions [86-88]. Bacteria species such as Bacillus, Pseudomonas, Streptomyces, Escherichia, Micrococcus, etc., have been tested for uptake metals or organics. Table 7 summarizes some of the important results of metal biosorption using bacterial biomasses, according to some published references [70, 89]. Heavy metal binding onto the surface of bacterial cell wall is generally, a two-stage process. The first stage involves the interaction between metal ions and reactive groups on cell surface and second stage includes deposition of successive metal species in greater concentrations.

9.2. Algae

Algae are considered one of the most promising materials in search for or in development of new biosorbents [102, 103]. This is mainly because of their high sorption capacity and being readily available in large quantities in seas and oceans [104]. Algae have low nutrient requirements, being autotrophic they produce a large biomass, and unlike other biomass and microbes, such as bacteria and fungi. They generally do not produce toxic substances.

There are three groups of algae:

- Micro-algal (green algae or fresh water algae)
- Macro-algal (brown algae or marine algae)
- Red algae

The metal uptake capacities exhibited by non-living biomass of micro-algal (green algae) and macro-algal species (brown algae) varies from 0.066 to 1.20 mmol/g and 0.65 to 1.21 mmol/g respectively [79, 105]. Using algae as bio sorbent material have been less compared with the other kinds of biomass, especially fungi (15%) and bacteria (85%) [71]. Binding of metal ions on algal surface depends on different conditions [106]:

- Ionic charge of metal ion
- Algal species, and
- Chemical composition of the metal ion solution.

Potential metal cationbinding sites of algal cell components include carboxyl, amine, imidazole, phosphate, sulphate, sulfhydryl, hydroxyl and chemical functional groups contained in cell proteins and sugars [107]. Microalgae can sequester heavy metal ions by the same biosorption mechanisms as other microbial biomass as well as by the formation of phytochelatins which they synthesize in response to toxic heavy metal stress.

The two principal mechanisms involved in biosorption by microalgae appear to be:

- Ion exchange wherein ions such as Na, Mg, and Ca become displaced by heavy metal ions
- Complexion between metal ions and various functional groups such as carboxyl, amino, thiol, hydroxyl, phosphate, and hydroxyl-carboxyl that can interact in coordinated way with heavy metal ions.

In general, the mechanism of biosorption is based on a number of metal-binding processes taking place with components of the algae cell wall. The algae cell walls can reversibly biosorb metals, and thus function in a similar way to an ion-exchange resin. Thus, the biosorption mechanism can be considered as being dependent on the composition of the algal cell wall. Algal cell walls can be made up with further polysaccharides: mannan, xylene, alginic acid, chitin, etc. These components, along with the proteins present, can provide acid binding sites such as amino, amine, hydroxyl, imidiazole, phosphate and sulphate groups [107].

Table 7. Biosorption data of different metals by bacteria

Biomass Type	Metal studied	pH	Temp (C°)	C_o (mg/l)	Wt (g/l)	Speed (rpm)	time (h)	q (mg/g) or % removal	References
Staphylococcus saprophyticus	Cr(VI) Pb Cu	2 4.5 3.5	27	193.6 100-150 105	0.2	150	3 4 2	24.1% 100% 14.5%	[90]
Sulphate-reducing bacteria (SRB)	Zn	6.0	30	10-200	1		24	5.6	[91]
Enterobacter cloacae	Pb Cu Cr(VI) Hg Cd	5 5 4 4 5	25	200 150 100 100 300 (25-350)	0.1	240	2	67.9% 78.9% 55.8% 43.23% 58.9%	[92]
Bacillus sp. Pseudomonas sp. Micrococcus sp.	Cu Cd Pb	5-9	30	25 25 25 (20-100)		100	24	69.34% 90.41% 84.27%	[93]
Pseudomonas sp.	Cr(VI) Cu Cd Ni	5.5	30	1-10 mmol/l	5	200		8.9-238 8.9-238 500 556	[94]
Thiobacillus thiooxidans	Zn Cu	6 5	30	50 25-250	0.25	786	2	95.24 39.84	[95]
Pseudomonas Aeruginosa	Cr(VI) Cu Zn		25	1000 1000 1000 (0-1000)				1.07 0.67 1.33	[96]

Table 7. (Continued)

Biomass Type	Metal studied	pH	Temp (C°)	C_o (mg/l)	Wt (g/l)	Speed (rpm)	time (h)	q (mg/g) or % removal	References
Bacillus cereus	Cu Pb	5.5	25	(5-100)	1.0		24	50.32 36.71	[97]
Geobacillus themodenitrificans	Cu Zn Pb	5 5 4	25	0.5 0.5 2.0 (0.5-3.0)		100	12	57 18 53	[68]
Geobacillus themocatenulatus	Cu Zn Pb	5 5 4	25	1.0 1.0 1.0		100	12	65 12.3 54	
Bacillus licheniformis	Cr(VI) Fe Cu	3.5 3.5 2.5	28	1200 1500 1500 (200-1500)		120	48	95% 52% 32%	[98]
Stenotrophomonas maltophilia	Cu Cd Pb	5.0	25	50 10 20 (0-140)	20	140	2	0.57 0.12 0.41	[99]
Actinomycete sp.	Cd Ni	6 5	30	(50-400)	5	150	24	32.63 36.55	[100]
Micrococcus sp.	Cr(VI) Ni	5	35	100 50		120	24	92% 90%	[101]

Table 8. Biosorption data of different metals by algae

Biomass type	Metal studied	pH	Temp. (C°)	C_o mg/l	Wt g/l	Speed rpm	Time H	Q mg/g	References
Calotropis procera (Ait)	Pb Cu	4 5	25	25-100	2	150	6	22.8 14.5	[108]
Bifurcaria bifurcata,	Cd	4.5		10-350	2.5	175	3	95	[109]
Oocystis	Cd Cu Pb Zn	7.5 5.5 5.5 5.5	28	0.5-35 0.5-25 10-30 2-12	28,51 4.4,60 16-80 0.5-157.5	60-80	72		[110]
Green algae	Zn	5	25	20-350	3		1	7.62	[111]
Sargassum filipendula	Cu Ni	4.5	25	250	5	175	6		[112]
Filamentous alga of pithpphoraspswas	Cd Cr(VI) Pb		25	2-30 2-30 2-30	0.17-14 0.13-4.9 0.12-10.3		9 days		[113]
Sargassum sp.	Cr(III)	4.0	30	92,162	2-5		6	68.9	[114]
Microalgae	Cu Zn		30	50-250	5	150		0.66 0.72 mmol/g	[115]
Green algae (Spirogyra spp)	Cr(VI)	4.0	30	1-25	1,2,3	180	3	265	
Brown algae (Sargassum sp.)	Cr Zn Ni Cu Cd Hg Fe	3 3 5 4 5 4 3	30 30 30 30 30 30 30			150 150 150 150 150 100 150		20.2 15.4 26.1 18.6 22.2 14.8 14.6	[116]
Schizosaccharomyces pombe	Cu	4	25	100-400 100	300 beads		96	70%	[117]
Sargassum tenerrimmum	Cd	6	25	100 20-500	4.0 0.05-8	150	24	0.4 mmol/g	[118]

Table 8. (Continued)

Biomass type	Metal studied	pH	Temp. (C°)	C_o mg/l	Wt g/l	Speed rpm	Time H	Q mg/g	References
Brown algae (Fucusvesiculous)	Cu Pb Ni Cd	5 5 5 6	25	10-150	0.25		2	0.97 1.04 0.80 1.12 Mmol/g	[119]
Brown marine macro algae	Cd Zn Pb	5	25	75-100	2.0		1	24-40 19-32 32-50	[120]
Sargassum glauceseens	Cd	5	25	0-150	2.5	150	2	84.7	[121]
Ascophylhum nodosum	Cd Ni Zn Cu Pb	6 6 6 5 5	25	10-150	0.5,1		2	114.9 50 53.2 70.9 204.1	[122]
Fucus vesiculosus	Cu	5.5	20	.01-.4 mmol/l	0.08			1.85 mmol/g	[123]
Cladophora fasicularis	Cd Hg Pb		25	20-80	1	180	2	20 20 20	[124]
Spirogyra hyalina	Cd Hg Pb AS Co		25	40 40 80 40 80	2.5	180	2	9.832 39.212 15.471 8.719 7.856	[125]
Sargassum Sp.	Cd	5	25	150	2.5	150	2	84.7	[125]
Sargassum Sp. Wightii (Greville)	Cr	4	25	50 50-400	8		12	85.3	[125]
Mixture of Chlorophyta and Cyanobacteria	Pb Cu Cd As	4	25	50 10-100	10	200	3	16.55 15.97 10.52 7.45	[126]

Table 9. Biosorption data of different metals by fungi

Biomass type	Metal studied	pH	Temp. (C°)	C_o (mg/l)	Wt (g/l)	Speed (rpm)	Time (h)	q (mg/g)	References
Rhizopus nigricans	Pb	5.5	25	300 (10-300)	25 25-200	225		80.8	[134]
Trichoderma longibrachiatum	Pb	7.0	25	100 (25-100)			0.33	71.0	[135]
Pleurotus ostreatus	Pb	5.5	25	1-25	2		3	4.84	[136]
Aspergillus cristatus	Cd	6.0	25	100	0.4	120	2	23.26	[137]
Aspergillus niger	Cd	4.75	25	5-10	0.7	125	6	13	[138]
Hydrilla verticillata	Cd	5.0	25	1-100	3-9	150	0.33	15	[139]
Aspergillus flavus I-V, Aspergillus fumigatus I-II	Hg	5.5	30	100 100-500	10	100	8	95.3%	[140]
p. chrysogenum	As	3-4	25	1-300	1	190		24.52	[141]
Waste tea fungal	As(III) As(V)		30	1.3 0.9	20	160	1.5		[142]
Waste tea fungal	As	6-8	22	1	1	160	1.5	0.004 mmol/g	[143]
Aspergillus niger	Ni	4.5	25	25-100	1	150	3	7.69	[144]
Pleurotus ostreatus fungal	Cu	5	25	50	2	125	0.16	4.0	[145]
A. lentulus	Cu	6	35	100	4	180	0.41		[146]
Fomes fasciatus	Cu	5.5	25	100	1.0	200	1	32.2	[147]
Mycelium of Aspergillus awamori	Cu	5	20	50 25-100	1	150	3	28.75	[148]
Penicillium canrscens	Cr(VI)	6	20	100 10-750	2	100	4	34.8	[149]
Living mycelium of phanerochaete chrysosporium	Cr(VI) Cu		27	100 20-500		100	2	48.6 90.6	[150]

Table 9. (Continued)

Biomass type	Metal studied	pH	Temp. (C°)	C_o (mg/l)	Wt (g/l)	Speed (rpm)	Time (h)	q (mg/g)	References
Mucor	Cr(VI)	5.5	35	50 50-400	0.0165				[151]
Trichoderma viride	Cr(VI)	6		175	3.75	150	0.75	4.66	[152]
Aspergillus niger	Cr(VI)	4.5	28	300	10	150	1	16.39	[153]
Pleurotus ostreatus	Cr(III)	4.5	25	14.35 4-20	2	150	3	1.97	[154]
Aspergillus niger	Cr(VI)	2	22-28	50 25-200	10	200	158	2.2	[155]
Mycelial biomass of Pleurotus Florida	Cd	5	25	10	1.0	9.76	125	2	[156]
Fungal Waste Biomass	Cd Pb	6 5.5	25	0-100	2	200	70 30	5 30	[157]

Table 10. Biosorption data of different metals by Yeast

Biomass Type	Metal studied					q (mg/g) or % removal	Speed (rpm)	Time (h)	References
		pH	T (C°)	C_o (mg/l)	Wt. (g/l)				
Saccharomyces cervisiae	Hg Cu Ni Zn Cd Co Cr Fe Al	7	25	25-200	2	76.2 29.9 14.1 11.8 12.3 8.2 6.2 5 4.1	100	2	[158]

Biomass Type	Metal studied					q (mg/g) or % removal	Speed (rpm)	Time (h)	References
		pH	T (C°)	C_o (mg/l)	Wt. (g/l)				
Candida spp	Hg Zn Cd Pb Fe Cu Cr	5	28	0.008 0.091 0.006 0.003 0.075 0.82 0.027	0.2	2.74 31.16 2.05 1.03 25.68 28.08 9.28	150	2	[159]
Saccharomyces cervisiae	Cd Cr(III) Cr(VI) Cu Pb Zn	5.8 5.2 5-6 6.5 5.2 5.8	25	5.0 mm/l 10-60	80	65.3% 55.3% 49.0% 50.7% 50.2% 47.6%	150	1	[160]
Saccharomyces cervisiae	Cr(III)	5.5 5.5	35 35	200 200	2.0 2.0	Static = 35 Aerobic = 86.3	120	6	[161]
Candida utilis	Cd	5.5	25	50 5-100	1.0	Dried = 28 Cell = 8.7 Nativ = 5.5	160	1	[162]
Candida pelliculosa	Cu	6.0	30	100 100-1000	13.397	95.04%	120	120	[163]
Schizosaccharomyces pombe	Cu	4.0	25	100 100-400		74.85		96	[117]
Spent yeast	Cu	4.0	35	20-80	63.33	0.014-0.037	200	8	[164]
Yeast biomass	Cr(VI)	4.0	310	120 10-250	40	86.95	180	0.75	[165]
Saccharomyces cervisiae	Cu Cd Pb Ni	5.0	28	50 50-500	0.005	0.64 0.66 0.60 0.68	150	48	[166]

Table 10. (Continued)

Biomass Type	Metal studied					q (mg/g) or % removal	Speed (rpm)	Time (h)	References
		pH	T (C°)	C_o (mg/l)	Wt. (g/l)				
Saccharomyces cervisiae	Cr Sn	6.5	25	--------	0.4	29% 33% removal	110	1	[167]
Saccharomyces cervisiae	Cu Ni Zn	6.0	25	5-50 5-200 5-50 (5, 100)	4.0	9.1 mmol/g 6.4 mmol/g 11.0 mmol/g	150	1	[168]
Saccharomyces cervisiae	Cr Cu Ni Zn	6.0	25	34 mmol/l	12		150	0.5	[169]
Mucorrouxii	Pb Cd Ni Zn	5	25	10		17.13 10.07 6.07 6.28	125	15	[170]
Saccharomyces cervisiae	Cu	4.0	30	25 10-250	1.0	120.7	200 200	3 3	[171]
Saccharomyces cervisiae	Cd	6	28	250 50-350	3	55 (dead cell) 36 (live cell)	120	24	[172]
Saccharomyces cervisiae	U(VI)	5	27	100 25-300	10	7.89		1.25	[173]

The biosorption mechanism has been described as not involving van der Waals' forces at the cellulose network of the cell walls, thus both ionic charge and covalent bonding are involved in the metal biosorption process. It is thought that the proteins and polysaccharides are the major components responsible for the biosorption. Covalent bonding could be expected with amino and carboxyl groups and ionic charge bonding with carboxyl and sulphate groups associated with these components. Studies with the micro alga Scenedesmusobfiquus indicated that the cell wall behaved like a weak acidic cation exchanger containing various cell wall ligands with different exchange capacities. Table 8 shows the biosorption data of different metals by algae.

9.3. Fungi

Fungi are one of the industrial fermentation waste biomass which is considered as excellent metal sorbents. Fungi are considered as an efficient and economical sorbent for sequestering heavy toxic metals from dilute aqueous solutions due to:

- It offers the advantage of having a high percentage of cell wall materials; large surface area.
- It shows excellent metal binding properties.
- It is available in large quantities from the antibiotic and food industries.
- It provides an eco-friendly environment.

Three groups of fungi have major practical importance: the molds, yeasts and mushrooms. Filamentous fungi and yeasts have been used in many wastewater treatments to bind metallic elements. The structures of fungi have wide range of morphologies from unicellular yeasts to polymorphic and filamentous fungi, many of them have complex macroscopic fruiting bodies.

Fungal cell walls are complex macromolecular structures mainly consisting of chitins, glucans, mannans and proteins, but also containing other polysaccharides, lipids and pigments, e.g., melanin [127-129]. In the fungal cell wall, several types of ionizable sites affect the metal uptake capacity: phosphate groups, carboxyl groups on uranic acids and proteins, and nitrogen- containing ligands on protein as well as on chitin or chitosan [130].

In yeasts, higher concentration of heavy metals can be accumulated by bioaccumulation process than biosorption [131]. However, general biosorption is responsible for the major uptake of heavy metals for many filamentous fungi. A wide range of biosorption capacities from 0.012 to 1.979 mmol/g for fungi has been reported [132]. Biosorption of heavy metals by macrofungi or mushroom is a known phenomenon but till now a little work has been done. Because mushrooms grow under natural habitats, the pollutants of heavy metals found in the soil or in the natural substrates are removed by the fruiting bodies of mushrooms and as a result, the metals are accumulated in the mycelia and sporocarps [133]. Table 9 shows the biosorption data of different metals by different types of fungi.

9.4. Yeast

Table 10 shows the biosorption data of different metals by yeast, while Table 11 shows the biosorption data of other different biomasses.

9.5. Other Bio Masses

Table 11. Biosorption data of different metals by different biomass

Biomass type	Metal studied	pH	Temp. (C°)	C_o mg/l	Wt g/l	Q mg/g	Speed rpm	Time h	References
Dry activated sludge (DAS)	Pb Hg Cr As	4	25	50	2.5	77.485 10.787 4.640 2.828	800	4-7	[174]
Mixing of dead anaerobic biomass	Pb Cr(III) Cd	4	35	50	0.5	54.92 34.78 29.99	200	4	[175]
Dead microorganism	Pb Ph	4	30		0.05-0.6	89.873 70.018	250	24	[175]

9.6. Mechanism of Biosorption

The realization of mechanisms by which biosorbents uptake pollutants is essential for the development of biosorption processes including the removal and recovery of heavy metal pollutants from aqueous solutions. Fundamentally, many types of biosorbents are derived from various forms of raw biomass, including bacteria, fungi, yeasts, and algae as was explained before. The complex structure of the raw biomass, which may contain several diversable function groups, implies that there are many ways, by which these biosorbents can remove various pollutants, but these are not yet fully understood. It is well understood that the mechanism of uptake of metal ions in such situation is very complicated process. Several factors are found to influence the mechanism of metal biosorption:

- The status of biomass
- Types of biomaterials
- Properties of metal solution chemistry, and
- Environmental conditions such as pH, temperature, etc.

Biosorption mechanisms may be classified based on location where biosorption occurs. In this concern, biosorption mechanisms are classified as extra cellular accumulation/ precipitation, cell surface sorption/precipitation and intra cellular accumulation. Biosorption of metals occurs mainly through several interactions such as physical adsorption, ion exchange, complexation, precipitation and entrapment in inner spaces. The metal ion uptake by living and dead cells can consist of two different modes. The mechanisms of uptake by living materials (bioaccumulation) and removal by dead ones (biosorption) are entirely different. The first uptake mode is independent of cell metabolic activity and involves surface binding of metal ions to cell walls and extracellular material. This is referred to biosorption or passive uptake. The second mode of metal uptake into the cell across the cell membrane is dependent on the cell metabolism, and is referred to intracellular uptake, active uptake or bioaccumulation. Metal removal by non-living cells is mainly in passive mode which is metabolism independent and proceeds rapidly by any one or a combination of metal binding mechanisms such as complexation, ion exchange, physical adsorption, etc. It is a dynamic equilibrium of reversible adsorption-desorption and is independent on energy. It has several advantageous over bioaccumulation

- There is no need of live cell growing
- Growth media is not required
- These materials are available as wastes or by-products.
- Absence of toxicity limitations, absence of nutrient requirements in the feed solution and reuse of regenerated biomass.

The sources of heavy metal ions as toxic materials in aqueous streams as well as the reason why they need to be removed from our environment have been well reviewed and presented [176]. Conventional methods of removal are expensive; hence the search of using low cost, friendly ecosystem material is needed. It is observed that biosorption is one of the most economical and eco-friendly method for removal of heavy metal from domestic as well as industrial wastewater. It is being an alternative to conventional methods for the removal of toxic heavy metals from industrial effluents. It offers several advantages including cost effectiveness, high efficiency, minimization of chemical/biological sludge, and regeneration of biosorbent with possibility of metal recovery. The process of biosorption has many attractive features including removal of metals over quit broad range of pH and temperature, its rapid kinetics of adsorption and desorption and low capital and operation cost.

Biological biomass can be regenerated relatively easily and used again. Most significantly, the metal removal capacity of biological biomass is a good or better than other conventional adsorbents. The new biological – based technologies need not necessarily replace conventional treatment approaches but may complement them. So, the biosorption technology becomes more beneficial and attractive than currently used technologies.

REFERENCES

[1] F. L. Slejko, Adsorption Technology: A step-by-step Approach to Process Evaluation and Application. *Chemical Industries*. Marcel Dekker, New York (1985) 19.

[2] K. Wark, C. F. Warner, *Air pollution (its origin and Control)*, New York (1976).

[3] H. G. Cassidy, Chromatography. *Adsorption*. Inter science Publishers (1951).

[4] M. Sustarsic, T. T. Nus, Wastewater treatment. Understanding the activated sludge process. Chemical Engineering Progress. *AIchE*. November (2009) 26-29.

[5] T. Asano, F. L. Burton, H. L. Leverenz, R. Tsuchihashi, G. *Tchobanoglous, Water Reuse Issues, Technologies and Applications*. The McGraw-Hill companies, ISBN 13 (2007).

[6] S. C. Bhatia, *Environment Pollution and Control in Chemical Process Industries*. Khanna publisher (2001) 250-260.

[7] J. D. Seader, *Henley, Separation Process: Principle*. Wiley, J. and Sons, Inc. (1999).

[8] J. R. Coulson, J. F. Richardson, J. R. Backhurts, J. H. Harker, *Coulson and Richardson Chemical Engineering*, vol. 2. 4 (1991).

[9] P. Sicdhar, Modeling of affinity separation by batch and fixed bed adsorption – A comparative study. *Chem. Eng. Technol*. 19 (1996) 357-363.

[10] R. S. Juang, R. L. Tseng, Role of micro porosity of activated carbons on their adsorption abilities for phenols and dyes. *Adsorption*, V. 7 (2001) 65-72.

[11] R. Qadeer, A. H. Rehan, A study of the adsorption of phenol by activated carbon from aqueous solution. *Turk. J. Chem*. 26 (2002) 357-361.

[12] B. Volesky, Z. R. Holan, Biosorption of heavy metals. *Biotechnol. Progr*. 11 (1995) 235-250.

[13] M. Ahemad, A. Malik, Bioaccumulation of heavy metals by zinc resistant bacteria isolated from agricultural soils irrigated with wastewater. *Bacteriology Journal* 2 (2011) 12-21.

[14] J. T. Richardson, *Principle of catalyst development*. Plenum press, New York (1989).

[15] D. Graham, *Chem. Eng. Prog. Symposium Series* 55 (24) (1959) 17-24.

[16] R. F. Treybel, *Mass Transfer Operation*. McGraw-Hill, New York 3 (1981).

[17] C. N. Sawyer, P. L. McCarty, *Chemistry for environmental engineering*. McGraw-Hill, New York 3 (1978).

[18] J. J. Kipling, *Adsorption from solutions of non-electrolytes*. Academic, New York (1965).

[19] C. N. *Satterfield, Heterogeneous catalysis in practice*. McGraw-Hill, New york (1980).

[20] A. Adamson, *Physical chemistry of surfaces*. John Wiley and sons, New York 4 (1982).

[21] A. Clark, *The theory of Adsorption and Catalysis*. Academic press, London (1970).

[22] J. C. Crittenden, R. R. Trusell, D. W. Hand, K. J. Howe, G. *Tchobanoglous, water treatment principles and design*. John Wiley and Sons, Inc. 2 (2005).

[23] B. Fonseca, H. Maio, C. Quintelas, A. Teixeira, T. Tavares, Rentention of Cr(VI) and Pb(II) on a loamy sand soil kinetics, equilibrium breakthrough curve. *J. Chem. Eng*. 152 (2009) 212-219.

[24] R. Ashkenazy, L. Gottlieb, S. Yannai, Characterization of acetone-washed yeast biomass functional groups involved in lead biosorption. *Biotechnol. Bioeng*. 55 (1997) 1-10.

[25] N. Kuyucak, B. Volesky, Biosorbents for recovery of metals from industrial solutions, *Biotechnol. Lett*. 10 (1988) 137-142.

[26] H. Eccles, Treatment of metal-contaminated wastes: Why select a biological process? *Biotechnology* 17 (1999) 462-465.

[27] C. R. Wilk, P. Chang, *Correlation of Diffusion Coefficient in Dilute Solutions (1955)*.

[28] D. O. Cooney, *Adsorption Design for Wastewater Treatment*. Lewis Publishers, F.L. Baco Raton (1999).

[29] P. Monneyron, C. Faur-Brasqet, Competitive Adsorption of Organic Micro -pollutant in the Aqueous Phase onto Activated Carbon Cloth. *Langmuir* 18 (2002) 5163-5168.

[30] H. T. Liao, C. Y. Shiau, Analytical solution to an axial Dispersion Model for the Fixed-Bed Adsorber. *J. AICHE* (2000) 1168-1176.

[31] I. T. Al Ewi, Preparation, characterization and evaluation of mesoporous activated carbons derived from agricultural by-products for adsorption of Methylene blue and 2,4,6-trichlorophenol. *Thesis*; University SainsMalasia (USM) (2008).

[32] G. Tchobanoglous, F. L. Burton, *Wastewater Engineering*. McGraw Hill, Inc. 3 (1991).

[33] I. A. Basheer, Predicting Dynamic Response of Adsorption Column with Neural Nets. *J. Computing in Civil Engineering* 10 (1) (1996) 31-39.

[34] W. L. McCabe, J. C. Smith, P. Hariot, *Unit operation of chemical engineering*. McGraw-Hill 6 (2001).

[35] P. A. Schweitzer, *Handbook of Separation Techniques for Chemical Engineers*. McGraw-Hill book Company 1-417.

[36] M. Bonnevie-Svendsen, *Noordhoff-Leyden, Sorption and Filtration Method for Gas and Water Purification*. 13 (1975) 13.

[37] J. Cazes, Handbook of affinity chromatography. *Chromatographic Science Series of Textbooks and References books* 2 (1995).

[38] R. T. Yang, *Gas Separation by Adsorption Processes*. New York (1987).

[39] T. Gu, G. J. Tsai, G. T. Taso, Multicomponent Adsorption and Chromatography with Uneven Saturation Capacity. *J. AIChE* 37 (9) (1991) 1333-1340.

[40] D. M. Ruthven, H. K. Yoshida, *Chem. Eng. Sci.* 39 (1984) 1489-1494.

[41] G. Belfort, Selective Adsorption of Organic Homologues onto Activated Carbon from Dilute Aqueous Solutions. *Environmental Science and Technology* 13 (8) (1979) 939-946.

[42] W. J. Weber, *Chem. Eng. Prog. Symposium Series* 74 (1967) 78-89.

[43] A. H. Sulaymon, B. A. Abid, Al-Najar, Removal of Lead and Copper onto granular activated carbon in batch and fixed adsorber. *J. Eng. Tesh.* 27 (2009) 12.

[44] D. M. Giusti, R. A. Conway, *JWPCF* 46 (1974) 947-965.

[45] D. Mohan, Jr. C. U. P, Arsenic removal from water/wastewater using adsorbents-A critical review. *J. Hazardous Materials* 142 (2007) 1-53.

[46] R. H. Perry, *Chilton, Perry's chemical Engineering Handbook*. McGrew-Hill, New york 5 (1984).

[47] R. H. Perry, D. W. Green, *Perry's chemical Engineering Handbook*. McGrew-Hill, New york 7 (1999).

[48] L. Warren, P. Harriot, Unit Operation of Chemical Engineering. McGrew-Hill. *J. AIchE* 1 (1993) 264-269.

[49] G. Kalyan, G. Babu Rao, B. V. Saradhi, Y. P. Kumar, Equilibrium and kinetic studies on biosorption of Znic onto Gallus Domesticus shell powder. *ARPN journal of Engineering and Applied Science* 4 (1) (2009) 39-49.

[50] G. A. Tito, L. H. G. Chaves, R. S. de Souza, Zinc adsorption in bentonite clay,particle size and pH influence. *Re vista caatinga* 21 (5) (2008) 1-4.

[51] O. Yavuz, Y. Altunkaynate, F. Guzel, Removal of Copper, Nickel, Cobalt and Manganese from Aqueous Solution by Kaolinite. *Wat. Res.* 37 (2003) 948-952.

[52] B. Crittenden, W. J. Thomas, *"Adsorption Technology and Design,"* Reed Education and Professional Publishing Ltd (1998).

[53] A. Fahmi, K. Munther, Competitive Adsorption of Nickel and Cadmium on Sheep Manure Waste, Experimental and Prediction Studies. *Separation Science and Technology* 38 (2) (2003) 483-497.

[54] J. Walter, J. R. Weber, *Adsorption processes*. The university of Mechigan,College of engineering Anes Arbor, Michigan 48104, US (1973).

[55] N. Muhammad, J. Parr, M. D. Smith, A. D. Whealty, Adsorption of Heavy Metals in Slow Sand Filters. *Proceeding of the 24th WEDC international Conference on Water Treatment*, Islamabad, Pakistan (1998) 346-349.

[56] D. Mohan, K. P. Singh, Single and Multi-Component Adsorption of Cadimum and Znic Using Activated Carbon Derived from Bagasse-an Agricultural Waste. *Wat. Res.* 36 2304-2318.

[57] C. J. Radk, J. M. Prausnitz, Adsorption of Organic Compounds from Dilute Aqueous Solution on Activated Carbon. *Ind. Eng. Chem. Fund.* 11 (1972) 445-451.

[58] L. Jossens, J. M. Prausnitz, W. Frits, *Thermodynamic of Multi-Solute (1978).*

[59] R. Sips, *J. Chem. Phys.* 16 (1984) 490-495.

[60] S. J. Jain, L. S. Vernon, Adsorption from Bisolute Systems on Activated Carbon. *JWPCF* 45 (12) 2463-2479.

[61] Wilson, E. J. Wilson, C. J. Geankoplis, Liquid Mass Transfer at Very Low Reynolds Numbers in Packed Beds. *Ind. Eng. Chem. Fundam.* 5 (9) (1966).

[62] M. L. Martinez, L. Moiraghi, M. Agnese, C. Guzman, Making and some properties of activated carbon produced from agricultural industrial residues from Argentina. J. *The Journal of the Argentine Chemical Society*. 91 (4/6) (2003) 103-108.

[63] R. Leyva-Ramos, C. J. Geankoplis, Model Simulation and Analysis of Surface Diffusion of Liquids in Porous Solids. *Chem. Eng. Sci.* 40 (5) (1985) 799-807.

[64] N. Ivars, Analysis of Some Adsorption Experiments with Activated Carbon. *Chem. Eng. Sci.* 31 (1976) 1029-1035.

[65] J. L. Wang, C. Chen, Biosorbents for heavy metals removal and their future. *Biotechnol. Adv*. 27 (2009) 195-226.

[66] J. L. Wang, C. Chen, Biosorption of heavy metals by Saccharomyces cerevisiae. *Biotechnol. Adv*. 24 (2006) 427-451.

[67] G. M. Gadd, Heavy metal accumulation by bacteria and oth microorganisms. *Experiential* 46 (1990) 834-840.

[68] L. Babak, P. Šupinova, M. Zichova, R. Burdychova, E. Vitova, Biosorption of Cu, Zn and Pb by thermophilic bacteria – effect of biomass concentration on biosorption capacity. *Acta universitatis agriculturae et silviculturae mendelianae brunensis LX* (5) (2012).

[69] A. Demirbas, Heavy metal adsorption onto agro-based waste materials. A review. *J. Hazard. Mater*. 157 (2008) 220-229.

[70] K. Vijayaraghavan, Y. S. Yun, Bacterial biosorbents and biosorption. *Biotechnol. Adv*. 26 (2008) 266-291.

[71] E. Romera, F. Gonzalez, A. Ballester, M. L. Blazquez, J. A. Munoz, Biosorption with algae. A statistical review. *Crit. Rev. Biotechnol.*, 26 (2006) 223-35.

[72] T. A. Davis, B. Volesky, A. Mucci, A review of the biochemistry of heavy metal biosorption by brown algae. *Water Res*. 37 (2003) 4311-4330.

[73] D. R. Lovley, J. D. Coates, Bioremediation of metal contamination. *Curr. Opin. Biotechnol*. 8 (1997) 285-289.

[74] F. Veglio, F. Beolchini, Removal of metals by biosorption: a review. *Hydrometallurgy* 44 (1997) 301–316.

[75] H. K. Alluri, S. R. Ronda, V. S. Settalluri, J. S. Bondili, V. Suryanarayana, P. Venkateshwar, Biosorption: An eco-friendly alternative for heavy metal removal. *African Journal of Biotechnology*, 6 (2007) 2924-2931.

[76] Y. Orhan, H. Buyukgungor, The removal of heavy metals by using agriculture wastes. *Wat. Set. Tech.* 28 (2) (1993) 247-255.

[77] A. I. Zouboulis, E. G. Rousou, K. A. Matis, I. C. Hancock. *Removal of toxic metals from aqueous mixtures.* Part 1: Biosorption (1999).

[78] A. Kapoor, T. Viraraghavan. Fungal biosorption — an alternative treatment option for heavy metal bearing wastewaters: a review. *Bioresour. Technol.* 53 (1995)195-206.

[79] B. Volesky, Advances in biosorption of metals: selection of biomass types. *FEMS Microbiol. Rev.* 14 (1994) 291-302.

[80] R. H. S. F. Vieira, B. Volesky, Biosorption: a solution to pollution? *Int. Microbiol.* 3 (2000) 17-24.

[81] N. Ahalya, T. V. Ramachandra, R. D. Kanamadi, Biosorption of Heavy Metals. *Research Journal of Chemistry and Environment* 7 (2003) 71-79.

[82] J. M. Modak, K. A. Natarajan, Biosorption of metals using nonliving biomass-a review. *Miner. Metall. Proc.* 12 (1995) 189-196.

[83] H. Mann, (1990) Removal and recovery of heavy metals by biosorption. In: B. Volesky (editor). *Biosorption of heavy metals. Boca Raton.* CRC press (1990) 93-137.

[84] J. S. Chang, J. Hong, Biosorption of mercury by the inactivated cells of Pseudomonas aeruinosa. *Biotechnol. Bioeng.* 44 (1994) 999-1006.

[85] A. Norberg, H. Persson, Accumulation of heavy metal ions by Zoogloearamigera. *Biotechnol. Bioeng.* 26 (1984) 239-245.

[86] Z. G. Eman, Production and characteristics of a heavy metals removing bioflocculant produced by Pseudomonas aeruginosa. *Polish Journal of Microbiology* 61 (2012) 281-289.

[87] H. Kinoshita, Y. Sohma, F. Ohtake, M. Ishid, Y. Kawai, H. Kitazawa, T. Saito, K. Kimura, Biosorption of heavy metals by lactic acid bacteria and identification of mercury binding protein. *Research in Microbiology* doi: 10.1016/j.resmic.04.004 (2013).

[88] M. M. Urrutia, General Bacterial Sorption Processes. In: J. Wase, C. Forster (editors), *Biosorbents for metal ions*, CRC Press, London, UK (1997) 39-66.

[89] S. S. Ahluwalia, D. Goyal, Microbial and plant derived biomass for removal of heavy metals from wastewater. *Bioresource Technology* 98 (2007) 2243-2257.

[90] S. Iihan, M. Nourbakhsh, S. Kilicarslan, H. Ozdag, Removal of chromium, Lead and Copper ions from industrial waste waters by Staphylococcus saprophyticus; *Turkish electronic Journal of Biotechnology* 2 (2004) 50-57.

[91] J. Jencarova, A. Luptakova, the Elimination of Heavy Metal Ions from Waters by Biogenic Iron Sulphides. *Chemical Eng. Transaction* 28 (2012) 205-210.

[92] J. Suriya, S. Bharathiraja, R. Rajasekaran, Biosorption Of Heavy Metals By Biomass of Enterobacter Cloacae Isolated From Metal-Polluted Soils; *International Journal of Chem. Tech. Research* 5 (3) (2013) 1329-1338.

[93] M. J. Rani, B. Hemambika, J. Hemapriya, V. Rajesh Kannan, Comparative assessment of heavy metal removal by immobilized and dead bacterial cells. A biosorption approach. *African Journal of Environmental Science and Technology* 4 (2) (2010) 077-083.

[94] H. Hussein, S. F. Ibrahim, K. Kandeel, H. Moawad, Biosorption of heavy metals from waste water using Pseudomonas sp.; *Electronic Journal of Biotechnology* 7 (1) April 15 (2004).

[95] H. Liu, B. Chenb, Y. Lana, Y. Chenga, Biosorption of Zn(II) and Cu(II) by the indigenous Thiobacillusthiooxidans; *Chemical Engineering Journal* 97 (2004) 195-201.

[96] V. Nagashetti, G. K. Mahadevaraju, T. S. Muralidhar, A. Javed, D. Trivedi, K. P. Bhusal, Biosorption of Heavy Metals from Soil by Pseudomonas Aeruginosa; *International Journal of Innovative Technology and Exploring Engineering (IJITEE)* ISSN 2 (6) (2013) 2278-3075.

[97] P. J. hua, L. R. Xia, T. H. Xiao, Surface reaction of Bacillus cereus biomass and its biosorption for lead and copper ions; *Journal of Environmental Sciences* 19 (2007) 403-408.

[98] D. P. Samarth, C. J. Chandekar, R. K. Bhadekar, Biosorption of Heavy Metals from Aqueous Solution using Bacillus Licheniformis; *Int. J. Pure Appl. Sci. Technol.* 10 (2) (2012) 12-19.

[99] M. M. Parungao, P. S. Tacata, C. R. G. Tanayan, L. C. Trinidad, Biosorption of Copper, Cadmium and Lead by Copper-Resistant Bacteria Isolated from Mogpog River, Mari nduque; *Philippine Journal of Science* 136 (2) (2007) 155-165.

[100] R. M. Karakagh, M. Chorom, H. Motamedi, Y. K. YuseKalkhajeh, S. Oustan, Biosorption of Cd and Ni by inactivated bacteria isolated from agricultural soil treated with sewage sludge; *Ecohydrology and Hydrobiology* 12 (3) (2012) 191-198.

[101] S. Congeevaram, S. Dhanarani, J. Park, M. Dexilin, K. Thamaraiselvi, Biosorption of chromium and nickel by heavy metal resistant fungal and bacterial isolates. *Journal of Hazardous Materials* 146 (2007) 270-277.

[102] R. Flouty, G. Estephane, Bioaccumulation and biosorption of copper and lead by a unicellular algae Chlamydomonasreinhardtii in single and binary metal systems: a comparative study. *Journal of Environmental Management* 111 (2012) 106-114.

[103] M. A. Trinelli, M. M. Areco, S. AfonsoMdos, Co-biosorption of copper and glyphosate by Ulva lactuca. *Colloids and Surfaces B: Biointerfaces*. 105 (2013) 251-258.

[104] N. Kuyicak, B. Volesky, Biosorption by fungal biomass. In: B. Volesky (editor), *Biosorption of heavy metals*. Florida. CRC press (1990) 173-98.

[105] T. C. Wang, J. C. Weissman, G. Ramesh, R. Varadarajan, J.R. Benemann, Heavy metal binding and re- moval by Phormidium. *Bull. Environ. Contamin. Toxicol.* 60 (1998) 739-744.

[106] P. S. Sheng, Y. P. Ting, J. P. Chen, L. Hong, Sorption of lead, copper, zinc and nickel by marine algal biomass; Characterization of biosorptive capacity and investigation of mechanisms. *J. Colloid Interface Sci.* 275 (2004) 131-141.

[107] R. H. Crist, K. Oberholser, K. Shank, M. Nguyen, Nature of bonding between metallic ions and algal cell walls. *Environ. Sci. Technol.* 15 (1981) 1212-1217.

[108] T. A. Oyedepo, Biosorption of Lead (II) and Copper (II) Metal Ions on Calotropisprocera (Ait.). *Science Journal of Purel and Applied Chemistry* (2011).

[109] P. Lodeiro, B. Cordero, J. L. Barriada, R. Herrero, M. E. Sastre, M. E. Vicente, Biosorption of cadmium by biomass of brown marine macroalgae. *Bioresource Technology*, 96 (16) (2005) 1796-1803.

[110] K. Y. Gina, Y. Z. Tanga, M. A. Aziz, Derivation and application of a new model for heavy metal biosorption by algae. *Water Research* 36 (2002) 1313-1323.

[111] D. Sheikha, I. Ashour, F. A. Abu Al-Rub, Biosorption of Zinc on Immobilized Green Algae. Equilibrium and Dynamics Studies. *The Journal of Engineering Research* 5 (1) (2008) 20-29.

[112] S. J. Kleinübinga, R. S. Vieiraa, M. M. Beppua, E. Guibalb, M. G. Carlos da Silvaar, Characterization and Evaluation of Copper and Nickel Biosorption on Acidic Algae SargassumFilipendula. *Materials Research*. 13 (4) (2010) 541-550.

[113] Brahmbhatt, N. H. Rinku, V. Patel, R. T. Jasrai, Removal of cadmium, chromium and lead from filamentous alga of Pithophora sp. of industrial wastewater. *International Journal of Environmental Sciences* 3 (1) (2012).

[114] E. S. Cossich, C. R. G. Tavares, T. M. K. Ravagnani, Biosorption of chromium(III) by Sargassum sp. Biomass. *EJB Electronic Journal of Biotechnology* 5 (2) 15 August (2002).

[115] R. Ramsenthil, R. Meyyappan. Single and Multi-component Biosorption of Copper and Zinc Ions using Microalgal Resin; *International Journal of Environmental Science and Development* 1 (4) October (2010).

[116] Saravanan, V. Brindha, S. Krishnan, Studies on the structural changes of the biomass sargassum sp. On metal adsorption. *Journal of Advanced Bioinformatics Applications and Research* 2 (3) (2011)193-196.

[117] S. Subhashini, S. Kaliappan, M. Velan (2011). Removal of heavy metal from aqueous solution using Schizosaccharomycespombe in free and alginate immobilized cells. *2nd International Conference on Environmental Science and Technology* IPCBEE IACSIT Press, Singapore 6 (2011).

[118] B. S. Prakash, S. V. Kumar, Batch Removal of Heavy Metals by Biosorption onto Marine Algae-Equilibrium and Kinetic Studies; *Int. J. Chem. Tech. Res.* 5 (3) (2013).

[119] J. Rincon, F. Gonzalez, A. Ballester, M. L. Blazquez, J. A. Munoz, Biosorption of heavy metals by chemically-activated alga Fucusvesiculosus. *Journal of Chemical Technology and Biotechnology* 80 (2005) 1403-1407.

[120] Olga M. M. Freitas, Ramiro J. E. Martins, Cristina M. Delerue-Matos4, Rui A. R. Boaventura, Removal of Cd(II), Zn(II) and Pb(II) from aqueous solutions by brown marine macro algae: Kinetic modeling. *J. Hazardous Materials*, 153 (1-2) (2008) 493-501.

[121] M. Hajar, Biosorption of Cadmium from Aqueous Solution using Dead Biomass of Brown Alga Sargassum Sp. mohamad.hajar@yahoo.com (M. Hajar).

[122] E. Romera, F. Gonza´lez, A. Ballester, M. L. Bla´zquez, J. A. Mun˜oz, Comparative study of biosorption of heavy metals using different types of algae. *Bioresource Technology* 98 (2007) 3344-3353.

[123] S. Ahmady-Asbchin, M. Mohammadi. Biosorption of Copper Ions by Marine Brown Alga Fucusvesiculosus; *J. Biol. Environ. Sci.*, 5 (15) (2011) 121-127.

[124] J. N. Al Kumar, C. Oommen, R. N. Kumar, Biosorption of Heavy Metals from Aqueous Solution by Green Marine Macroalgae from Okha Port, Gulf of Kutch, India; *American-Eurasian J. Agric. and Environ. Sci.* 6 (3) (2009) 317-323.

[125] J. I. N. Kumar, C. Oommen. Removal of heavy metals by biosorption using freshwater alga Spirogyra hyaline; *J. Environ. Biol.* 33 (2012) 27-31.

[126] Sulaymon, A. A. Mohammed, T. J. Al-Musawi. Column bioaorption of lead, cadmium, copper and arsenic ions onto algae. *J. Bioprocess Biotech.* 3 (2013) 128.

[127] G. M. Gadd. Interactions of fungi with toxic metals. *New Phytol.* 124 (1993) 25-60.

[128] G. M. Gadd, A. J. Griffiths, Effect of copper on morphology of Aureobasidium pullulans. *Trans. Brit. Mycol. Soc.* 74 (1980) 387-392.

[129] G. M. Gadd, J. L. Mowll, Copper uptake by yeast-like cells. Hyphae and chlamydospores of Aureobasidium. *Exp. Mycol.* 9 (1985) 230-240.

[130] J. Remacle, The cell wall and metal binding. In: B. Volesky (Ed.). *Biosorption of Heavy Metals.* CRC Press. Boca Raton. Florida. US (1990) 83-92.

[131] A. H. Sulaymon, S. E. Ebrahim, S. M. J. Al-MusawiTAbdullah, Removal of lead, cadmium, and mercury ions using biosorption. *Desalination and water treatment* 24 (2010) 344-352.4.

[132] C. Huang, C. P. Application of Aspergillus oryzae and Rhizopusoryzae for Cu (II) removal. *Wat. Res.* 30 (1996) 1985-1990.

[133] V. D. pullulans, R. Vimala, P. Kartika, *Biosorption of heavy metals-An overview* (2008).

[134] Kogej, A. Pavko, Laboratory experiments of lead biosorption by self-immobilized Rhizopusnigricans pellets in the batch stirred tank reactor and the packed bed column; *Chem. Biochem. Eng. Q.* 15 (2) (2001) 75-79.

[135] L. M. Jr. Barros, G. R. Macedo, M. M. L. Duarte, E. P. Silva, A. K. C. L. Lobato, Biosorption of cadmium using the fungus Aspergillus niger, Brazilian. *J. Chem. Eng.* 20 (3) (2003) 229-239.

[136] M. S. Osman, M. Bandyopadhyay, Biosorption of lead (II) ions from wastewater by using a fungus p. Ostreatus. *Journal of Civil Engineering. Bangladish.* 27 (2) (1999) 193-196.

[137] S. W. Hassan, H. Y. El-Kassas, Biosorption of Cadmium from Aqueous Solutions using A local Fungus Aspergillus cristatus (Glaucus Group). *African Journal of Biotechnology* 11 (9) (2012) 2276-2286.

[138] L. M. B. Júnior, G. R. Macedo, M. M. L. Duarte, E. P. Silva, A. K. C. L. Lobato, Biosorption of cadmium using the fungus Aspergillus niger. *Braz. J. Chem. Eng.* 20 (3) (2003). http://dx.doi.org/10.1590/S0104-66322003000300003.

[139] S. Bunluesin, M. Kruatrachue, P. Pokethitiyook, S. Upatham, G. R. Lanza, Batch and Continuous Packed Column Studies of Cadmium Biosorption by Hydrillaverticillata Biomass. *J. of Bioscience and Bioenginerring* 103 (6) (2007) 509-513.

[140] V. M. Ju´arez, J. F. C. Gonz´alez, M. E. T. Bouscoulet, A. Rodr´ıguez, Biosorption of Mercury (II) from Aqueous Solutions onto Fungal Biomass. *Bioinorganic Chemistry and Applications* (1-5) doi:10.1155/2012/156190 (2012).

[141] M. X. Loukidou, K. A. Matis, A. I. Zouboulis, K. M. Liakopoulou, Removal of As(V) from wastewaters by chemically modified fungal biomass. *Water Res.* 37 (2003) 4544-4552.

[142] G. S. Murugesan, M. Sathishkumar, K. Swaminathan, Arsenic removal from groundwater by pretreated waste tea fungal biomass. *Bioresource Technology* 97 (2006) 483-487.

[143] S. Mamisahebei, Gh. R. J. Khaniki, A. Torabian, S. Nasseri, K. Naddafi, Removal of arsenic from an aqueous solution by pretreated waste tea fungal biomass. *Iran. J. Environ. Health. Sci. Eng.* 4 (2) (2007) 85-92.

[144] A. Shoaib, A. Naureen, F. Tanveer, N. Aslam, Removal of Ni(II) Ions from Substrate using Filamentous Fungi; *Int. J. Agric. Biol.* 14 (2012) 831-833.

[145] C. Tay, H. Liew, S. Yong, S. Surif, G. Redzwan, S. Abdul-Talib, Cu(II) Removal onto Fungal Derived Biosorbents: Biosorption Performance and the half Saturation Constant Concentration Approach. *Int. J. Res. Chem. Environ.* 2 (3 July 2012) 138-143.

[146] S. Jha, S. N. Dikshit, G. Pandey, Comparative study of Growing/Immobilized biomass verses resting biomass of A. lentulus for the effect of pH on Cu2+ metal removal. *Research Journal of Pharmaceutical. Biological and Chemical Sciences. RJPBCS* 3 (3) (2012) 421-427.

[147] C. Sutherland, C. Venkobachar, A diffusion-chemisorption kinetic model for simulating biosorption using forest macro-fungus, fomesfasciatus. *International Research Journal of Plant Science* 1 (4) (2010) 107-117.

[148] Z. Velkova, M. Stoytcheva, V. Gochev, Biosorption of Cu (II) onto chemically modified waste mycelium of Aspergillus awamori. Equilibrium, kinetics and modeling studies. *J. BioSci. Biotech.* 1 (2) (2012) 163-169.

[149] R. Say, N. Yilmaz, A. Denizli, Removal of chromium (VI) ions from synthetic solutions by the fungus Penicilliumcanescens, *The European Journal of Mineral Processing and Environmental Protection* 3 (1) (2003) 36-41.

[150] M. Nikazar, L. Davarpanah, F. Vahabzadeh, Biosorption of aqueous chromium (VI) by living mycelium of phanerochaetechrysosporium, http://www.aidic.it/IBIC2008/webpapers/27Nikazar.pdf.

[151] B. K. Sethi, S. Kanungo, J. R. Rout, P. K. Nanda, S. L. Sahoo, Effect of Chromium on Mucor species and optimization of growth conditions. *Nature and Science*. 8 (4) (2010) 29-32.

[152] H. Y. El-Kassas, E. M. El-Taher, Optimization of Batch Process Parameters by Response Surface Methodology for Mycoremediation of Chrome-VI by a Chromium Resistant Strain of Marine Trichoderma Viride. *American-Eurasian J. Agric. and Environ. Sci.* 5 (5) (2009) 676-681.

[153] S. Chhikara, R. Dhankhar, Biosorption of Cr(VI) ions from electroplating industrial effluent using immobilized Aspergillus niger biomass. *J. Environ. Biol.* 29 (2008) 773-778.

[154] A. Javaid, R. Bajwa, Biosorption of Cr (III) ions from tannery wastewater by Pleurotusostreatus. *Mycopath.* 5 (2007) 71-79.

[155] P. Donghee, Y. Yeoung-sang, and P. M. Jong, Use of dead fungal biomass for the detoxification of hexavalent chromium: screen and kinetics. *Process Biochemistry* 40 (2007) 2559.

[156] N. Das, D. Charumathi, R. Vimala, Effect of pretreatment on Cd2+ biosorption by mycelial biomass of Pleurotusflorida. *African Journal of Biotechnology* 6 (22) (2007) 2555-2558.

[157] S. Abbas, T. Mostafa, A. Sulaymon, I. Ismail, A Comparative Experimental and Theoretical of Removal Pb (II) and Cd (II) from Wastewater Using Different Adsorbents, *International Journal of Chemical and Environmental Engineering* 5 (4) (2014) 195-205.

[158] H. Yavuz, A. Denizli, H. G¨ung¨unes, M. Safarikova, I. Safarik, Biosorption of mercury on magnetically modified yeast cells. *Separation and Purification Technology* 52 (2006) 253-260.

[159] A. Anaemene, the Use of Candida sp in the Biosorption of Heavy Metals from Industrial Effluent. *Euro. J. Exp. Bio.* 2 (3) (2012) 484-488.

[160] M. Mapolelo, N. Torto. Trace enrichment of metal ions in aquatic environments by Saccharomyces cerevisiae. *Talanta* 64 (2004) 39-47.

[161] I. S. Abd-Elsalam, Factorial design for some parameters affecting on chromium (III) uptake by saccharomyces cerevisiae; *International Journal of Applied Biology and Pharmaceutical Technology* 2 (1) (2011) 33-40.

[162] P. Kujan, A. Prell, H. Safar, M. Sobotka, T. Rezanka, P. Holler. Use of the industrial yeast candida utilis for cadmium sorption, *André Folia Microbiol.* 54 (4) (2006) 257-260.

[163] A. Arakaki, L. Vandenberghe, V. Soccol, R. Masaki, E. Filho, A. Gregório, C. Soccol, Optimization of Biomass Production with Copper Bioaccumulation by Yeasts in Submerged Fermentation. *Brazilian archives of biology and technology* 54 (5) 1027-1034.

[164] M. Apinthanapong, M. Phensaijai, Biosorption of copper by spent yeast immobilized in sodium alginate beads. *Kasetsart Journal Natural Sciences* 43 (5) (2009) 326-332.

[165] N. Lokeshwari, K. Joshi, Biosorption of Heavy Metal (Chromium) Using Biomass. *Global Journal of Environmental Research* 3 (1) (2009) 29-35.

[166] E. A. F. Hosam, B. Hamuda, Nikolett Tóth. Functioning of divalent alkaline metal on yeast multiplication in heavy metal contaminated soil. *Tájökológiai Lapok* 10 (2) (2012) 195-208.

[167] C. S. V. Wyk, Removal of heavy metals from metal-containing effluent by yeast biomass. *African Journal of Biotechnology* 10 (55) (2011) 11557-11561.

[168] M. D. Machado, S. Janssens, H. M. V. M. Soares, E. V. Soares, Removal of heavy metals using a brewer's yeast strain of Saccharomyces cerevisiae: advantages of using dead biomass. *Journal of Applied Microbiology* 106 (2009) 1792-1804.

[169] D. M. Machado, E. V. Soares, M. Soares, Removal of heavy metals using a brewer's yeast strain of Saccharomyces cerevisiae: Chemical speciation as a tool in the prediction and improving of treatment efficiency of real electroplating effluents. *Journal of Hazardous Materials* 180 (2010) 347-353.

[170] G. Yan, T. Viraraghavan, Effect of pretreatment on the bioadsorption of heavy metals on Mucorrouxii. *Water SA* 26 (1) (January, 2000) 119-124.

[171] Y. Göksungur, S. Üren, U. Güvenç, Biosorption of copper ions by caustic treated waste baker's yeast biomass. *Turk. J. Biol.* 27 (1) (2003) 23-29.

[172] M. T. El-Sayed, Removal of lead (II) by Saccharomyces Cerevisiae AUMC 3875. *Ann. Microbiol.* 63 (2013) 1459-1470.

[173] R. Dhankhar, A. Hooda, R. Solanki, P. A. Sainger, Saccharomyces cerevisiae: a potential biosorbent for biosorption of uranium. *International Journal of Engineering Science and Technology (IJEST)* 3 (6) (2011) 5397-5407.

[174] A. H. Sulaymon, S. A. Yousif, M. M. Al-Faize, Single-multicomponent biosorption of lead, mercury, chromium and arsenic onto activated sludge in batch and fixed-bed adsorber. *Desalination and Water Treatment* 53 (13) (2015) 3499-3512.

[175] A. H. Sulaymon, D. W. Abbood, A. H. Ali, Competitive biosorption of phenol and lead from synthetic wastewater onto live and dead microorganisms. *Desalination and water treatment* 45 (2012) 331-342.

[176] S. Abbas, I. Ismail, T. Moustafa, A. Sulaymon, Biosorption of Heavy Metals: A Review. *Journal of Chemical Science and Technology* 3 (4) (2014) 74-102.

In: Heavy Metals
Editor: Deepak Pathania
ISBN: 978-1-63484-740-7

Chapter 6

TOXICITY AND ADVANCED DETECTION STRATEGY OF HEAVY METALS

***Debabrata Nandi*[1,*] *and Arjun Maity*[1,2,†]**
[1]Department of Civil and Chemical Engineering,
University of South Africa (UNISA), South Africa
[2]DST/CSIR, National Centre for Nanostructured Materials,
Materials Science and Manufacturing, Council for Scientific
and Industrial Research, Pretoria, South Africa

ABSTRACT

Though, certain heavy metals (e.g., Fe, Mn, Cu, Zn) are nutritionally indispensable for a healthy life, beyond permissible limit, these affect molecular structure of proteins or else inhibit enzyme through forming bonds especially with ligands of biotic matter comprising oxygen, nitrogen, and sulphur atom. These chemical/physical interactions may germinate toxicological effects of heavy metals such as those affecting the kidneys or liver (Cd^{2+}, Cu^{2+}, Pb^{2+}, Hg^{2+}); the central nervous system (Pb^{2+}, Hg^{2+}, As^{3+}); skin, bones, or teeth (As^{3+}, As^{5+}, Ni^{2+}, Cu^{2+}, Cd^{2+}, Zn^{2+}, Cr^{3+}, Cr^{6+}). As heavy metals are non-biodegradable in nature, they remain in food chain as well as ecological systems indefinitely, displaying top-level predators to break out high levels of contamination. Hence, safe limits of contaminant levels have been demarcated for drinking water by various organizations throughout the world. Based on toxicology data and scientific reports, World Health Organization (WHO) and Environmental Protection Agency (EPA) recommended standard permissible limits for heavy metals in drinking water (Table 1). According to chemical recognition along with size (macro to nano), shape, morphology and porosity (macro to nano), the selective affinity of heavy-metal ions with the specific species/receptor customarily encompasses noncovalent interaction, viz van der Waals forces, hydrogen bonding, coordination, π-π interactions, hydrophilic/phobic interaction, electrostatic and electromagnetic effects. Efforts are utilizing to develop prompt and inexpensive techniques for heavy-metal detection in different chemical systems,

* Corresponding author address: E-Mail: nandidebabrata87@gmail.com (D. Nandi), amaity@csir.co.za, Phone: +2712 841 2658, Fax: +2712 841 3553.
† maityarjun@gmail.com (A. Maity).

including living species in ecosystem. From lab bench to field scale, the standard sophisticated analytical techniques (even at ppt and ppq level) such as atomic absorption spectroscopy (AAS), mass spectroscopy (MS), inductively coupled plasma-mass spectrometry (ICPMS), X-ray fluorescence spectroscopy (R-FS), potentiometric methods have been adapted. Still scientists are burning the midnight oil to improve strategies for early detection of trace heavy metals. Therefore, this chapter envisioned a universal knowledge on toxic/carcinogenic effects of various heavy metals in living systems in account of their bonding interactions and the leading-edge strategies in the development of heavy-metal-detection devices/systems constructed on chemical recognition and specific affinity of receptor.

Keywords: abundance, pourbiax diagram, toxicity, detection strategy

Table 1. Permissible limit of heavy metals in drinking water recommended by EPA and WHO

Heavy Metal	EPA (mg/L)	WHO (mg/L)
Hg	0.002	0.001
Cd	0.005	0.003
Pb	0.015	0.010
As	0.010	0.010
Ni	0.040	0.070
Cu	1.300	2.000
Zn	5.000	3.000

1. INTRODUCTION

The uninterrupted discharge of copious amounts of heavy metal rich domestic and industrial effluents onto surface water and soil, bereft of appropriate treatment, fouls not only the surface water and soil but the groundwater also [1]. The accumulation of heavy metals and metalloids through emissions from the rapidly expanding industrial areas, mine tailings, disposal of high metal wastes, leaded gasoline and paints, animal manures, sewage sludge, pesticides, wastewater irrigation, land application of fertilizers, coal combustion residues, spillage of petrochemicals, and atmospheric depositions pose several severe threats to environmental health [2-3]. Heavy metals set up an ill-defined group of inorganic chemical hazards, and those most commonly reported at contaminated sites are lead (Pb), chromium (Cr), arsenic (As), zinc (Zn), cadmium (Cd), copper (Cu), mercury (Hg), and nickel (Ni) [4]. Soils and surface water are the chief sinks for heavy metals released into the environment by aforementioned anthropogenic activities and unlike organic contaminants which are oxidized to carbon (IV) oxide by microbial action, most metals do not undergo microbial or chemical degradation [5], and their total concentration in soils continues for a long term after their introduction [6]. Modifications in their chemical forms (speciation) and bioavailability are, however, possible. The presence of toxic metals in surface can rigorously inhibit the biodegradation of organic contaminants [7]. Heavy metal contamination may carry risks and hazards to humans and the ecosystem through: direct ingestion or contact with contaminated soil, the food chain (soil-plant-human or soil-plant-animal-human), drinking of contaminated ground water, falling in food quality (safety and marketability) via phytotoxicity, decrease in land usability for agricultural production causing food insecurity, and land tenure complications [8-10]. The adequate protection and restoration of ecosystems contaminated by heavy metals necessitate their characterization and remediation. Contemporary legislation respecting environmental protection and public health, at both national and international levels, are established on data that describe chemical properties of environmental phenomena, especially those that reside in our food chain [11]. While its characterization can ventilate an insight into heavy metal speciation and bioavailability, attempt at remediation of heavy metal contaminated soils/water would demand knowledge of the source of contamination, basic chemistry, and environmental and associated health risks of heavy metals. Risk assessment is a real scientific tool which enables decision makers to manage sites so contaminated in a cost-effective manner while preserving public and ecosystem health [12]. Despite their cost-effectiveness and environment friendliness, field applications of these technologies have only been testified in developed countries. With greater awareness by the governments and the public of the implications of contaminated surface on human and animal health, there has been increasing interest amongst the scientific community in the development of technologies to remediate contaminated sites [13]. In developing countries with great population density and scarce funds accessible for environmental restoration, low-cost and ecologically sustainable remedial options are mandatory to restore contaminated lands so as to reduce the associated risks, make the land resource accessible for agricultural production, enhance food security, and scale down land tenure problems. Here, scattered literature is utilized to review the possible sources of contamination, basic chemistry, and the associated environmental and health hazards of prior heavy metals (Pb, Cr, As, Zn, Cd, Cu, Hg, and Ni) which can offer insight into heavy metal speciation, bioavailability, and hence selection of appropriate remedial options.

2. Sources of Heavy Metals

Heavy metals observe naturally in the environment from the pedogenetic processes of weathering of parent materials at levels that are regarded as trace and rarely toxic [11, 13]. Due to the disturbance and acceleration of nature's slowly occurring geochemical cycle of metals by man, rural and urban environments may accumulate one or more of the heavy metals above demarcated background values high enough to cause hazards to human health, plants, animals, ecosystems, or other media [14]. The heavy metals become pollutants in the environments because (i) their rates of generation via man-made cycles are more rapid relative to natural ones, (ii) they become transferred from mines to random environmental locations where higher potentials of direct exposure occur, (iii) the concentrations of the metals in discarded products are relatively high compared to those in the receiving environment, and (iv) the chemical form (species) in which a metal is found in the receiving environmental system may render it more bioavailable [14]. A simple mass balance of the heavy metals in the soil can be communicated as follows [15-16]:

$$M_{total} = (M_p + M_f + M_a + M_{ag} + M_{ow} + M_{ip}) - (M_{cr} + M_l) \quad (1)$$

where "M" is the heavy metal, "p" is the parent material, "f" is the fertilizer sources, "a" is the atmospheric deposition, "ag" are the agrochemical sources, "ow" are the organic waste sources, "ip" are other inorganic pollutants, "cr" is crop removal, and "l" is the losses by leaching, volatilization, and so forth. Metal containing solids at contaminated sites can create from a wide variety of anthropogenic sources in the form of metal mine tailings, disposal of high metal wastes in inappropriately protected landfills, leaded gasoline and lead-based paints, land application of fertilizer, animal manures, biosolids (sewage sludge), compost, pesticides, coal combustion residues, petrochemicals, and atmospheric deposition [17] are described hereunder.

2.1. Fertilizers

Historically, agriculture was the first major human effect on the surface [18]. To grow and complete the lifecycle, plants must attain not only macronutrients (N, P, K, S, Ca, and Mg), but also indispensable micronutrients. Some surfaces are deficient in the heavy metals (such as Co, Cu, Fe, Mn, Mo, Ni, and Zn) which are essential for healthy growth [19]. Large amounts of fertilizers are frequently added to surface soil/water in intensive farming systems to provide adequate N, P, and K for crop growth. The compounds used to supply these elements encompass trace amounts of heavy metals (e.g., Cd and Pb) as impurities. Metals, such as Cd and Pb, have no recognized physiological activity. Use of certain phosphate fertilizers inadvertently increases Cd and other potentially toxic elements to the soil, including F, Hg, and Pb [20].

2.2. Pesticides

Common pesticides were utilised extensively in agriculture and horticulture in the past contained substantial concentrations of metals. For example in the recent past, about 10% of the chemicals have permitted for use as insecticides and fungicides in UK were containing Cu, Hg, Mn, Pb, or Zn. Lead arsenate was applied in fruit orchards for many years to control some parasitic insects. Arsenic-containing compounds were also applied extensively to control cattle ticks and to control pests in banana in New Zealand and Australia, timbers have been conserved with formulations of Cu, Cr, and As (CCA). In comparison with fertilizers, the use of such materials has been more confined, being restricted to particular sites or crops [9].

2.3. Biosolids and Manures

The application of copious biosolids (e.g., livestock manures, composts, and municipal sewage sludge) to land inadvertently tips to the accumulation of heavy metals such as As, Cd, Cr, Cu, Pb, Hg, Ni, Se, Mo, Zn, Tl, Sb, and so forth, in the soil [17]. Certain animal wastes such as poultry, cattle, and pig manures made in agriculture are commonly used to crops and pastures either as solids or slurries [21]. While most manure are seen as valuable fertilizers, in the pig and poultry industry, the Cu and Zn added to diets as growth promoters and As contained in poultry health products may cause metal contamination of the surface [21- 22]. Heavy metals most frequently observed in biosolids are Pb, Ni, Cd, Cr, Cu, and Zn, and the metal concentrations are administered by the nature and the intensity of the industrial activity, as well as the type of process utilized during the biosolids treatment [23]. Current studies on some New Zealand surfaces treated with biosolids have revealed increased concentrations of Cd, Ni, and Zn in drainage leachates [24-25].

2.4. Wastewater

The use of municipal and industrial wastewater and related effluents to land dates back 400 years and now is a common repetition in many of arable land are irrigated with waste water. Agriculture oriented wastewater irrigation accounted for 50 percent of the vegetable supply to urban areas [26]. Farmers usually are not worried for environmental hazards and are primarily interested in maximizing their yields and profits.

2.5. Metal Mining, Milling Processes and Industrial Wastes

Mining and milling of metal ores combined with industries have willed many countries, the legacy of wide distribution of metal contaminants in soil/water. Extensive Pb and Zn ore mining and smelting have resulted in contamination of soil/water that poses risk to human and ecological health. Other materials are produced by a variety of industries such as textile, tanning, petrochemicals from accidental oil spills or utilization of petroleum- based products, pesticides, and pharmaceutical facilities and are highly inconstant in composition. In addition, many are potentially unsafe because of heavy metals (Cr, Pb, and Zn).

2.6. Airborne Sources

Airborne sources of metals comprise stack or duct emissions of air, gas, or vapour streams, and fugitive emissions such as dust from storage areas or waste piles. Metals from airborne sources are usually released as particulates in the gas stream. High-temperature processing can also volatilize some metals such as As, Cd, and Pb. These metals will transform to oxides and condense as fine particulates unless a reducing atmosphere is maintained [27]. Stack emissions can be circulated over a wide area by natural air currents until dry and/or wet precipitation mechanisms remove them from the gas stream. All solid particles in smoke from fires and in other releases from factory chimneys are ultimately deposited on land or sea. For example, very high concentration of Cd, Pb, and Zn has been observed in plants and soils adjacent to smelting works. Zn and Cd may also be used in surface adjacent to roads, the sources being tyres, and lubricant oils [28].

3. Potential Risks of Heavy Metals

The most abundant heavy metals found at contaminated sites, are Pb, Cr, As, Zn, Cd, Cu, and Hg [28]. Those metals are imperative since they are capable of decreasing crop production due to the risk of bioaccumulation and biomagnification in the food chain. There's also the jeopardy of superficial and groundwater contamination. Awareness of the basic chemistry, environmental and associated health effects of these heavy metals is essential in understanding their speciation, bioavailability, and remedial options. The fate and transport of a heavy metal in soil be subject to the chemical form and speciation of the metal. Once in the surface, heavy metals are adsorbed primarily in fast reactions (minutes, hours), followed by slow adsorption reactions (days, years) and are, therefore, rearranged into different chemical forms with varying bioavailability, mobility, and toxicity [29-30].

3.1. Arsenic

Arsenic is a group VA metalloid in period 4 of the periodic table that occurs in an extensive variety of minerals, mainly as As_2O_3, and can be recovered from processing of ores comprising typically Cu, Pb, Zn, Ag and Au. Common minerals contain the two sulfides realgar (As_4S_4) and orpiment (As_2S_3) and the oxidized form arsenolite (As_2O_3). The arsenides of Fe, Co and Ni and their mixed sulfides produce another set of minerals, e.g., safforlite (CoAs), loellingite (FeAs2), niccolite (NiAs), arsenopyrite (FeAsS), rammelsbergite ($NiAs_2$), cobaltite (CoAsS), gerdorffite (NiAsS) and the quaternary sulfide glaucodot [(Co,Fe)AsS]. As is found on an industrial scale by smelting [31] $FeAs_2$ or FeAsS at 650-700°C in the absence of air and condensing the sublimed element:

$$FeAsS \longrightarrow FeS + As(g) \longrightarrow As(s)$$

Arsenic encompasses the following properties: atomic mass 75, atomic number 33, density 5.72 g cm^{-3}, melting point 817°C, and boiling point 613°C, and exhibits fairly complex chemistry and can be present in several oxidation states (−III, 0, III, V) [27]. In aerobic environments, As (V) is predominant, usually in the form of arsenate (AsO_4^{3-}) in various protonation states: H_3AsO_4, $H_2AsO_4^-$, $HAsO_4^{2-}$, and AsO_4^{3-}. Arsenate and other anionic forms of arsenic act as chelates and can precipitate when metal cations are existing [32]. Metal arsenate complexes are stable merely under certain conditions. Arsenic (V) may also coprecipitate with or adsorb onto iron oxyhydroxides under acidic and moderately reducing conditions. Coprecipitates are immobile under these conditions, but its mobility increases as pH increases [27]. Under reducing conditions As(III) predominates, surviving as arsenite ($AsO3^{3-}$), and its protonated forms H_3AsO_3, $H_2AsO_3^-$, and $HAsO_3^{2-}$. Arsenite may adsorb or coprecipitate with metal sulfides and has a high affinity towards other sulfur compounds. Elemental As and arsine (AsH_3) may exist under extremely reducing conditions. Biotransformation (via methylation) of arsenic generates methylated derivatives of arsine, such as dimethyl arsine $HAs(CH_3)_2$ and trimethylarsine $As(CH_3)_3$ which are extremely volatile. Since arsenic is often occur in anionic form, it does not form complexes with simple anions such as Cl^- and SO_4^{2-}. Arsenic speciation also contains organometallic forms such as methylarsinic acid $(CH_3)AsO_2H_2$ and dimethylarsinic acid $(CH_3)_2AsO_2H$. As cycle is shown below in Figure 1.

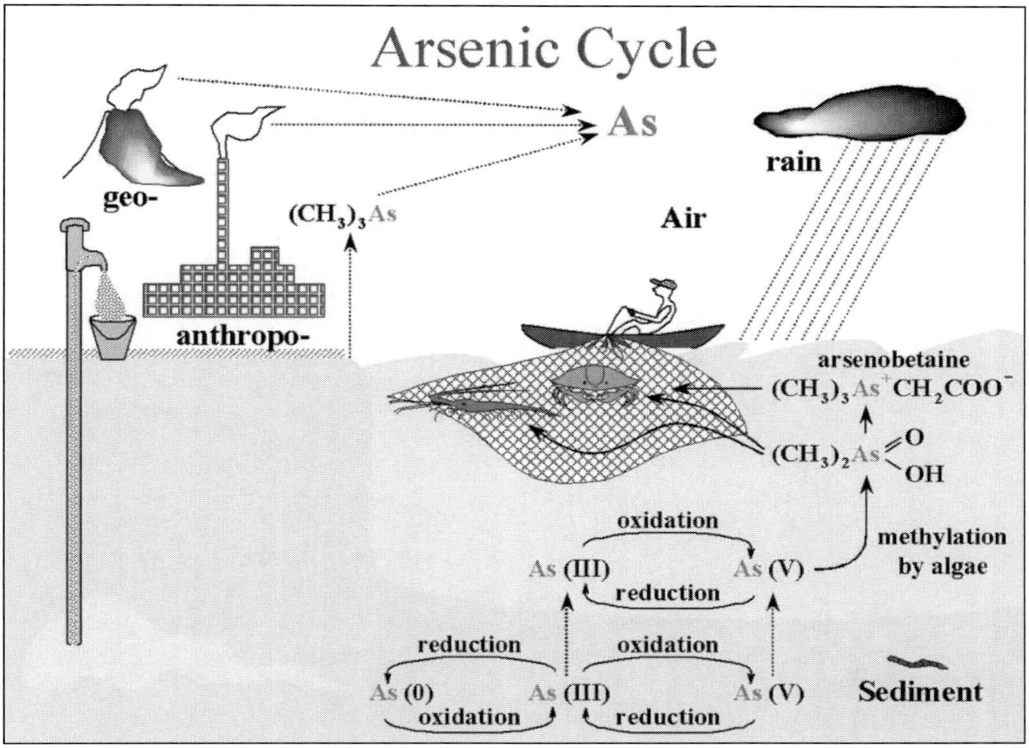

Figure 1. Proposed arsenic cycle with variable oxidation states.

Arsenic content in igneous rocks are generally low ranging from 1.5-1.8 $mg.kg^{-1}$, in metamorphic rocks is 5 $mg.kg^{-1}$, in sedimentary rocks it is in the range of 5-10 $mg.kg^{-1}$. Arsenic concentration in coal and bituminous deposites are found in the range 3500 $mg.kg^{-1}$ and 100-900 $mg.kg^{-1}$, respectively. Some iron stone and iron rich rocks are also relatively rich in arsenic with value up to 2900 $mg.kg^{-1}$ [33]. Although, the dominant source of arsenic in soils is geogenic, the additional imputes may be derived locally from industrial sources and fossil fuel combustion products and agricultural sources such as pesticides and weed killers. Arsenic can occur in nature in several oxidation states (-3, 0, +3 and +5), but in natural water it is mostly found in inorganic forms as oxyanions of trivalent and pentavalent arsenic. However, which species predominate in the medium depends on the redox potential and pH.

Figure 2 shows that the arsenic (V) remains almost entirely as undissociated H_3AsO_4 in strongly acidic medium ($pH \leq 2.0$) at high Eh. The pK values of H_3AsO_4 are $pK_1 = 2.19$, $pK_2 = 6.94$ and $pK_3 = 11.5$. In acidic medium (pH = 2.0 to 7.0), pentavalent arsenic mainly exists as $H_2AsO_4^-$ where as in neutral or moderately alkaline pH medium (pH = 7.0 to 11.5), the predominant species is $HAsO_4^{2-}$. AsO_4^{3-} only exists in highly alkaline pH (pH > 11.5). Thus, the commonest species in natural water is $HAsO_4^{2-}$, which is stable under neutral to mildly alkaline medium at negative Eh values. Arsenic in trivalent state exists as undissociated H_3AsO_3 over a wide range of pH under the reducing environment. The pK_a

values of H_3AsO_3 are $pK_1 = 9.20$, $pK_2 = 14.22$ and $pK_3 = 19.22$. Under reducing condition at pH less than 9.2, the dissociation of H_3AsO_3 is highly unlikely and it remains undissociated. Thus arsenic is the most problematic in the environment because of its relative mobility over a wide range of redox conditions. Organic form of arsenic is usually minor in surface as well as in ground water. The dominant organic forms of arsenic are mono and dimethyl arsenic acid. In both the cases arsenic present is in pentavalent state. Organo arsenic is formed by methylation of inorganic arsenic catalysed by microbial activity.

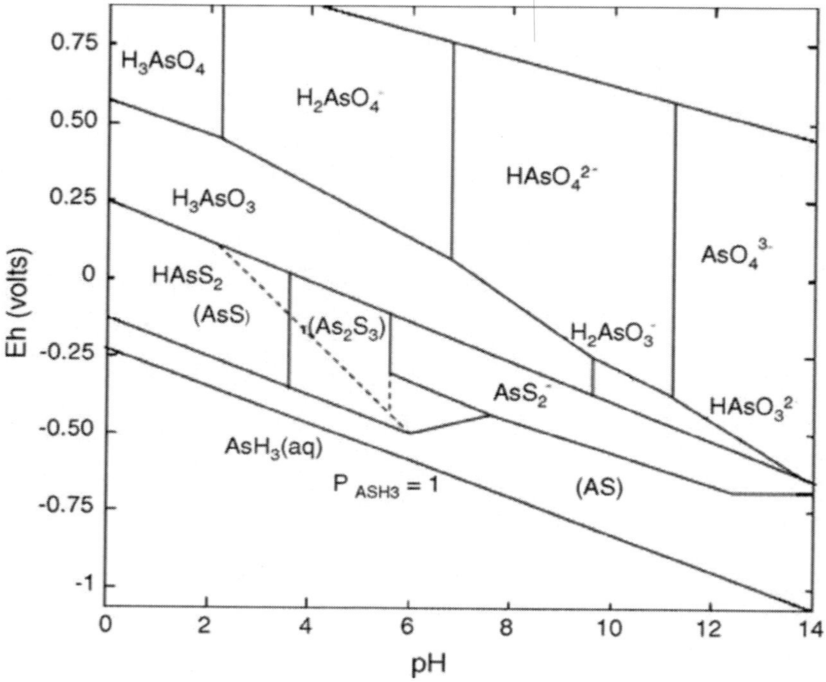

Figure 2. Eh-pH diagram for arsenic species at 25°C and 101.3 kPa.

Arsenic in the environment above tolerance limit is a worldwide problem. Arsenic pollution had been reported in the USA, China, Chile, Bangladesh, Taiwan, Mexico, Argentina, Poland, Canada, Hungary, New Zealand, Vietnam, Romania, Japan, and India [34]. Arsenic is supposed to derive worldwide from various sources, including natural dissolution/desorption reactions, geo thermal trigger and mining activity (Figure 3).

Arsenic is recognized as a micronutrient essential to life up to a certain level (10 $\mu g.L^{-1}$). But when the desired level exceeds it shows physicochemical and toxicological effect [35]. The carcinogenic effect of arsenic inhibits replication of DNA and interrupts the repair mechanism possibly through the linkage with thiol groups of DNA proteins causing headache, nervous disorder, respiratory disorders, loss of nails and hairs. High dose of arsenic cause black-foot disease, caratosis, melanosis, and muscular weakness, loss of appetite, nausea, and long term drinking water exposure causes skin, lung, bladder, kidney cancer. Arsenic also has phytotopic effects on certain plants as it is capable of replacing phosphorous in plant cells causing biological disorders (Figure 4). Looking onto the entire spectrum of arsenic, it becomes apparent that arsenic is extremely hazardous to both plants and animals.

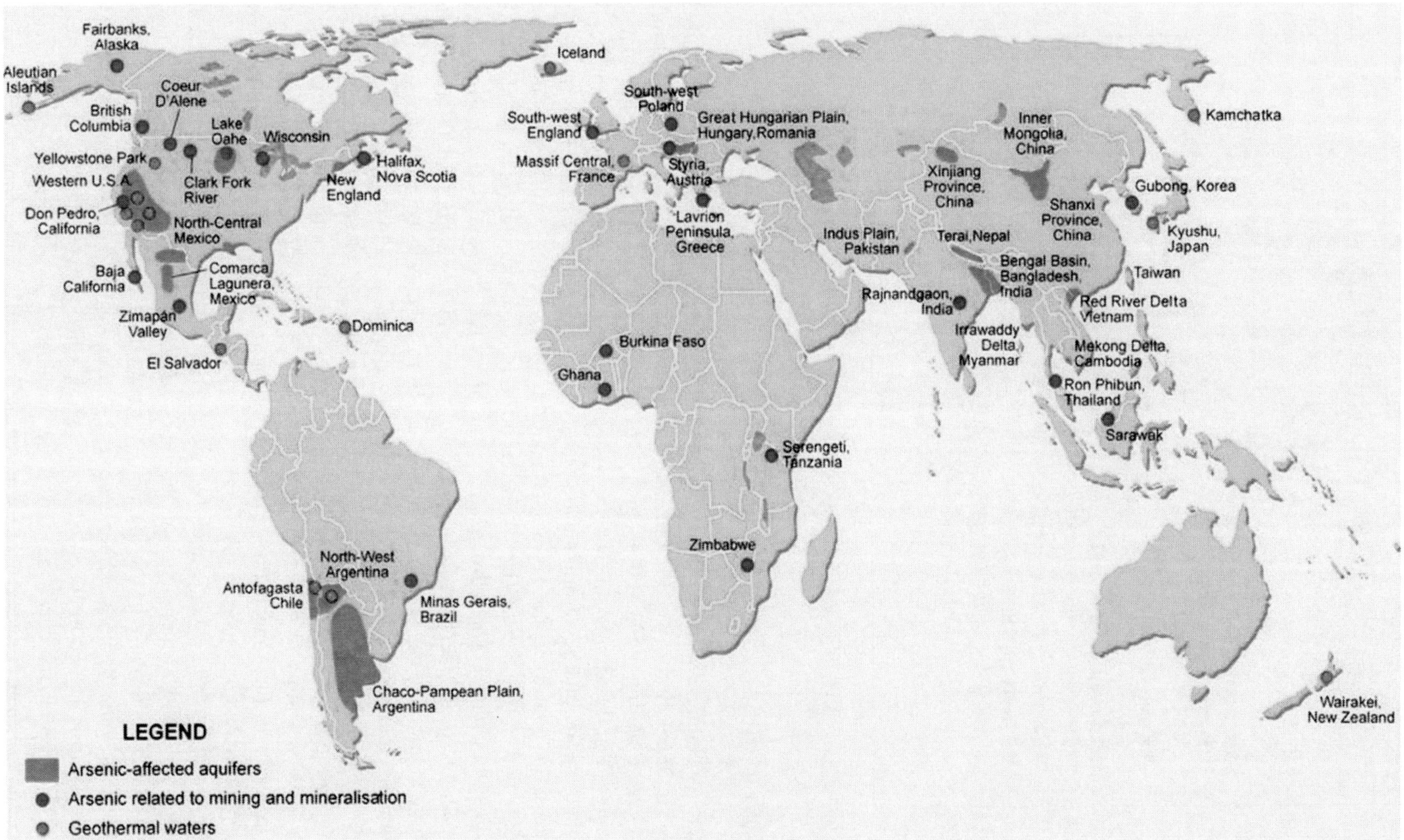

Figure 3. Worldwide arsenic sources.

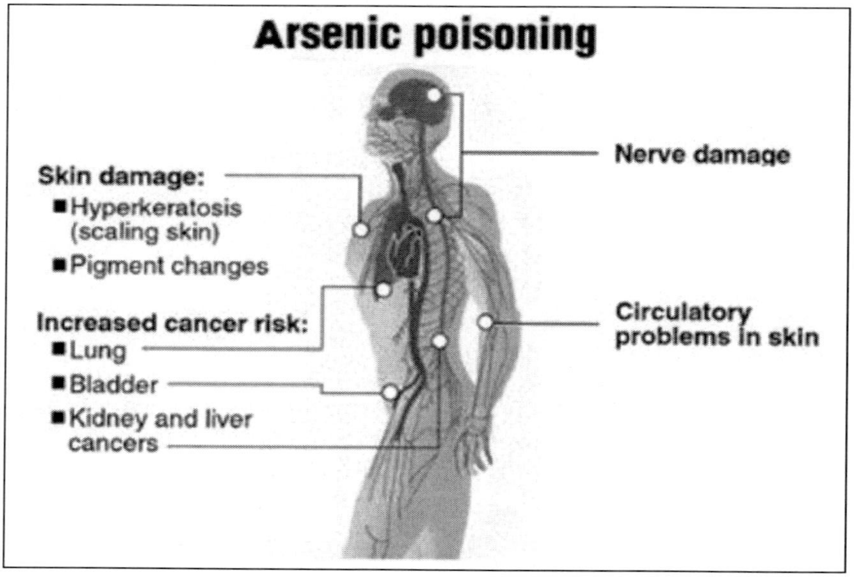

Figure 4. Arsenic poisoning effect on human health.

3.2. Mercury

Mercury belongs to same group with Zn and Cd in the periodic table. It is the only liquid metal at stp. It has atomic weight 200.6, atomic number 80, density 13.6 g cm^{-3}, melting point −13.6°C, and boiling point 357°C and is customarily recovered as a byproduct of ore processing [27]. Discharge of Hg from coal combustion is a foremost source of Hg contamination. Discharges from manometers at pressure-measuring stations along gas/oil pipelines also contribute to Hg contamination. After release to the environment, Hg commonly occurs in mercuric (Hg^{2+}), mercurous (Hg_2^{2+}), elemental (Hg^0), or alkylated form (methyl/ethyl mercury). The redox potential and pH of the system determine the stable forms of Hg. Mercurous and mercuric forms are more stable under oxidizing conditions. When a bit reducing conditions exist, organic or inorganic Hg may be reduced to elemental Hg, which may then be transformed to alkylated forms by biotic or abiotic processes. Mercury is maximum toxic in its alkylated forms which are soluble in water and volatile in air [27]. Mercury(II) can forms strong complexes with a variety of both inorganic and organic ligands, creating it very soluble in oxidized aquatic systems [32]. Sorption to soils, sediments, and humic materials is a key mechanistic pathway for the removal of Hg from solution. Sorption is pH dependent and increases as pH rises.

Diagram shows that Quicksilver (Liquid Hg) is the most stable, as we get reducing condition, mono and bi mercury irons becomes soluble. The "Cinnabar (HgS)" phases are now one of the solid phase in the diagram. Mercury may also be separate from solution by coprecipitation with sulphides. Under anaerobic conditions, both organic and inorganic forms of Hg may be transformed to alkylated forms by microbial activity, such as by sulfurreducing bacteria. Elemental mercury may also be produced under anaerobic conditions by demethylation of methyl mercury, or by reduction of Hg(II). Acidic conditions ($pH < 4$) also facilitate the formation of methyl mercury, whereas higher pH values favour precipitation of HgS(s) [27]. The mobilization of Hg is shown below (Figure 6).

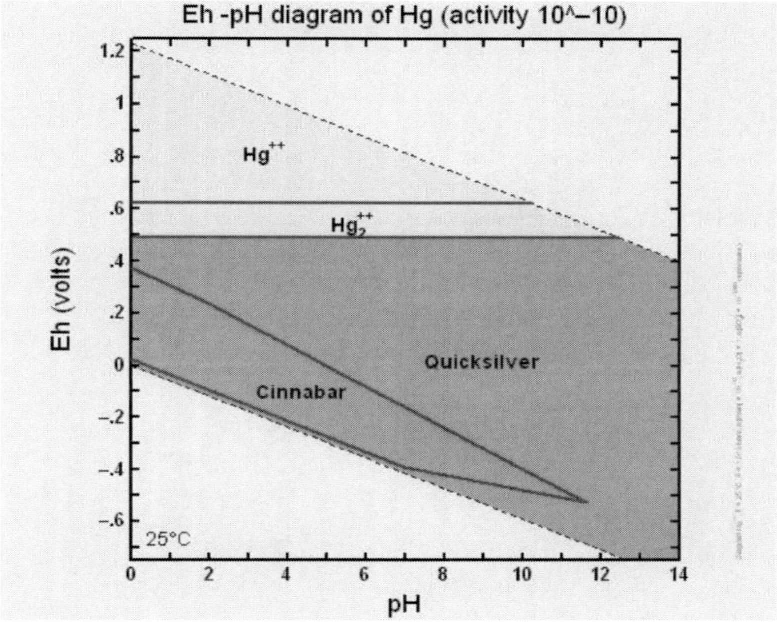

Figure 5. Eh-pH diagram for Hg species at 25°C and of 1mM concentration.

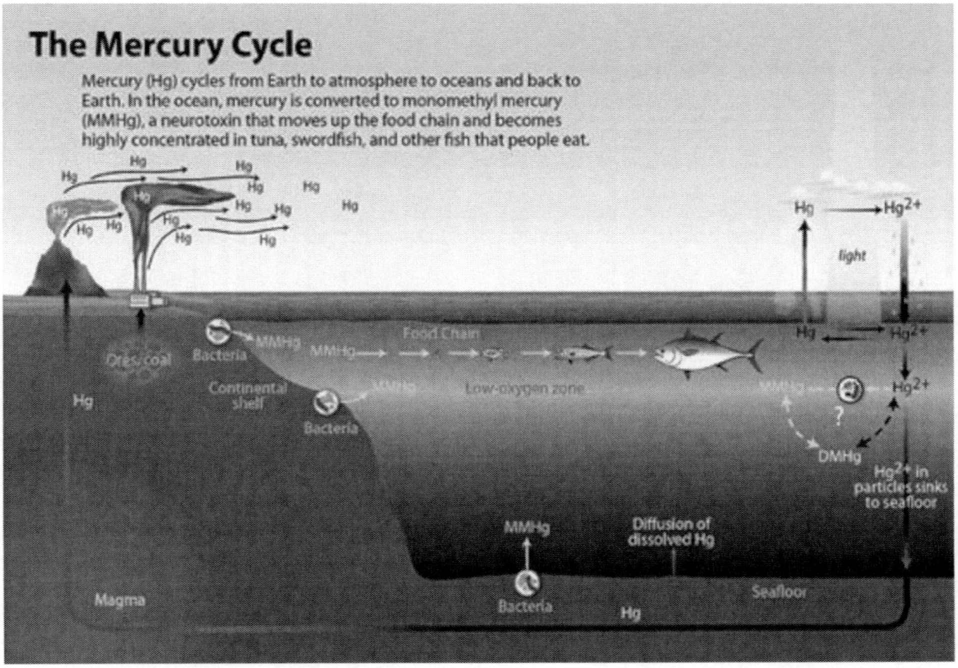

Figure 6. Mercury cycle with variable oxidation states.

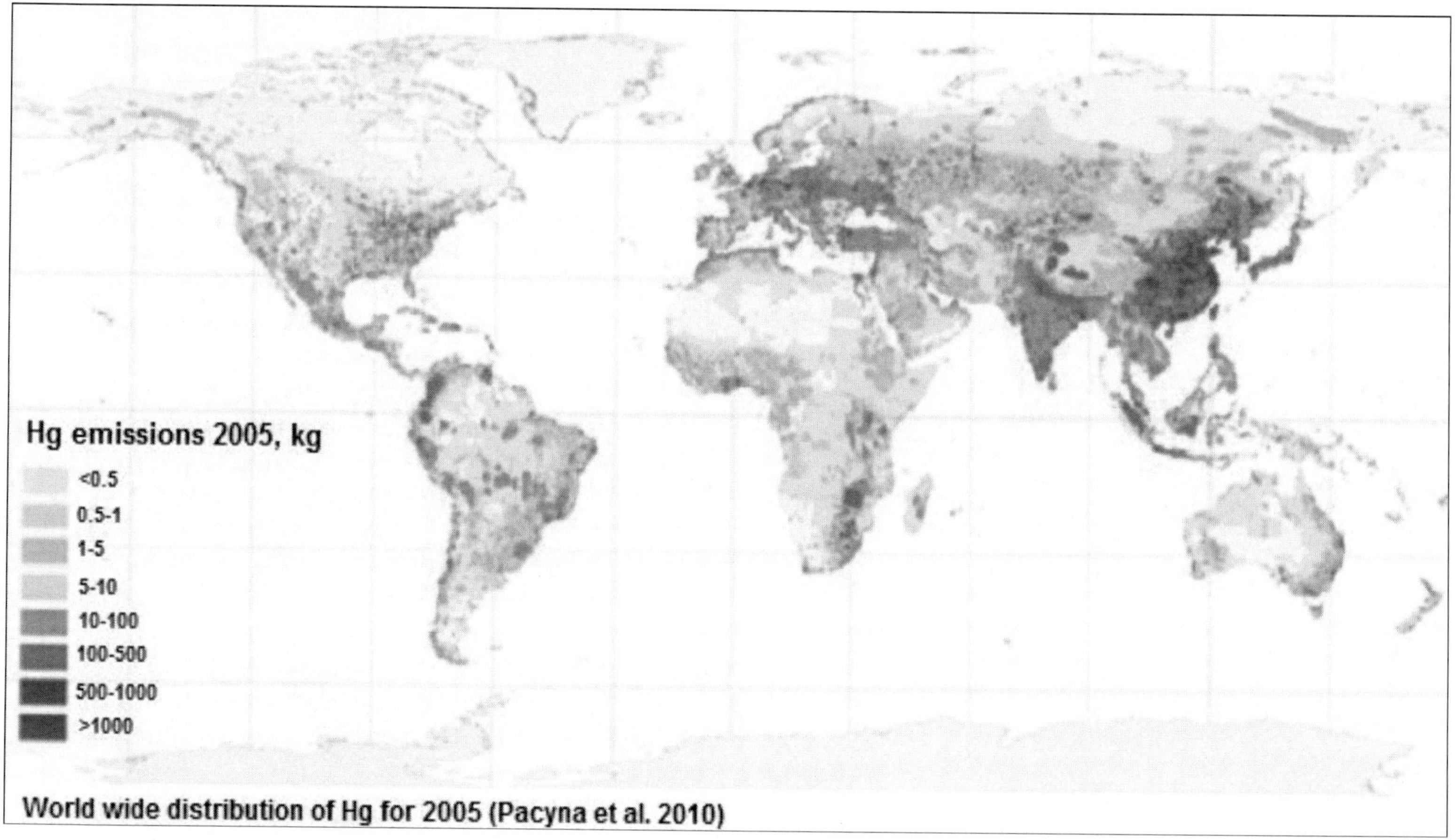

Figure 7. Worldwide abundance of mercury.

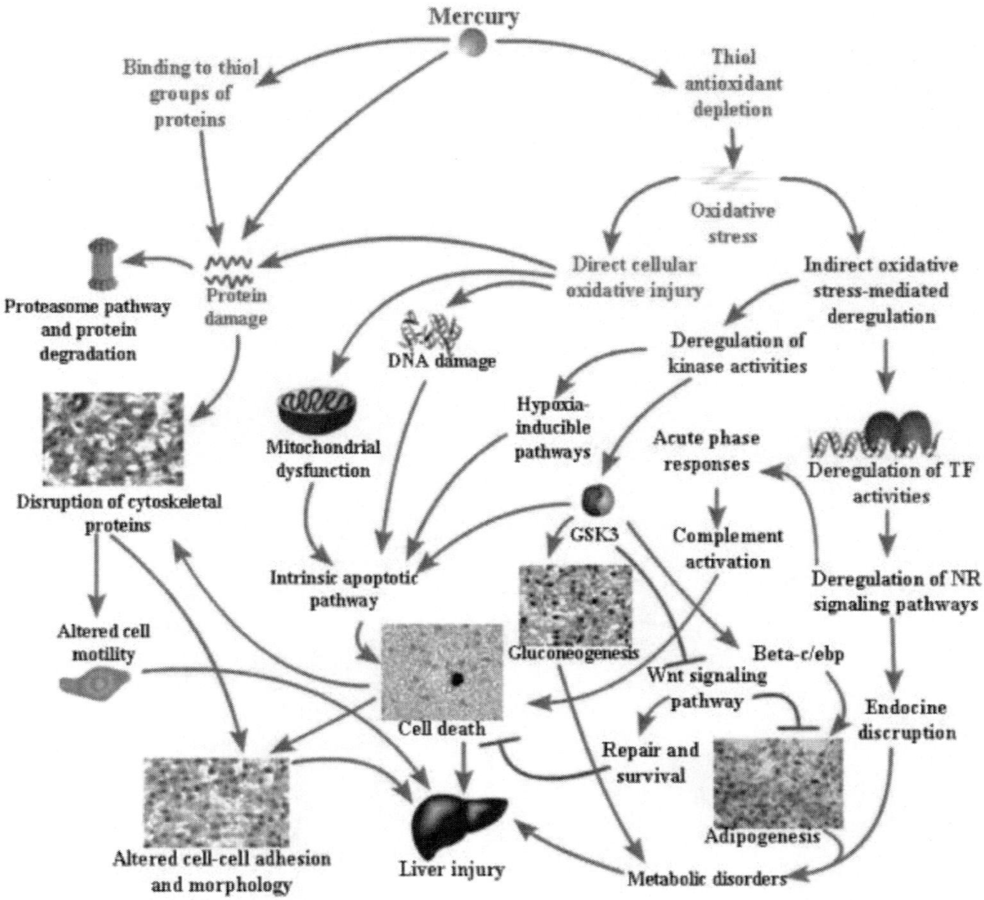

Figure 8. Toxic effects of Hg.

Mercury is present in various forms in nature but the primary forms that affect humans beings are methylmercury, elemental mercury, and to a minor degree a variety of inorganic mercury-based salts. Methylmercury exposure most commonly occurs when you consume fish and shellfish that have been contaminated with methylmercury formed by various marine microorganisms (these little critters convert inorganic mercury substances into methylmercury). On the other hand, elemental mercury exposure typically occurs when a device that containing mercury (e.g., thermometers, barometers, thermostat switches, dental fillings, etc.) breaks and spills some of its contents. The spilled mercury subsequently evaporates into a toxic vapor that is both odorless and colorless. This poisonous vapor is inhaled, particularly in poorly ventilated spaces, and the elemental mercury then builds up in your body and can potentially cause a variety of symptoms [36]. Hg in the environment above tolerance limit is a worldwide problem. Hg pollution had been reported in the USA, India, Australia, China, Chile, Bangladesh, Taiwan, Mexico, Argentina, Poland, Canada, Hungary, New Zealand, Vietnam, Romania, Japan, and Germany as shown below (Figure 7).

To make a generalization, you can think of mercury intoxication as a problem that affects the nervous system (a.k.a. your brain, kidney, spinal cord, and nerves). Furthermore, the type of mercury compound causing the intoxication makes a big difference as it will affect the severity of symptoms and whether any other parts of your body will be affected. The model is anticipated based on our transcriptome analysis, phenotype anchoring and targeted gene validation. Words in red and blue fonts are direct and indirect upstream processes induced by mercury, respectively. Black words are downstream processes that are induced or suppressed by either direct or indirect effects of mercury. Please see text for detail description of the indicated mechanism (Figure 8) [37].

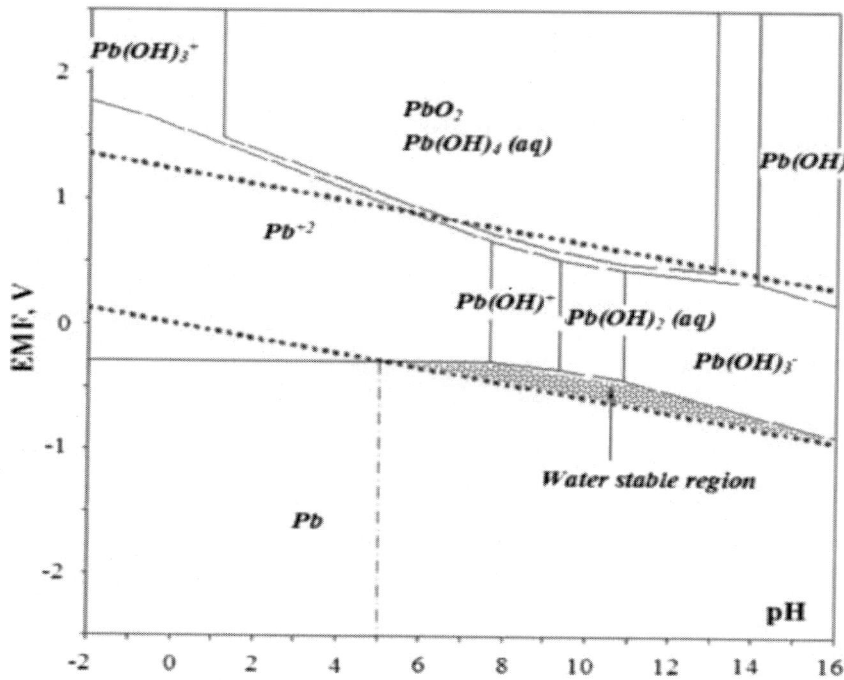

Figure 9. Eh-pH diagram for Pb species at 25°C and of 1mM concentration.

3.3. Lead

Lead metal belongs to group IV and period 6 of the periodic table with atomic mass 207.2, atomic number 82, density 11.4 g cm^{-3}, melting point 327.4°C, and boiling point 1725°C. It is a naturally abundant, bluish-gray metal usually found as a mineral combined with other elements, such as sulphur (i.e., PbS, $PbSO_4$), or oxygen ($PbCO_3$), and ranges from 10 to 30 mg kg^{-1} in the earth's crust. Typical mean Pb concentration for surface soils worldwide be around 32 mg kg^{-1} and ranges from 10 to 67 mg kg^{-1} [11]. Lead ranks fifth behind Fe, Cu, Al, and Zn in industrial production of metals. About half of the Pb used in the U.S. drives for the manufacture of Pb storage batteries. Other usages consist of solders, bearings, cable covers, ammunition, plumbing, pigments, and caulking. Metals normally alloyed with Pb are antimony (in storage batteries), silver (Ag) (for solder and anodes), calcium (Ca) and tin (Sn) (in maintenance-free storage batteries), strontium (Sr) and Sn (as anodes in electro winning processes), tellurium (Te) (pipe and sheet in chemical installations and nuclear shielding), Sn (solders), and antimony (Sb), and Sn (sleeve bearings, printing, and high-detail castings). Ionic lead, Pb(II), lead oxides and hydroxides, and lead-metal oxyanion complexes are the common forms of Pb that are released into the soil, groundwater, and surface waters. The most established forms of lead are Pb(II) and lead-hydroxy complexes. Lead(II) is the most common and responsive form of Pb, forming mononuclear and polynuclear oxides and hydroxides. The dominant insoluble Pb compounds are lead phosphates, lead carbonates (form when the pH is above 6), and lead (hydr)oxides. Pourbiax diagram is of lead is appended below (Figure 9).

Lead sulfide (PbS) is the utmost stable solid form within the soil matrix and forms under reducing conditions, when increased concentrations of sulphide are found. Under anaerobic conditions a volatile organolead (tetramethyl lead) can be observed due to microbial alkylation. Lead (II) compounds are dominantly ionic (e.g., Pb^{2+} $SO4^{2-}$), whereas Pb(IV) compounds incline to be covalent (e.g., tetraethyl lead, $Pb(C_2H_5)_4$). Some Pb (IV) compounds, such as PbO_2, are strong oxidants. Lead forms several basic salts, such as $Pb(OH)_2 \cdot 2PbCO_3$, which was once the most extensively used white paint pigment and the source of considerable chronic lead poisoning to children who ate peeling white paint. Many compounds of Pb(II) and a few Pb(IV) compounds are useful. The two most usual of these are lead dioxide and lead sulphate, which are participants in the reversible reaction that takes place during the charge and discharge of lead storage battery. In addition to the inorganic compounds of lead, there are a numurous organolead compounds such as tetraethyl lead. The toxicities and environmental effects of organolead compounds are particularly worth mentioning because of the former widespread use and distribution of tetraethyllead as a gasoline additive. Although more than 1000 organolead compounds have been made, those of commercial and toxicological meaning are largely limited to the alkyl (methyl and ethyl) lead compounds and their salts (e.g., dimethyldiethyllead, trimethyllead chloride, and diethyllead dichloride). Inhalation and ingestion are the two routes of exposure, and the effects from both are the identical. Pb accumulates in the body organs (i.e., brain), which may lead to harming (plumbism) or even death. The gastrointestinal tract, kidneys, and central nervous system are also affected by the attendance of lead. Children exposed to lead are at danger for impaired development, lower IQ, shortened attention span, hyperactivity, and mental deterioration, with children under the age of six being at a more considerable risk. Adults frequently experience decreased reaction time, loss of memory, nausea, insomnia, anorexia, and weakness of the joints when exposed to lead. Lead is not an essential element. It is well recognized to be toxic and its effects have been more extensively studied than the effects of other trace metals. Lead can cause severe injury to the brain, nervous system, red blood cells, and kidneys. Exposure to lead can effect in a wide range of biological belongings depending on the level and duration of exposure. Various effects arise over a broad range of doses, with the developing young and infants being more sensitive than adults. Lead poisoning, which is so severe as to cause marked illness, is now very rare. Lead executes no known essential function in the human body, it can merely do harm after uptake from food, air, or water. Lead is a particularly hazardous chemical, as it can accumulate in individual organisms, but also in entire food chains. The most thoughtful source of exposure to soil lead is through direct ingestion (eating) of contaminated soil or dust. Commonly, plants do not absorb or accumulate lead. However, in soils testing high in lead, it is probable for some lead to be taken up. Studies have revealed that lead does not readily accumulate in the fruiting parts of vegetable and fruit crops (e.g., corn, beans, squash, tomatoes, strawberries, and apples).

Higher concentrations are more likely to be observed in leafy vegetables (e.g., lettuce) and on the surface of root crops (e.g., carrots). Since plants seldom take up large quantities of soil lead, the lead levels in soil considered safe for plants will be much higher than soil lead levels where eating of soil is a concern (pica). In general, it has been considered safe to use garden produce grown in soils with total lead levels less than 300 ppm. The risk of lead poisoning through the food chain escalating as the soil lead level rises above this concentration. Even at soil levels above 300 ppm, maximum risk is from lead contaminated soil or dust deposits on the plants rather than from uptake of lead by the plant.

Pb in the environment above tolerance limit is a worldwide problem. Pb pollution had been reported in USA, Australia, Germany, Bangladesh, Argentina, Peru, Indonesia, Vietnam, South Africa, Kenya, Romania, Hungary, Japan, Pakistan and India (Figure 10).

Common sources of lead include house paint, kajol eyeliners, pipes, bullets, lead figures, toys, jewellery, pottery, household items etc. Signs of continual lead exposure include abdominal pain, abdominal cramps, aggressive behavior, constipation, sleep problems, headaches, irritability, loss of developmental skills in children, loss of appetite, fatigue, high blood pressure, memory loss, anaemia, kidney dysfunction, encephalopathy, which manifests as confusion, coma, seizures etc., (Figure 11).

3.4. Chromium

Chromium is a first-row d-block transition metal of VIB group in the periodic table with the ensuing properties: atomic mass 52, atomic number 24, density 7.19 g cm^{-3}, melting point 1875°C, and boiling point 2665°C. It is one of the less common elements and does not observe naturally in elemental form, but only in compounds. Chromium is mined Major sources of Cr-contamination include releases from electroplating processes and the disposal of Cr containing wastes [27]. Chromium (VI) is majorly found at contaminated sites. Chromium can also observe in the +III oxidation state, based on pH and redox conditions (Figure 12).

Chromium(VI) is the predominant form of Cr in shallow aquifers where aerobic conditions exist. Chromium(VI) can be reduced to Cr(III) by means of S^{2-}, soil organic matter, and Fe^{2+} ions under anaerobic conditions often encountered in deeper groundwater. Major Cr(VI) species contain chromate (CrO_4^{2-}) and dichromate ($Cr_2O_7^{2-}$) which precipitate promptly in the presence of metal cations (especially Pb^{2+}, Ba^{2+} and Ag^+). Chromate and dichromate also adsorb on soil surfaces, exclusively iron and aluminium oxides. Chromium(III) is the predominant oxidation state of Cr at low pH (<4). Cr^{3+} makes complexes with NH_3, CN−, OH^-, Cl^-, F^-, SO_4^{2-}, and soluble organic ligands. Chromium (VI) is the more noxious form of chromium and is also more mobile. Chromium(III) mobility is reduced by sorption to clays and oxide minerals below pH 5 and low solubility above pH 5 due to the formation of $Cr(OH)_3(s)$. Chromium mobility is subject to sorption characteristics of the soil, including clay content, iron oxide content, and the amount of organic matter present. Chromium may be transported by surface overflow to surface waters in its soluble or precipitated form. Soluble and un-adsorbed chromium complexes may leach from soil into groundwater. The leachability of Cr(VI) increases with increasing pH. Most of Cr released into natural waters is particle linked, however, and is eventually deposited into the sediment [27]. Chromium is associated with allergic dermatitis in humans [18]. Worldwide abundance of Cr is shown in Figure 13.

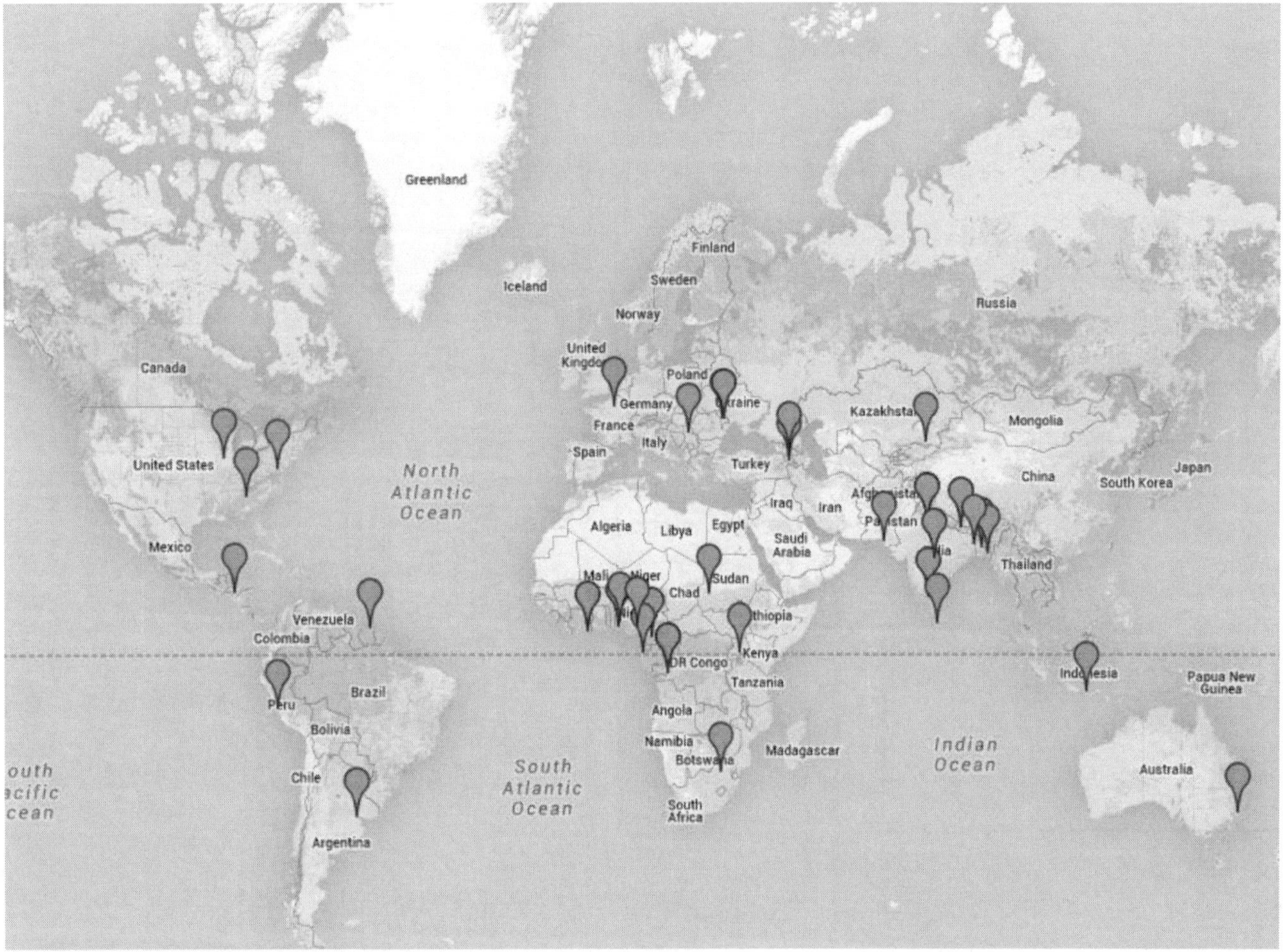

Figure 10. Worldwide abundance of Pb.

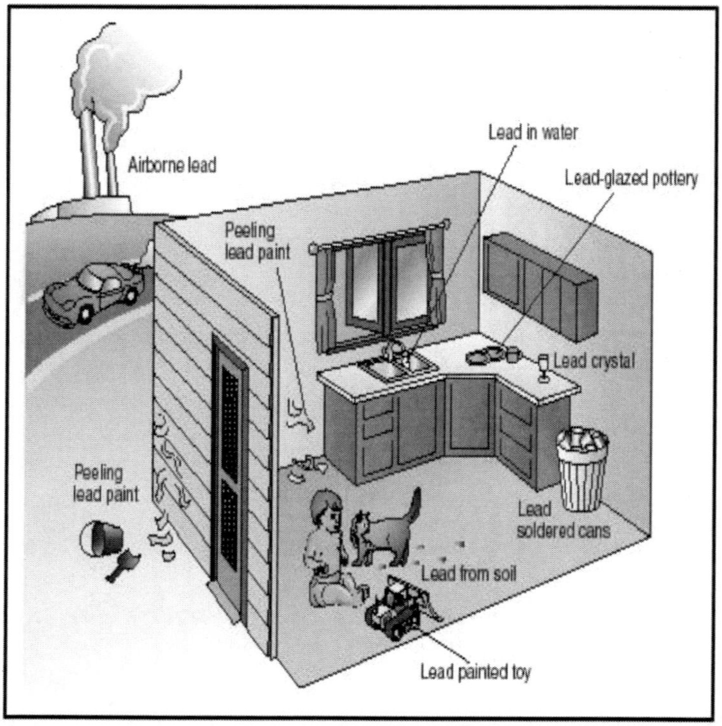

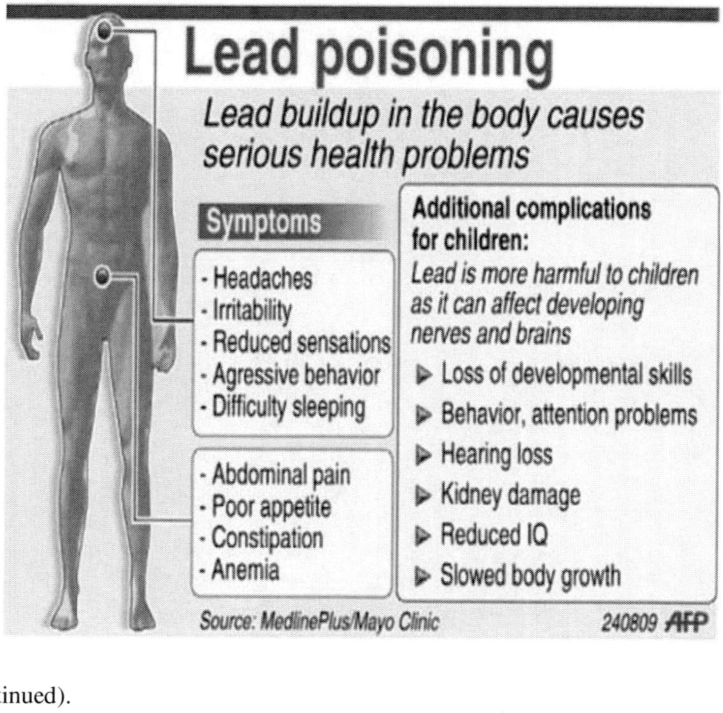

Figure 11. (Continued).

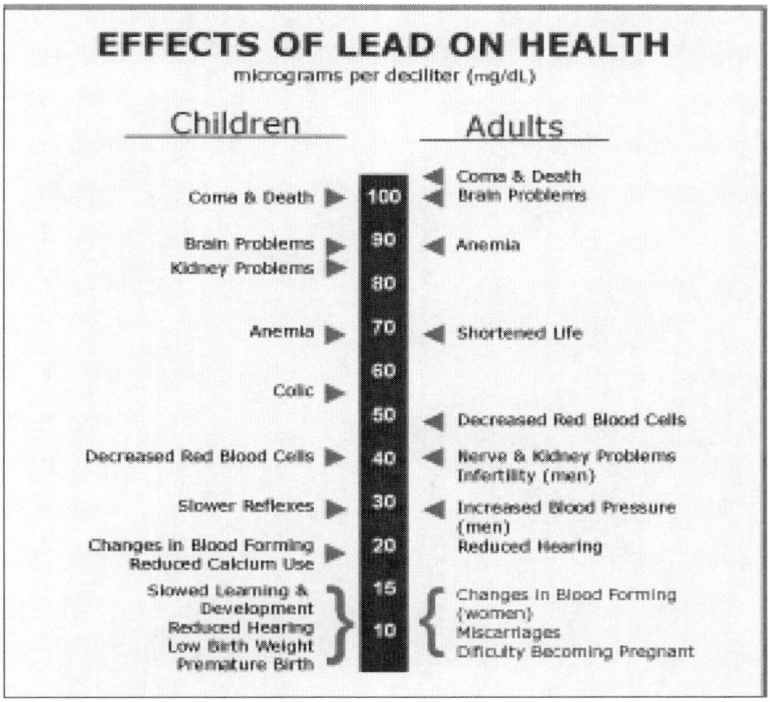

Figure 11. Lead source and poisoning.

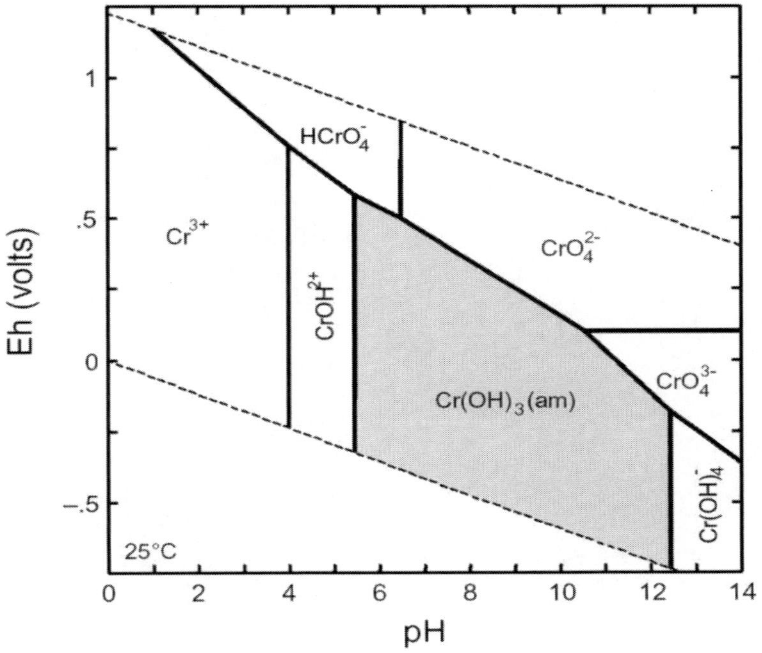

Figure 12. Eh-pH predominance diagram for chromium species at 25°C ($Cr_{tot} = 1\ \mu M$).

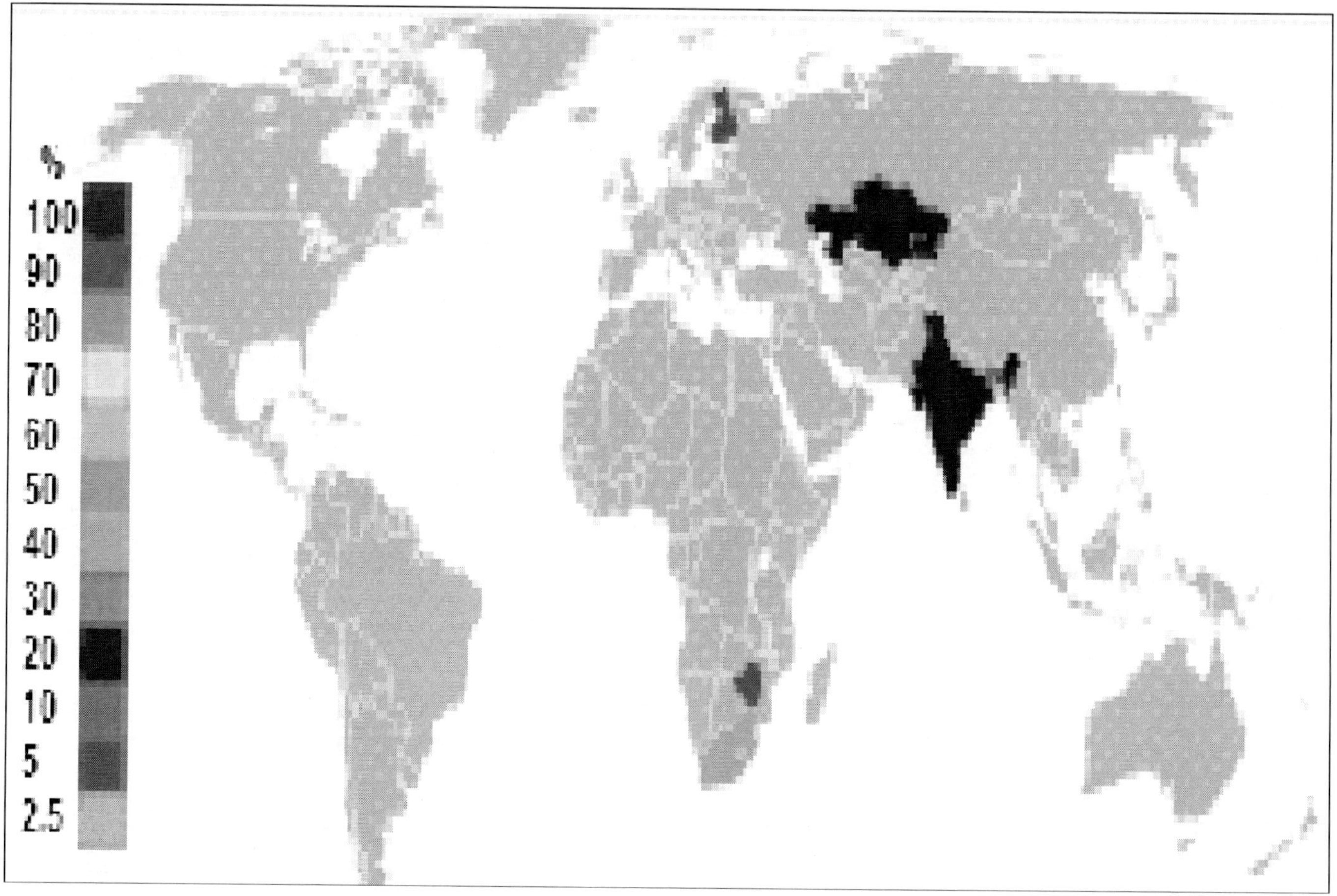

Figure 13. Worldwide abundance of Cr.

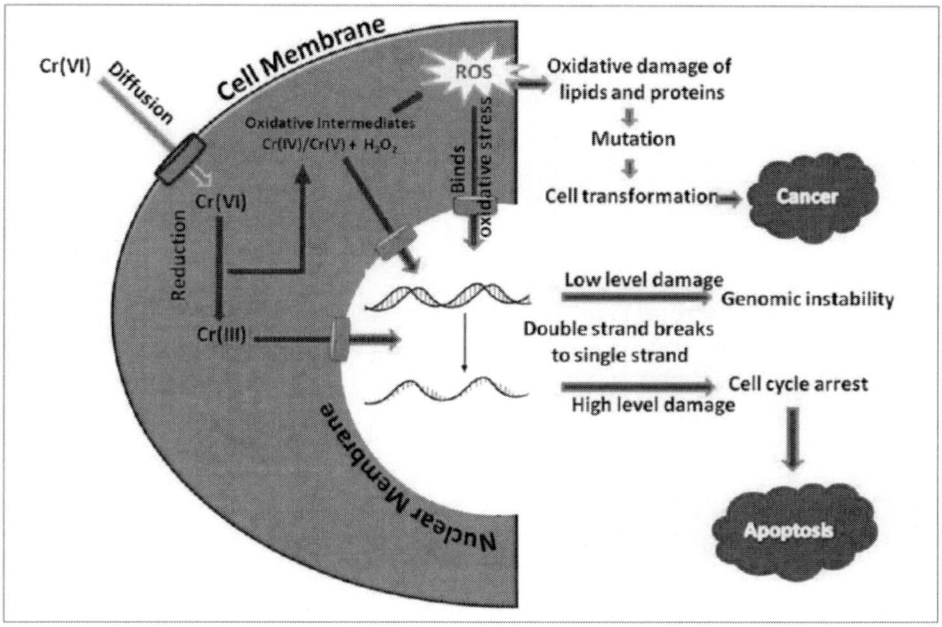

Figure 14. Toxicity of Cr in cell.

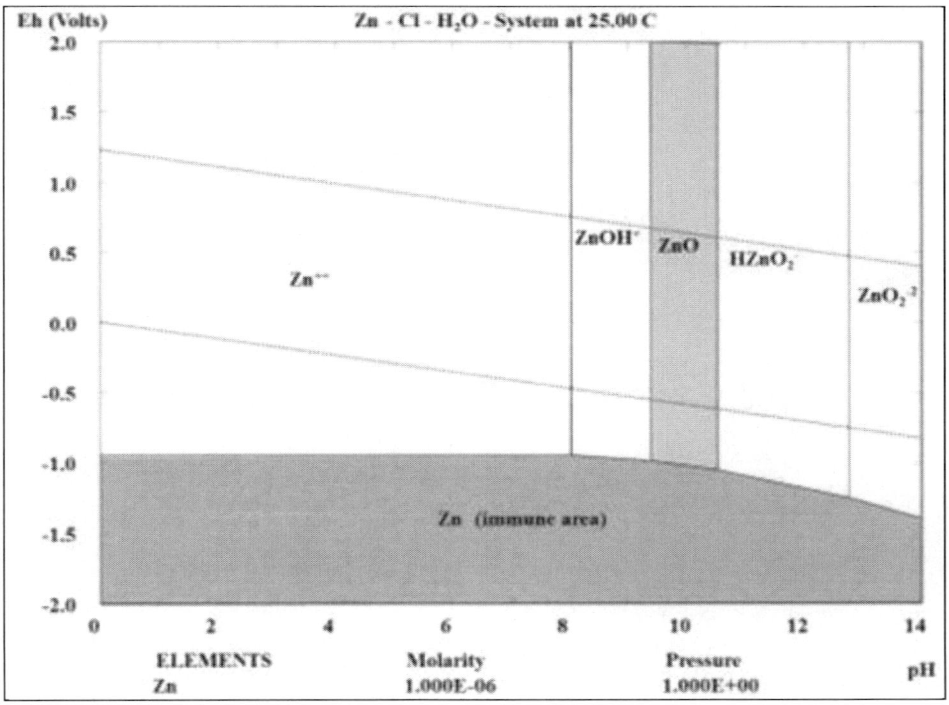

Figure 15. Eh-pH diagram for Zn species at 25°C and of 1mM concentration.

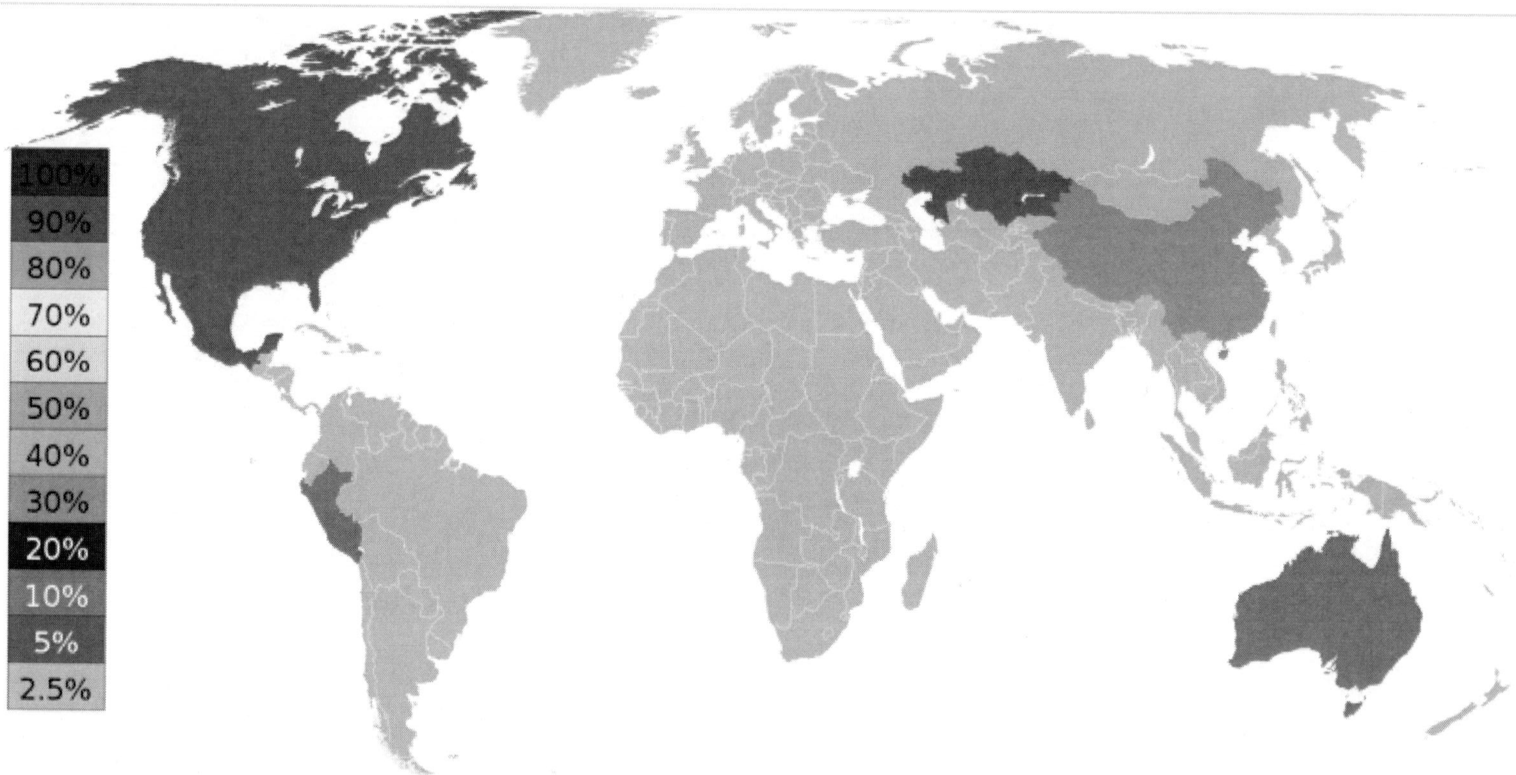

Figure 16. Worldwide abundance of Zn percentage.

Chromium (VI) is carcinogenic to human via mutagenesis. Hexavalent chromium is readily taken up by cells through sulfate transporters, due to the structural similarity of hexavalent chromium to the tetrahedral sulfate and phosphate anions (Figure 14). Inside the cell, chromium (VI) readily undergoes a series of reduction reactions to yield pentavalent, tetravalent, and ultimately the thermodynamically stable chromium (III). Many potential enzymes as well as non-enzymatic cellular reductants enable to reduce hexavalent chromium existing in cell. These reductants contain glutathione, ascorbate, cysteine, lipoic acid, NADH, fructose, and ribose. Following this intracellular reduction, several possible mechanisms leading to mutagenicity can occur. Chromium (VI) itself does not interact directly with DNA. However, the products of its reduction within the cell (pentavalent, tetravalent, and trivalent chromium) have all been shown to be DNA reactive. Chromium (VI) is reduced by glutathione to yield pentavalent chromium and thiyl radicals, which can react with other thiol molecules to produce superoxide radicals. Both chromium (V) and chromium (III) can participate in Fenton reactions, generating hydroxyl radicals. All of these species can cause DNA single- and double-strand breaks, base modifications, and lipid peroxidation, which may lead to mutations if not adequately repaired. Chromium (III) is the ultimate product of the intracellular reduction of chromium (VI). Trivalent chromium is capable of interacting directly with DNA, forming stable coordination complexes with nucleic acids and peptides. In particular, chromium (III) is capable of forming ternary complexes with DNA and an intracellular reducer, such as ascorbate, glutathione, or cysteine, as well as crosslinking DNA and proteins, and forming intrastrand DNA-DNA crosslinks. These chromium-DNA complexes, as well as DNA-protein and DNA-DNA crosslinks, all have the capability of causing DNA single- and double-strand breaks, which, if not adequately repaired, could lead to cell death, or if misrepaired, could result in mutation. Thus, once inside the cell, hexavalent chromium, through reduction to its pentavalent, tetravalent, and trivalent forms, is capable of inducing a wide range of mutagenic and genotoxic damage, including the formation of DNA adducts, DNA-protein and DNA-DNA crosslinks, mutations, DNA single and double-strand breaks, abasic sites, oxidized DNA bases, chromosomal aberrations, sister chromatid exchanges, and micronuclei [38].

3.5. Zinc

Zinc is a transition metal with the ensuing characteristics: period 4, group IIB, atomic mass 65.4, atomic number 30, density 7.14 g cm^{-3}, melting point 419.5°C, and boiling point 906°C. Zinc occurs naturally in soil (about 70 mg kg^{-1} in crustal rocks), but Zn concentrations are rising unnaturally, due to anthropogenic additions. Most Zn is found during industrial activities, such as mining, coal, and waste combustion and steel processing. Many foodstuffs contain definite concentrations of Zn. Drinking water also holds certain amounts of Zn, which may be higher when it is stored in metal tanks. Industrial sources or toxic waste sites may increase the concentrations of Zn in drinking water to reach levels facilitating health problems. Pourbiax diagram of zinc is shown below (Figure 15).

Zinc is an essentially trace element for human health. Zinc shortages can cause birth defects. The world's Zn production is still rising which indicating more Zn ends up in the environment. Water is contaminated with Zn, due to the presence of large quantities present in the wastewater of industrial plants. A consequence is that Zn polluted sludge is frequently being deposited by rivers on their banks. Zinc may also raise the acidity of water. Some fish can accumulate Zn in their bodies, due to Zn-contaminated waterways. When Zn enters the bodies of these fish, it enable to biomagnify up the food chain. Water-soluble zinc of soils can contaminate groundwater. Plants frequently have a Zn uptake that their systems cannot handle, due to the accumulation of Zn in soils. Finally, Zn can interrupt the activity in soils, as it negatively influences the activity of microorganisms and earthworms, thus retarding the breakdown of organic matter. Worldwide abundance% of Zinc is revealed in the following (Figure 16). Zinc in excess and deficiency both are severe threat for human health hazard as appended below by means of brain, respiratory track, and gastrointestinal track, prostate, thymus, skin, reproductive system etc. (Figure 17).

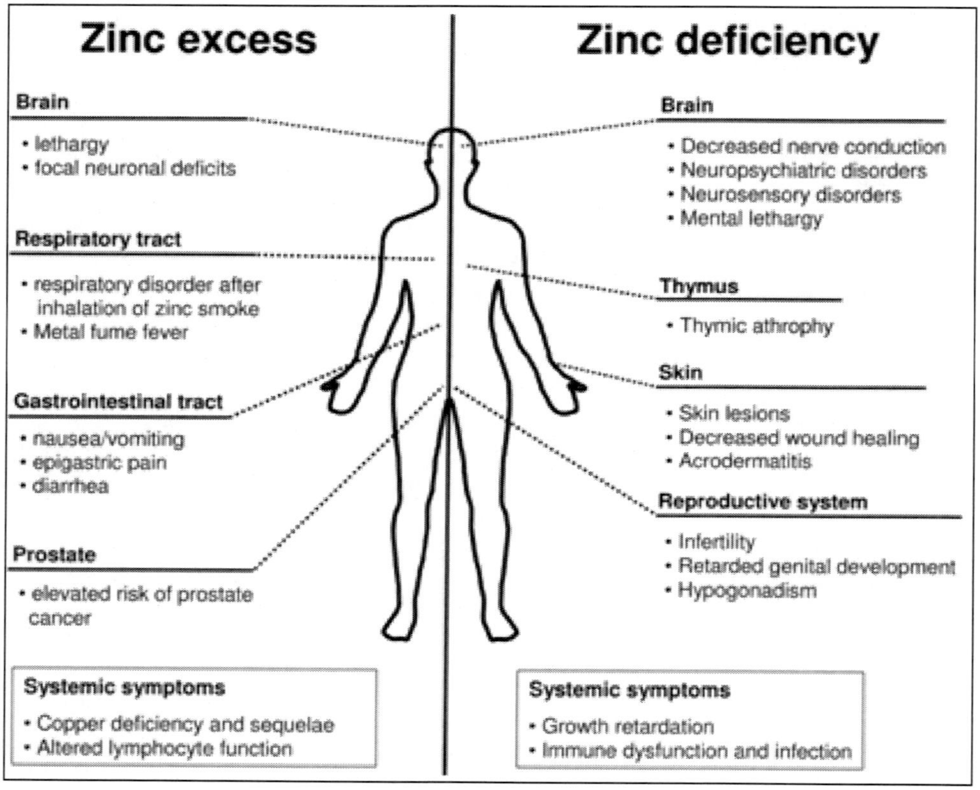

Figure 17. Toxic effects of Zn both in excess and deficiency.

3.6. Cadmium

Cadmium is placed at the end of the second row of transition elements with atomic weight 112.4, atomic number 48, density 8.65 g cm^{-3}, melting point 320.9°C, and boiling point 765°C. Together with Hg and Pb, Cd is one of the massive three heavy metal poisons and is not responsible for essential biological function. In its compounds, Cd observes as the divalent Cd(II) ion (Figure 18).

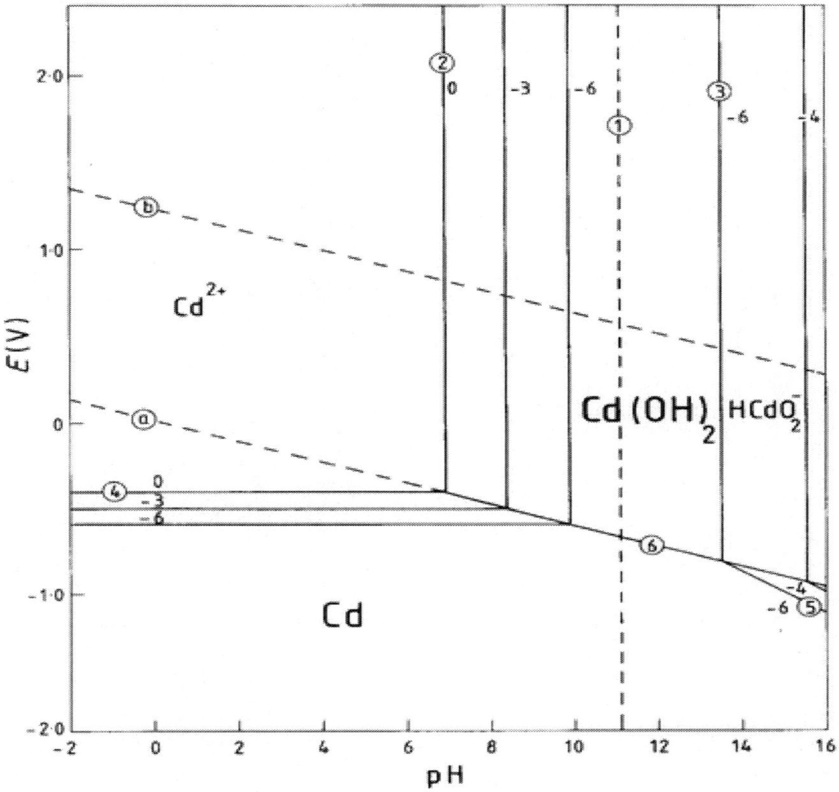

Figure 18. Eh-pH diagram for Cd species at 25°C and of 1mM concentration.

Cadmium is directly underneath Zn in the periodic table and has a chemical similarity to that of Zn, an essential micronutrient for plants and animals. This may describe in part for Cd's toxicity; because Zn being an essential trace element, its substitution by Cd may cause the malfunctioning of metabolic processes. The most noteworthy use of Cd is in Ni/Cd batteries, as rechargeable or secondary power sources demonstrating high output, long life, low maintenance, and high tolerance to physical and electrical stress. Cadmium coatings deliver good corrosion resistance coating to vessels and other vehicles, particularly in high-stress environments such as marine and aerospace. Other usages of cadmium are as pigments, stabilizers for polyvinyl chloride (PVC), in alloys and electronic compounds. Cadmium is also occurring as an impurity in several products, including phosphate fertilizers, detergents and refined petroleum products. In addition, acid rain and the resulting acidification of soils and surface waters have increased the geochemical mobility of Cd, and as a result its surface-water concentrations are apt to increase as lake water pH decreases. Cadmium is yielded as an inevitable by-product of Zn and occasionally lead refining. The use of agricultural inputs such as fertilizers, pesticides, and biosolids (sewage sludge), the disposal of industrial wastes or the deposition of atmospheric contaminants increases the total concentration of Cd in soils, and the bioavailability of this Cd determines whether plant Cd uptake happens to a significant degree. Cadmium is very bio-persistent but has few toxicological properties and, once absorbed by an organism, remains resident for many years.

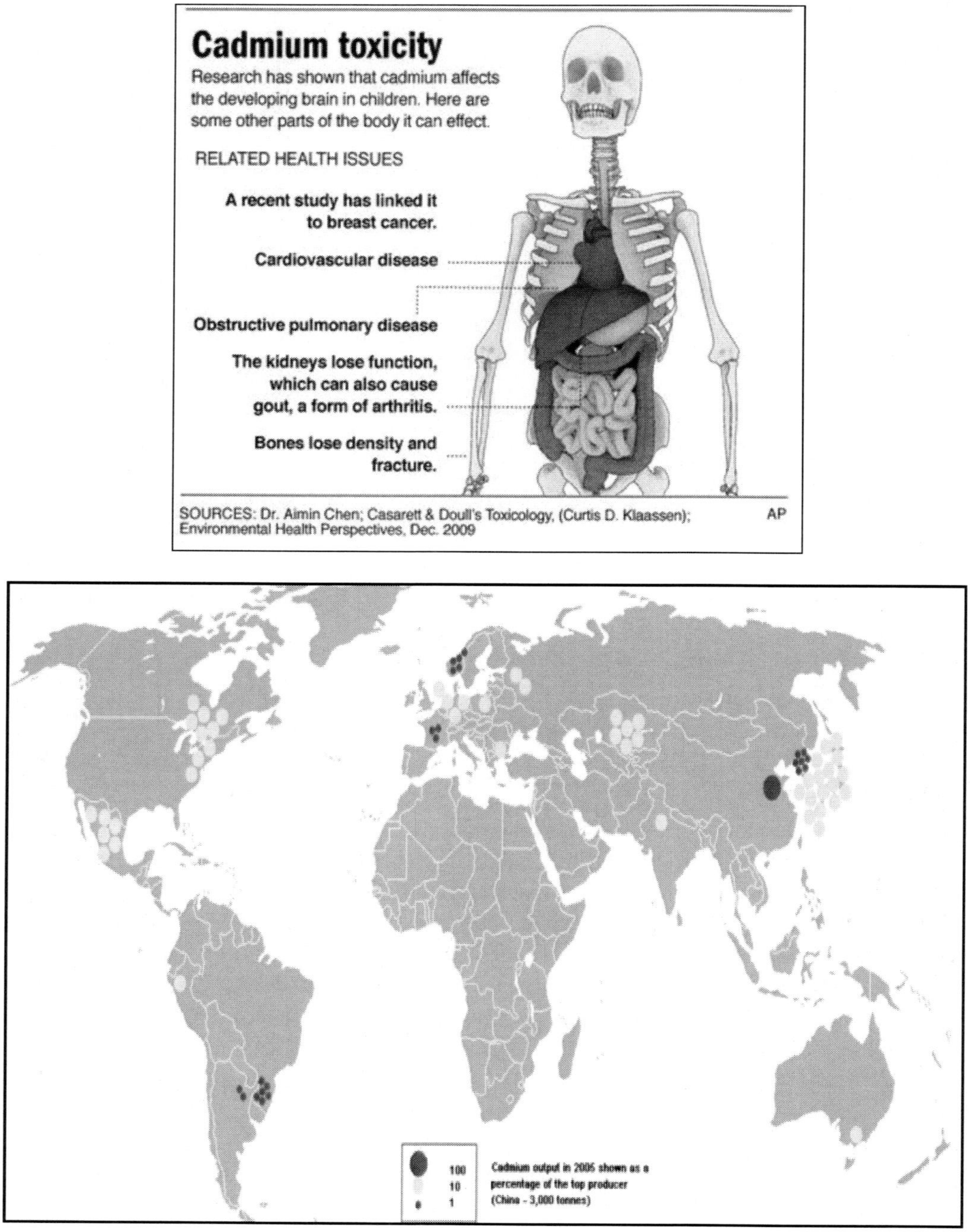

Figure 19. Cd toxicity in human health and its worldwide abundance.

Since the 1970s, there has been sustained interest in possible exposure of humans to Cd through their food chain, for example, via the consumption of certain species of fish or vegetables. Concern regarding this latter route (agricultural crops) directed to research on the possible consequences of relating sewage sludge (Cd-rich biosolids) to soils used for crops meant for human consumption, or of using cadmium-enriched phosphate fertilizer. This research has directed to the stipulation of highest permissible concentrations for a number of food crops [9]. Cadmium in the body is acknowledged to affect several enzymes. It is supposed that the renal damage that results in proteinuria is the result of Cd adversely affecting enzymes responsible for reabsorption of proteins in kidney tubules. Cadmium also decreases the activity of delta-aminolevulinic acid synthetase, arylsulfatase, alcohol dehydrogenase, and lipoamide dehydrogenase, whereas it improves the activity of delta-aminolevulinic acid dehydratase, pyruvate dehydrogenase, and pyruvate decarboxylase. The most remarkable and publicized occurrence of cadmium poisoning resulted from dietary intake of cadmium by people in the Jintsu River Valley, near Fuchu, Japan. The victims had been affected by ouch, ouch in Japanese. The symptoms are the consequence of painful osteomalacia (bone disease) combined with kidney malfunction. Cadmium poisoning in the Jintsu River Valley was ascribed by irrigated rice contaminated from an upstream mine producing Pb, Zn, and Cd. The major hazard to human health is chronic accumulation in the kidneys leading to kidney dysfunction. Food intake and tobacco smoking are the key routes by which Cd enters the body. Cadmium toxicity to the human health is shown below (Figure 19).

3.7. Copper

Copper is a transition metal which is situated in the period 4 and group IB of the periodic table with atomic weight 63.5, atomic number 29, density 8.96 g cm^{-3}, melting point 1083°C and boiling point 2595°C. Its average density and concentrations in crustal rocks are 8.1 × 103 kg m^{-3} and 55 mg kg^{-1}, respectively. It is the third most utilized metal in the world. Copper is an essential micronutrient for growth of both plants and animals. In humans, it has assistance in the production of blood haemoglobin. In plants, Cu is especially imperative in seed production, disease resistance, and regulation of water. Copper is indeed essential, but in high doses it can originate anaemia, liver and kidney damage, and stomach and intestinal irritation. Copper normally observes in drinking water from Cu pipes, as well as from additives designed to control algal growth. While the interaction of Cu with the environment is complex, research displays that most Cu introduced into the environment is, or rapidly becomes, stable and results in a form which does not carriage a risk to the environment. In fact, unlike some man-made materials, Cu is not expanded in the body or bioaccumulated in the food chain. In the soil, Cu strongly complexes to the organic inferring that only a small fraction of copper will be found in solution as ionic copper, Cu(II). The solubility of Cu is considerably increased at pH 5.5, which is rather close to the ideal farmland pH of 6.0–6.5 (Figure 20).

Copper and Zn are two vital elements for plants, microorganisms, animals, and humans. The correlation between soil and water contamination and metal uptake by plants is determined by many chemical and physical soil factors as well as the physiological properties of the crops. Soils polluted with trace metals may pose both direct and indirect threats: direct, through negative effects of metals on crop growth and yield, and indirect, by entering the human food chain with a potentially negative impact on human health especially on liver, kidney, intestine, eye, skin etc. (Figure21). Even a reduction of crop yield by a few percent could lead to substantial long-term loss in production and income. Some food importers are now specifying permissible maximum contents of metals in food, which might limit the possibility for the farmers to export their contaminated crops [26].

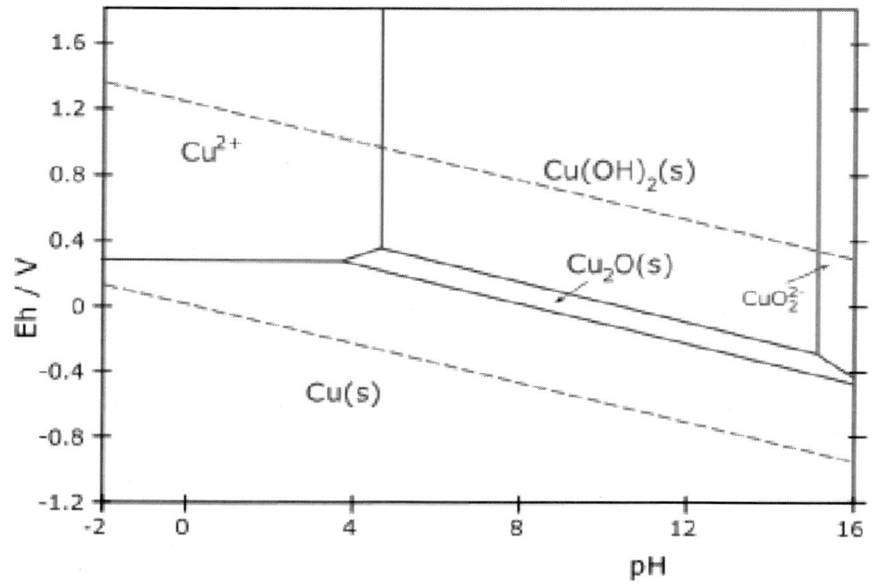

Figure 20. Eh-pH diagram for Cu species at 25°C and of 1mM concentration.

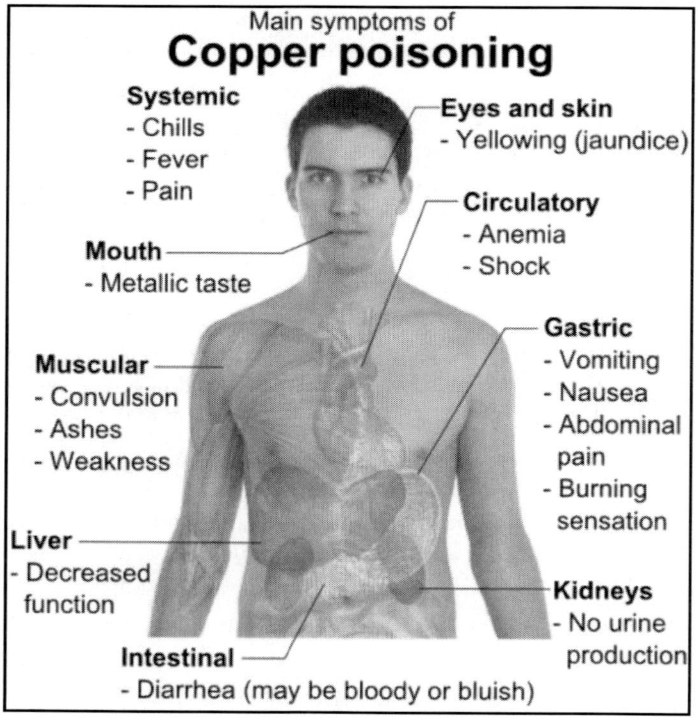

Figure 21. (Continued).

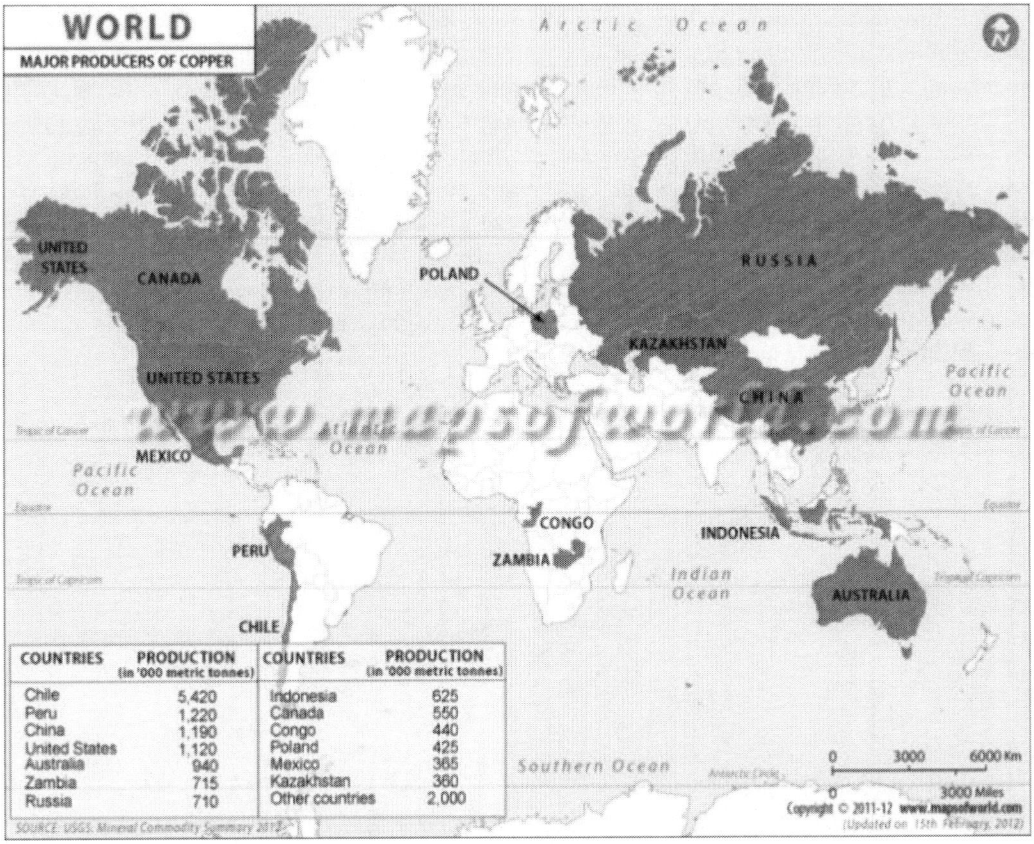

Figure 21. Cu toxicity in human health and its worldwide abundance.

3.8. Nickel

Nickel is a transition element with atomic weight 58.69, atomic number 28, density 8.908 g cm^{-3}, melting point 1455°C and boiling point 2730°C. In low pH regions, the metal occurs in the form of the nickelous ion, Ni(II). In neutral to slightly alkaline solutions, it precipitates as stable nickelous hydroxide, $Ni(OH)_2$. This precipitate readily dissolves in acid solutions forming Ni(III) and in high alkaline medium; it forms nickelite ion, $HNiO_2$, which is soluble in water. In very oxidizing and alkaline conditions, nickel exists in form of the stable nickelo-nickelic oxide, Ni_3O_4, which is soluble in acid solutions. Other nickel oxides such as nickelic oxide, Ni_2O_3, and nickel peroxide, NiO_2, are unstable in alkaline solutions and decompose by liberating oxygen. In acidic regions, however, these solids dissolve generating Ni^{2+}. Pourbiax diagram of nickel is in the following (Figure 22).

Nickel is an element that observes in the environment only at very low levels and is essential in small doses, but it can be dangerous when the maximum tolerable amounts are exceeded. This can cause various kinds of cancer on different sites within the bodies of animals, mainly of those that live proximate refineries. The most civic application of Ni is an ingredient of steel and other metal products. The main resources of nickel contamination in the soil are metal plating industries, combustion of fossil fuels, and nickel mining and electroplating. It is discharged into the air by power plants and trash incinerators and settles to the ground after undergoing precipitation reactions. It typically takes a long time for nickel to be removed from air. Nickel may also end up in surface water when it is a part of wastewater streams. The larger part of all Ni compounds that are discharged to the environment will adsorb to sediment or soil particles and become immobile as a result. In acidic soils, however, Ni becomes more mobile and frequently leaches down to the adjacent groundwater. Microorganisms are also suffering from growth decline due to the presence of Ni, but they usually develop resistance to Ni after a while. Nickel is not known to accumulate in plants or animals and as a result Ni has not been set up to biomagnify up the food chain. For animals Ni is an important foodstuff in small amounts.

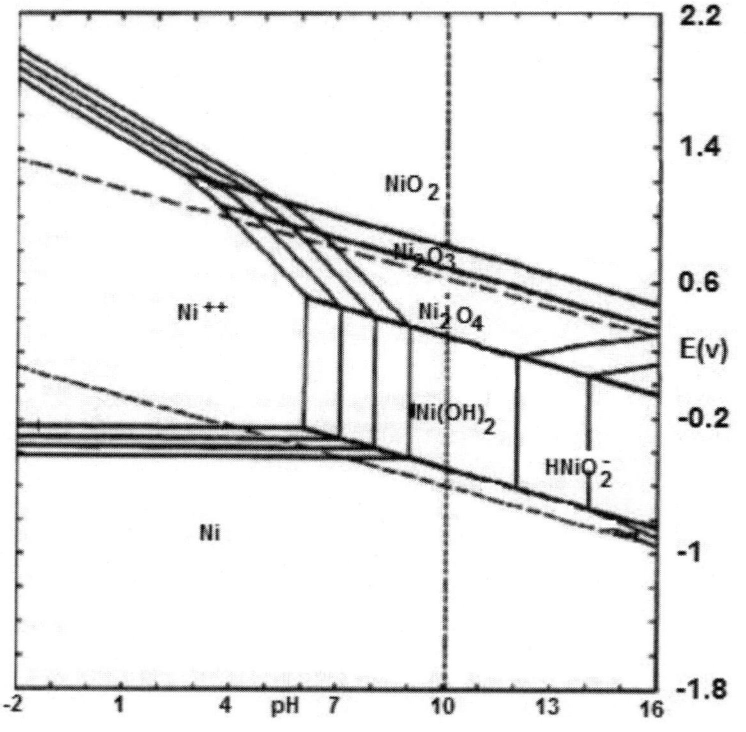

Figure 22. Eh-pH diagram for Ni species at 25°C and of 1mM concentration.

Nickel sulphide fume and dust are supposed to be carcinogenic (Figure 23). Nickel carbonyl, [Ni(CO)4], is an exceedingly toxic gas. Sensitized individuals may display an allergy to nickel, harming their skin, known as dermatitis. Sensitivity to nickel may also be found in patients with pompholyx. Nickel is an important reason of contact allergy, partly due to its use in jewellery intended for pierced ears. Nickel allergies affecting pierced ears are a lot marked by itchy, red skin. Many earrings are now made nickel-free owing to this problem. The amount of nickel allowed in products that come into contact with human skin is now regulated by the European Union. Reports also disclosed that both the nickel-induced activation of hypoxia-inducible factor (HIF) and the up-regulation of hypoxia-inducible genes are due to depleted intracellular ascorbate levels. The addition of ascorbate to the culture medium improved the intracellular ascorbate level and reversed both the metal-induced stabilization of HIF and HIF-α dependent gene expression.

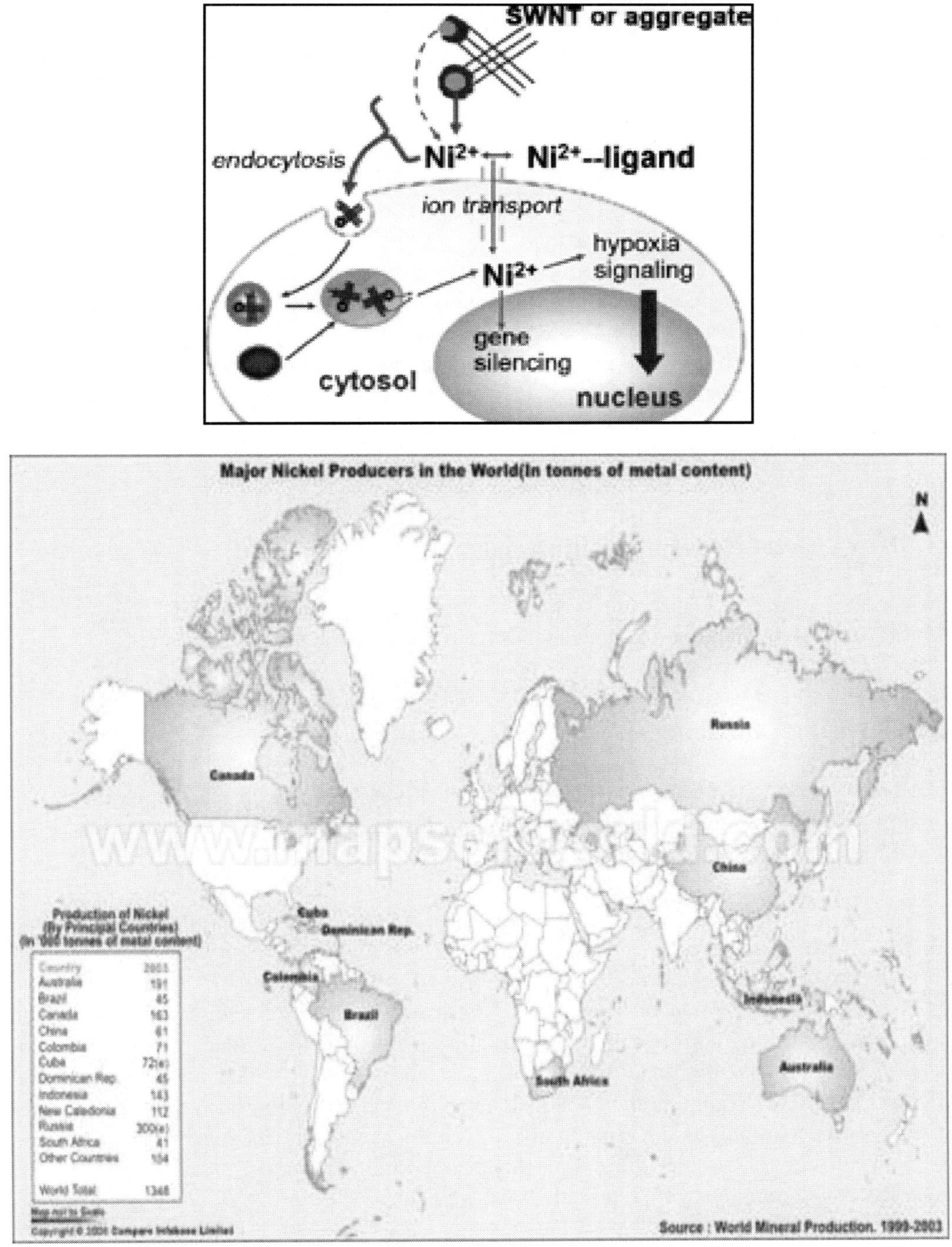

Figure 23. Ni toxicity and its worldwide abundance.

4. Detection Strategy

There is an on-going quest to develop sensing strategies for trace heavy metal for the early detection of pollution in various chemical systems, including living systems and the whole environment. Hence, this review is focused on a general overview on the latest trends in the development of heavy-metal-detection devices and systems. As for other sensing systems, here two important parts are considered: the receptor and the immobilization/transducing platform [4]. While the first part, the receptor (i.e., heavy-metal ionophores, biological receptors, etc.), is the accountable for the device selectivity, the second one is the answerable for the device stability as well as the sensitivity, which will depend upon the technique utilised (i.e., optical, electrochemical). The chemical recognition of heavy-metal ions is the base for the mentioned techniques to enable an appropriate sensing through specific substances/species i.e., receptors of the detection system. The selective interaction of heavy-metal ions with the specific receptor customarily includes noncovalent bonding, such as hydrogen bonding, metal coordination, hydrophobic forces, van der Waals forces, π-π interactions, and electrostatic and/or electromagnetic effects. The specific recognition demarcated as the different affinity of a specific ligand for different metal ions is of special concern in this chapter.

4.1. Heavy Metal Recognition Materials

Various types of inorganic, organic, biological, and combined materials for specific/selective recognition of heavy metals have been reported in the literature [4].

4.2. Alloys and Amalgams

Nontoxic metal solid electrodes (M SE, where M is Ag, Au, Cu, Ir, etc.) for example, can in many cases successfully substitute mercury electrodes and, in some special cases, offer possibilities not available to mercury electrodes.

4.3. Synthetic Receptors

The specific recognition of heavy metals by using synthetic or natural receptors called ionophores or carriers described elsewhere. Various ionophores have been described so far, including simple molecules, macromolecules, or caged molecules, which can monitor by two different ways of recognition: chemical affinity, cavity entrapment, or both ways simultaneously. Not only is the binding affinity significant for the chemical sensing but also the chemical reply of the ionophore after the analyte binding. Some typical commercial and non-commercial ionophores are reported elsewhere [4].

4.4. Biological Receptors

The biggest challenge in heavy-metal-sensing research is the design and synthesis of specific receptors with strong metal binding capabilities. The interaction of heavy-metal ions with biological molecules such as proteins, antibodies, or nucleic acids offers in this field remarkable advantages in terms of selectivity and limits of detection.

4.5. Immobilization Platforms

The key factor in the design of heavy-metal-sensing systems is the immobilization of the specific receptors onto a suitable platform. This immobilization can be done by different methods, based upon the nature of the receptor and the material that establishes the sensing platform. This subdivision described the different macro-, micro-, and nanomaterial sensing platforms, including the respective receptor immobilization strategies that apply to heavy-metal detection.

4.6. Conventional Platforms

Various approaches for the immobilization of synthetic heavymetal receptors onto conventional platforms (i.e., micro- or macrometallic electrodes, synthetic/polymeric platforms) are used [39]. Sol-gel encapsulation [40-41] polymer entrapment [42] polymerization of the receptor itself onto the electrode surface, physical adsorption (electrostatic and/or hydrophobic adhesion), or covalent bonding governing the formation of irreversible covalent bonds [43-44] are the most reported. Physical adsorption of receptors onto polymer supports is rarely used as weak bonding clues to a loss of the receptor.

4.7. Nanoparticles as Immobilization Platforms

Nowadays, the use of metal nanoparticles (MNPs) for sensing applications is growing interest due to their strong emission and stability to photo bleaching and also acknowledgments to their possibility to be easily functionalized by specific receptors to afford water compatibility and increase target specificity [4]. The physical features of the MNPs can be controlled during the reaction step by various parameters, such as the precursor concentration, the nature of the solvent, the acting of the reducing agent, and the capping agent employed. Though attaining homogeneous MNPs is of high importance for the sensibility of the sensing techniques, the most important step to accomplish easy-to-functionalize MNPs is the stabilization mechanism. Two well accepted methods are to enable the stabilization of MNPs: electrostatic stabilization and steric stabilization.

The electrostatic interaction comprises of the formation of an electric double layer (EDL) around the MNPs with a Coulombic interaction between oppositely charged ions. In reality, the deposition of a layer of another inorganic substance (inorganic core-inorganic shell composites), the attachment of certain organic capping agents, and the covalent bonding with biomolecules (i.e., DNA, antibodies) are the most distinctive surface derivatizations of MNPs [4]. On the other hand, the steric stabilization method is constructed on the use of organic ligands to form a protective layer. This is skilled by the use of polymers, surfactants, or ligands. Concretely, surfactants such as sodium bis(2-ethylhexyl)sulfosuccinate, cetyltrimethylammonium bromide, or tetraoctylammonium bromide (AOT, CTAB, or TOAB, respectively) propose great colloid stabilization, avoiding the Van der Waals interparticle attractions that lead to agglomeration [4]. The stabilization is attained when the polar group of the surfactant adsorbs on the surface of MNPs and the organic apolar "tail" faces the peripherals of the resulting assembly, thereby experiencing steric repulsion.

4.8. Electrochemical Detection

Electrochemical measurements have publicized numerous advantages for trace heavy-metal detection, counting rapid analysis, good selectivity, and sensitivity. There are several reported electrodes for this detection such as Chemically and Biochemically Modified Electrodes, Nanostructurated Electrodes, Metal Nanoparticle Modified Electrodes, Carbon Nanotube Modified Electrodes, Nanomaterials Combined with Synthetic, Receptors Modified Electrodes, Nanobiomodified Electrodes etc.

4.9. Optical Detection

Optical detection is an alternative way to the electrochemical detection methods. These epitomize attractive analytical tools whenever continuous monitoring and real-time information is desired. There are several reported electrodes for this detection such as Light Absorption Techniques, Nonmodified Metal Nanoparticles, Synthetic Receptor Modified Metal Nanoparticles, Biomodified Metal Nanoparticles [4].

4.10. Light Emission Fluorescence Techniques

Fluorescence has been widely used for the determination of heavy metals in environmental amongst light emission techniques. Fluorescence belongs to the ubiquitous luminescence family in which susceptible molecules excited to high energy levels decay to lower levels, emitting transition-allowed radiation. An significant feature of the sensing systems based upon fluorescence is their highly sensitive detection compared to absorbance-based techniques [45]. Fluorescence characteristically takes place from aromatic molecules due to the rigid conjugated structure and the high ring density of π electrons. The complexation of metal ion exhibits either enhanced fluorescence, i.e., chelation-enhanced fluorescence (CHEF), or in a decreased fluorescence-chelation-enhanced quenching (CHEQ). Mechanisms commonly follow explaining the response of fluorescent sensors to heavy metals is electron transfer (ET) and charge transfer (CT). These two general categories consist of photoinduced electron transfer (PET), intramolecular charge transfer (ICT), and photoinduced charge transfer (PCT).170 In addition to fluorescence techniques, chemiluminescence and electrochemiluminescence are also being utilized for heavy metal ion detection.(Qian et al. 2010; Li et al. 2010).

Conclusion

This chapter exhibited natural abundance, physical/chemical properties, worldwide existence, and toxic effects of heavy metals along with their pH dependant species in account of pourbiax diagram. Beyond permissible limit their influences on kidneys or liver (Cd^{2+}, Cu^{2+}, Pb^{2+}, Hg^{2+}); the central nervous system (Pb^{2+}, Hg^{2+}, As^{3+}); skin, bones, or teeth (As^{3+}, As^{5+}, Ni^{2+}, Cu^{2+}, Cd^{2+}, Zn^{2+}, Cr^{3+}, Cr^{6+}) were addressed. As heavy metals are non-biodegradable in nature, they remain in food chain as well as ecological systems indefinitely, displaying top-level predators to break out high levels of contamination. Hence, safe limits of contaminant levels have been demarcated for drinking water by various organizations throughout the world. Based on toxicology data and scientific reports, World Health Organization (WHO) and Environmental Protection Agency (EPA) recommended standard permissible limits for heavy metals in drinking water. According to chemical recognition along with size (macro to nano), shape, morphology and porosity (macro to nano), the selective affinity of heavy-metal ions with the specific species/receptor customarily comprehends noncovalent interaction, viz van der Waals forces, hydrogen bonding, coordination, π-π interactions, hydrophilic/phobic interaction, electrostatic and electromagnetic effects. Efforts are utilizing to develop prompt and inexpensive techniques for heavy-metal detection in different chemical systems, including living species in ecosystem. Hence, this chapter envisioned a universal knowledge on toxic/carcinogenic effects of various heavy metals in living systems in account of their bonding interactions and the leading-edge strategies in the development of heavy-metal-detection devices/systems constructed on chemical recognition and specific affinity of receptor.

References

[1] Nandi, D., Basu, T., Debnath, S., Ghosh, A.K., De, A., Ghosh, U.C., (2013). Mechanistic insight for the sorption of Cd(II) and Cu(II) from aqueous solution on magnetic Mn- doped Fe(III) oxide nanoparticle implanted graphene. *J. Chem. Eng. Data,* 58, 2809- 2818.

[2] Khan, S., Cao, Q., Zheng, Y. M., Huang, Y. Z., Zhu, Y. G., 2008. Health risks of heavy metals in contaminated soils and food crops irrigated with wastewater in Beijing, China. *Environ. Pollut.* 152, 686–692.

[3] Zhang, M., Liu Z., Wang, H., 2010. Use of single extraction methods to predict bioavailability of heavy metals in polluted soils to rice. *Commun. Soil Sci. Plant Anal.* 41, 820–831.

[4] Aragay, G., Pons, J., Merkoci, A., 2011. Recent trends in macro-, micro-, and nanomaterial-based tools and strategies for heavy-metal detection. *Chem. Rev.* 111, 3433-3458.

[5] Kirpichtchikova, T. A., Manceau, A., Spadini, L., Panfili, F., Marcus, M. A., Jacquet, T., 2006. Speciation and solubility of heavy metals in contaminated soil using X-ray,microfluorescence, EXAFS spectroscopy, chemical extraction, and thermodynamic modelling. *Geochim. Cosmochim. Acta*, 70, 2163–2190.

[6] Adriano, D.C., 2003. Trace elements in terrestrial environments: Biogeochemistry, bioavailability and risks of metals. Springer, New York, NY, USA, 2nd edition.

[7] Maslin, P., Maier, R., 2000. Rhamnolipid-enhanced mineralization of phenanthrene in organic-metal co-contaminated soils. *Bioremediation J.* 4, 295–308.

[8] McLaughlin, M. J., Hamon, R., McLaren, R., Speir, T., Rogers, S., 2000. Review: a bioavailability-based rationale for controlling metal and metalloid contamination of agricultural land in Australia and New Zealand. *Australian J. Soil Res.* 38, 1037-1086.

[9] McLaughlin, M. J., Zarcinas, B. A., Stevens, D. P., Cook, N., 2000. Soil testing for heavy metals. *Commun. Soil Sci. Plant Anal.* 31, 1661–1700.

[10] Ling, W., Shen, Q., Gao, Y., Gu, X., Yang, Z., 2007. Use of bentonite to control the release of copper from contaminated soils. *Australian J. Soil Res.* 45, 618–623.

[11] Kabata-Pendias A., Pendias, H., 2001. Trace metals in soils and plants, CRC Press, Boca Raton, Fla, USA, 2nd edition.

[12] Zhao, Q., Kaluarachchi, J. J., 2002. Risk assessment at hazardous waste-contaminated sites with variability of population characteristics. *Environ. Intern.* 28, 41–53.

[13] Bolan, N. S., Ko, B. G., Anderson, C.W.N., Vogeler, I., 2008. Solute interactions in soils in relation to bioavailability and remediation of the environment," in proceedings of the 5th international symposium of interactions of soil minerals with organic components and microorganisms, Pucón, Chile.

[14] D'Amore, J. J., Al-Abed, S. R., Scheckel, K. G., Ryan, J. A., 2005. Methods for speciation of metals in soils: a review. *J. Environ. Qual.* 34, 1707–1745.

[15] Alloway, B. J., 1995. Heavy metals in soils, Blackie Academic and Professional, London, UK, 2nd edition.

[16] Lombi, E., Gerzabek, M. H., Determination of mobile heavy metal fraction in soil: results of a pot experiment with sewage sludge. *Commun. Soil Sci. Plant Anal.* 29, 2545–2556.

[17] Basta, N., Ryan, J., Chaney, R., 2005. Trace element chemistry in residual- treated soil: key concepts and metal bioavailability. *J. Environ. Qual.* 34, 49–63.

[18] Scragg, A., 2006. Environ. Biotechnol. Oxford University Press, Oxford, UK, 2nd edition.

[19] Lasat, M., 2000. Phytoextraction of metals from contaminated soil: a review of plant/soil/metal interaction and assessment of pertinent agronomic issues. *J. Hazard. Subs. Res.* 2, 1–25.

[20] Raven, P. H., Berg, L. R., Johnson, G. B., 1998. Environment, Saunders College Publishing, New York, NY, USA, 2nd edition.

[21] Sumner, M. E., 2000. Beneficial use of effluents, wastes, and biosolids. Commun. *Soil Sci. Plant Anal.* 31, 1701–1715.

[22] Chaney, R. L., Oliver, D. P., 1996. Sources, potential adverse effects and remediation of agricultural soil contaminants. Contaminants and the Soil Environments in the Australia-pacific.

[23] Mattigod, S. V., Page, A. L., 1983. Assessment of metal pollution in soil. Appl. Environ. Geochem. Academic Press, London, UK, 355–394.

[24] Keller, C., McGrath, S. P., Dunham, S. J., 2002. Trace metal leaching through a Soil grassland system after sewage sludge application. *J. Environ. Qual.* 31, 1550– 1560.

[25] McLaren, R. G., Clucas, L. M., Taylor, M. D., Hendry, T., 2004. Leaching of macronutrients and metals from undisturbed soils treated with metal-spiked sewage sludge. *Australian J. Soil Res.* 42, 459– 471.

[26] Bjuhr, J., 2007. Trace metals in soils irrigated with waste water in a Periurban area downstream Hanoi City, Vietnam, Seminar Paper, Institutionen för markvetenskap, Sveriges lantbruksuniversitet (SLU), Uppsala, Sweden.

[27] Smith, L. A., Means, J. L., Chen, A., 1995. Remedial options for metals-contaminated sites, Lewis Publishers, Boca Raton, Fla, USA.

[28] USEPA, 1996. Report: recent developments for in situ treatment of metals contaminated soils, U.S. Environmental Protection Agency, Office of Solid Waste and Emergency Response.

[29] Shiowatana, J., McLaren, R. G., Chanmekha, N., Samphao, A. 2001. Fractionation of arsenic in soil by a continuous-flow sequential extraction method. *J. Environ. Qual.* 30, 1940–1949.

[30] Buekers, J., 2007. Fixation of cadmium, copper, nickel and zinc in soil: kinetics, mechanisms and its effect on metal bioavailability, Ph.D. thesis, Katholieke Universiteit Lueven.

[31] Greenwood, N., Earnshaw, A., 2007. Chemistry of the elements, Butterworth-Heinemann Linacre House, Jordan Hill, Oxford OX2 8DP. 225 Wildwood Avenue, Woburn, MA 01801-2041.

[32] Bodek, I., Lyman, W. J., Reehl, W. F. Rosenblatt, D. H., 1988. Environmental Inorganic Chemistry: Properties, processes and estimation methods, Pergamon Press, Elmsford, NY, USA.

[33] Boyle, R. W., Jonasson, I. R., 1973. The geochemistry of As and its use as an indicator element in geochemical prospecting. *J. Geochemical. Explor.* 2, 251-296.

[34] http://web.worldbank.org/wbsite/external/countries/southasiaext/0,contentmdk:22392781~pagepk:146736~pipk:146830~thesitepk:223547,00.html.

[35] Fawzy, E. M., 2008. Soil remediation using in situ immobilisation techniques. *Chemistry and Ecology*. 24, 147–156.

[36] http://www.myhousecallmd.com/the-mad-hatter-fish-the-truth-behind-mercurylevels/#sthash.fzzfqlWH.dpuf.

[37] Ung, C. H. et al. 2010. Mercury-induced hepatotoxicity in zebrafish: in vivo mechanistic insights from transcriptome analysis, phenotype anchoring and targeted gene.

[38] http://flipper.diff.org/app./items/info/5530.

[39] Svancara, I., Vytras, K., Kalcher, K., Walcarius, A., Wang, J., 2009. Carbon Paste Electrodes in Facts, Numbers, and Notes: A Review on the occasion of the 50-years jubilee of carbon paste in electrochemistry and electroanalysis. *Electroanal.* 21, 7-28.

[40] Ilangovan, R.; Daniel, D.; Krastanov, A.; Zachariah, C.; Elizabeth, R., 2006. Enzyme based biosensors for heavy metal ions determination. *Biotechnol. Biotechnol. Equip.* 20, 184-189.

[41] Pinheiro, S. C.; Raimundo, I. M.; Moreno-Bondi, M. C.; Orellana, G., 2010. Simultaneous determination of copper, mercury and zinc in water with a tailored

fluorescent bipyridine ligand entrapped in silica sol-gel. *Anal. Bioanal. Chem*398, 3127-3138.

[42] Guo, D.; Li, J.; Yuan, J.; Zhou, W.; Wang, E., 2010. Nafion Film Immobilized Nano Ag-Hg Amalgam Glassy Carbon Electrode Used for Simultaneous Determination of Lead, Cadmium and Copper. Electroanal. 22, 69-71.

[43] Fan, L.; Chen, J.; Zhu, S.; Wang, M.; Xu, G., 2009. Determination of Cd^{2+} and Pb^{2+} on glassy carbon electrode modified by electrochemical reduction of aromatic diazonium salts. *Electrochem Commun.* 11, 1823-1825.

[44] Combellas, C.; Kanoufi, F.; Pinson, J.; Podvorica, F. I. 2008. Sterically hindered diazonium salts for the grafting of a monolayer on metals. *J. Am. Chem. Soc.* 130, 8576- 8577.

[45] Zang, L., Liu, R., Holman, M. W., Nguyen, K. T., Adams, D. M., 2002. A single-molecule probe based on intramolecular electron transfer. *J. Am. Chem. Soc.* 124, 10640-10641.

[46] Qian, X.; Xiao, Y.; Xu, Y.; Guo, X.; Qian, J.; Zhu, W., 2010. "Alive" dyes as fluorescent sensors: fluorophore, mechanism, receptor and images in living cells. *Chem Commun.* 46, 6418-6436.

[47] Li, T.; Wang, E.; Dong, S., 2010. Lead(II)-induced allosteric G-quadruplex DNAzyme as a colorimetric and chemiluminescence sensor for highly sensitive and selective Pb^2 detection. *Anal. Chem.* 82, 1515-1520.

In: Heavy Metals
Editor: Deepak Pathania
ISBN: 978-1-63484-740-7

Chapter 7

THE EFFECT OF HEAVY METALS POLLUTION, SOURCES AND REMEDIAL MEASURE TECHNIQUE

Vaneet Kumar**, *Rakesh Kumar and Saruchi
Environmental Chemistry Lab, Department of Chemistry,
CT Group of Institution Jalandhar, India
*Dr. B.R Ambedkar National Institute of Technology Jalandhar,
Punjab, India

ABSTRACT

Metals are inseparable part of the biological system. The concentration of metals mainly depends on geological and biological processes. With the improvement of the worldwide economy, both type and content of heavy metals in the environment caused by human activities have continuously expanded in recent years, which have resulted in serious environment deterioration and degradation. Heavy metals are exceptionally harmful to the nature and organic entities. When the soil suffers from heavy metal contamination, it is hard to be remediated. Scattered writing is outfit to critically survey the possible sources, science, potential biohazards and best accessible remedial systems for a various number of overwhelming metals (lead, chromium, arsenic, zinc, cadmium, copper, mercury and nickel) commonly found in contaminated soils. The standards, favorable circumstances advantages and disadvantages of immobilization, soil washing and phytoremediation techniques are recorded many times with the help of the advancements for cleaning up heavy metal contaminated sites. Remediation of substantial metal contaminated soils is important to diminish the associated risks, to improve the area for horticultural creation, and upgrade food security and scale down land tenure issues which emerge from changes in the area utilization design.

* Corresponding author: vanet2106@gmail.com.

INTRODUCTION

Soils may be contaminated by the accumulation of heavy metals and metalloids through discharges from quickly expanding industrial regions, mine tailings, disposal of high metal wastes, leaded fuels and paints, land application of fertilizers, animal manures, sewage sludge, pesticides, wastewater watering system, coal combustion residues, spillage of petrochemicals and atmospheric deposition [1, 2]. Heavy metals do not constitute a well characterized gathering of inorganic chemical hazards, and those, which are most normally found at contaminated sites are lead (Pb), chromium (Cr), arsenic (As), zinc (Zn), cadmium (Cd), copper (Cu), mercury (Hg) and nickel (Ni) [3]. Soils are the significant sink for heavy metals discharge into the nature by aforementioned anthropogenic activities. This is dissimilar to natural contaminants which are oxidized to carbon (IV) oxide by microbial activity, most metals don't undergo microbial or chemical degradation [4] and their aggregate concentration in soils persist a long after their presentation [5]. However, changes in their compound structures and bioavailability are possible. The presence of harmful metals in soil can seriously inhibit the biodegradation of natural contaminants [6]. Heavy metal contamination of soil may pose danger and risk to human and the biological community through: direct ingestion or exposure to contaminated soil, the food chain, drinking of contaminated ground water, reduction in food quality by means of phytotoxicity, reduction in land convenience for agricultural production aggravating food insecurity, and land tenure problems [7–9].

The satisfactory assurance and restoration of soil ecosystems contaminated by heavy metals require their characterization and remediation. Modern-day legislation regarding environmental insurance and public health, at both national and global levels, are in view of information that describe chemical properties of environmental phenomena, particularly in our evolved way of life [10]. Soil characterization would provide us an understanding into substantial metal speciation and bioavailability, endeavor at remediation of heavy metal contaminated soils. It would involve learning the source of contamination, their fundamental science, environmental and related health effects of these heavy metals. Hazard assessment is effective scientific device to empower decision makers to manage contaminated in a cost-effective manner while protecting health, people and ecosystem [11].

Immobilization, soil washing and phytoremediation procedures are usually recorded among the best demonstrated available innovations (BDATs) for remediation of heavy metal-contaminated sites [3]. Their exposed adequacy and environment friendliness, field utilizations of these innovations have been accounted for in developing nations. In most developing nations, these are yet to end up commercially available nations perhaps because of lacking consciousness of their points of interest and standard. The governments and the public on the implications of contaminated soils on humans and other creature have raised the interest amongst the mainstream researchers in the advancements to remediate contaminated sites [12]. In developing nations with great population density and rare funds availability for ecological rebuilding and biologically manageable choices are required to decrease related damages and make the area accessible for agricultural production, food security to scale down land residing issues.

In this chapter, brief literature reviewed the possible sources of contamination, basic chemistry and the related natural and health dangers of heavy metals (Pb, Cr, As, Zn, Cd, Cu, Hg, and Ni) which can give knowledge of substantial metal speciation, bioavailability, and consequently choice of suitable therapeutic alternatives. The standards, preferences and impediments of immobilization, soil washing and phytoremediation systems as choices as options for soil cleanup are additionally exhibited in this chapter.

SOURCES OF HAZARDOUS METALS AND MINERALS

Natural contamination from hazardous metals and minerals can emerge from natural as well as anthropogenic sources. The natural sources are: drainage from rocks into water, volcanic movement, forest fires, etc. With fast growing industrialization and consumerist way of life, anthropogenic sources of environmental pollution have expanded. The pollution happens both at the level of industrial production as well as utilization of the items and keep running off. These toxic elements enter the human body basically through food and water and to a lesser extent by inhaling polluted air, utilization of cosmetics, drugs, low quality herbal formulations and even things like toys which have paints containing lead.

Heavy Metals and Aluminium

Table 1 lists the industrial sources of heavy metals.

Table 1. Sources of heavy metals

Metal	Industry
Chromium (Cr)	Mining, industrial coolants, chromium salts manufacturing, leather tanning
Lead (Pb)	lead acid batteries, paints, E-waste, Smelting operations, coal- based thermal power plants, ceramics, bangle industry
Mercury (Hg)	Chlor-alkali plants, thermal power plants, fluorescent lamps, hospital waste (damaged thermometers, barometers, sphygmomanometers), electrical appliances, etc.
Arsenic (As)	Geogenic/natural processes, smelting operations, thermal power plants, fuel burning
Copper (Cu)	Mining, electroplating, smelting operations
Vanadium (Va)	Spent catalyst, sulphuric acid plant
Nickel (Ni)	Smelting operations, thermal power plants, battery industry
Cadmium (Cd)	Zinc smelting, waste batteries, e-waste, paint sludge, incinerations & fuel Combustion
Molybdenum (Mo)	Spent catalyst
Zinc (ZN)	Smelting, electroplating

Other than the industrial sources, lead exposure also occurs through gasoline additives, food can solder, ceramic glazes, drinking water framework system, beautifying agent, folk remedies, and battery/plastic recycling industry (Table 1) [13]. According to some research work carried out at DPSAR University, New Delhi various brands of cosmetics like talcum powder, lipsticks, shampoos, 'kajal' and hair colours contain heavy metals. Ash dumps from thermal power plants, contain numerous polluting metals and complexes, which are conveyed to close by water bodies and ground water. Volatile complexes such as those from Uranium enter the atmosphere via chimney emissions. U content of the coal may be as low as 0.2 ppm, yet considering the huge amount of coal that is burnt it is an important pollutant. In recent years the utilization of energy-saving CFL bulbs has gone up enormously. Thus, along these lines, as indicated by recent report the production of CFL bulbs has increased from 19 million in 2002 to 500 million in 2010. Every bulb contains 3-12 mg of mercury. There is no framework to recover these bulbs and safe transfer or disposal; these can turn to be a major health hazard. Some of the metal contaminated sites in India are given in table 2.

Information of CPCB New Delhi demonstrate that Gujarat, Maharashtra and Andhra Pradesh contribute to 80% of hazardous waste (including heavy metals) in India.

Aside from commercial ventures, roadways and automobiles considerably contribute to the ecological burden of substantial heavy metal, since particulate matter in traffic emissions incorporate heavy metals like lead, cadmium and arsenic. Exposure to exhaust emissions, particularly diesel fumes may enhance asthma, allergen responsiveness and inflammation, prompty to atherosclerotic vascular disease [14, 15]. With the utilization of petrol, the burden of lead has diminished. Aluminium contamination is connected with bauxite mining. With expanding, increase in demand for aluminium in India, its anthropogenic pressure has increased. The position of India is 6th in bauxite mining and 8th in aluminium generation whereas condition of Orissa is exceedingly bad.

Significant concentration of hexavalent chromium is observed in many wells situated in nearby region of Ranipet, in Tamil Nadu. The total aggregation of chromium in these wells varies between 3.1 to 246 mg/L while the concentration of hexavalent chromium varies between 2.1 to 214 mg/L which far exceed the concentration of 0.05 mg/L recommended under Indian standards specification for drinking water quality. The ground water in these ranges is therefore, extremely contaminated with hexavalent chromium. Taking into account the definite research studies and techno – economic assessment, an in-situ bioremediation (biotransformation) alternatives were recommended by NEERI for execution of bio-remediation of contaminated ground water in the critically dirty regions. Town Khanpur in Rania zone of Kanpur Dehat also revealed high amount of hexavalent chromium in groundwater was ranging from 1.05 to 35.34 ppm.

Fluoride

Natural sources add the greater part to the ecological load of fluoride and arsenic. In India, 19 out of 35 states and Union domains have ground water highly contaminated with fluoride, with levels exceeding 1.0 mg/L and setting off up to 48 mg/L [16-17].

Table 2. Major heavy metals contaminated sites in India

Chromium	Lead	Mercury	Arsenic	Copper
Ranipet,	Ratlam,	Kodaikanal,	Tuticorin,	Tuticorin,
Tamil Nadu	Madhya Pradesh	Tamil Nadu	Tamil Nadu	Tamil Nadu
Kanpur,	Bandalamottu	Ganjam,	West Bengal	Singbhum Mines,
Uttar Pradesh	Mines,	Orissa	Ballia	Jharkhand
Vadodara,	Andhra Pradesh Vadodara, Gujarat	Singrauli,	UP	Malanjkahnd,

In states like Andhra Pradesh, Gujarat and Rajasthan, 70-100% districts contain high fluoride levels in food and water. As per Susheela, black rock salt (CaF_2) ordinarily utilized as flavouring agent in road side, as well as processed and home-cooked foods contributes significantly to the ingestion of fluoride. It contains 157 ppm Flouride, so public awareness in this regard is needed.

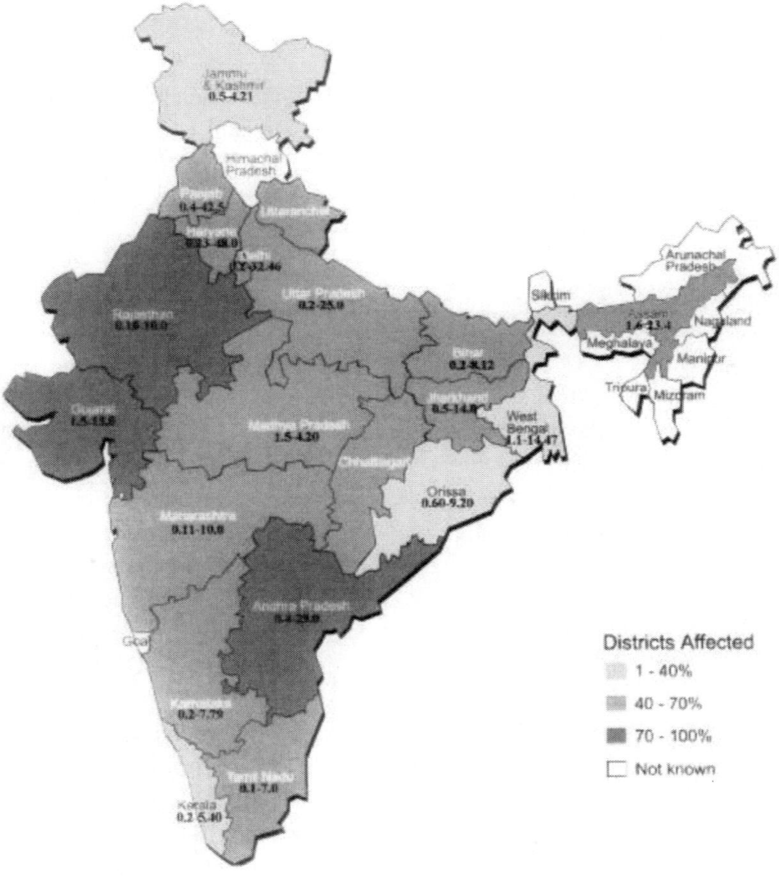

Dental items, anti-depressant and anti - cholesterol medications utilized for long term treatment are essential sources of fluoride. Industries which are using fluoride salts/hydrofluoric acid pollute the work environment are a hot spot for high ingestion/inhalation of fluoride dust and fumes by the Industrial workers.

Arsenic

High content of arsenic in water was found in West Bengal, Bihar, UP, Assam and Chhatisgarh in India, whereas, West Bengal being by far the worst affected. Ground water of 9 out of 16 districts of West Bengal was heavily contaminated with arsenic, affecting 26 million people [18]. In a U.P. Jal Nigam/IITR survey of 66671 samples of water from hand pumps in 20 districts of U.P., 42% were found to contain > 10 ppb arsenic of which 2610

(4%) had > 50 ppb arsenic. High burden of environmental pollution in developed countries like India is due to high level of resource consumption, and it is also due to outdated technologies, over exploitation of natural resources and weak environment regulations and enforcement.

TOXICITY DUE TO METALS AND MINERALS

In general, heavy metal toxicity can cause chronic, degenerative conditions. General symptoms include: headache, short-term memory loss, mental confusion, sense of unreality, distorted perception, pain in muscles and joints, and gastro-intestinal upsets, food intolerances, allergies, vision problems, chronic fatigue, fungal infections, etc. Sometimes the symptoms are vague and difficult to diagnose.

Lead (Pb)

Absorption of Pb from diverse sources is subject to the measure of Pb introduced to portals per unit time and the physical and chemical state in which Pb is displayed. It is additionally affected by the variables such as age and physiological status. In adults, almost 20-30% and in children almost 50% lead is accumulated through the GI track. Depending on the molecular size, lead can enter through lungs. While natural lead is well absorbed through the skin, whereas, inorganic lead is not. Since lead is chemically similar to calcium, body handles it like calcium. In the body lead is dispersed throughout-bone, teeth, liver, lung, brain and spleen and bone being the major accumulator. Lead can cross blood brain boundry as well as placental barrier. Excretion occurs through urine and faeces. Doses and length of time dependent genotoxic effects have been found [19].

Nutritional deficiency of iron raises Pb poisonous quality, increases concern that pregnant women and youngsters in whom iron deficiency anaemia is high, may be more susceptible to Pb toxicity [20]. Amount of Pb absorption increases considerably with fasting or in persons whose diet is deficient in calcium, iron, phosphorous or zinc. In children the blood lead level (BLL) above 10 μg/dl is labeled as poisonous. Latest studies from Hyderabad indicate unusual intellectual capacities in youngsters at levels > 10 μg/dl. A study done in Hyderabad indicated high blood lead levels in neonates and mothers in general population [21, 22].

Table 3. General signs and symptoms of lead toxicity

Fatigue	Motor neuropathy
Irritability	Encephalopathy
Lethargy	Cerebral edema
Paresthesis	Seizures
Myalgias	Coma
Abdominal pain	Severe abdominal cramping

Table 4. Range of lead-induced effects in humans at different blood levels

Blood lead levels	Adults	Children
10 μg/dL	Hypertension may occur	Crosses placenta Impairment IQ, growth Partial inhibition of heme synthesis
20 μg/dL	Inhibition of heme synthesis Increased erythrocyte protoporphyrin	Beginning impairment of nerve conduction velocity
30 μg/dL	Systolic hypertension Impaired hearing	Impaired vitamin D metabolism
40 μg/dL	Infertility in males Renal effects Neuropathy Fatigue, headache, abd pain	Haemoglobin synthesis inhibition
50 μg/dL	Anaemia, gastro-intestinal disorder, headache, tremor	Colicky abdominal pain, neuropathy
100 μg/dL	Lethargy, seizures, encephalopathy	Encephalopathy, anemia, nephropathy, seizures

In general Pb is discharged very slowly from the body. Its biological half-life is about 10 years, encourges accumulation in the body. Right around 90% lead is bound to red blood cells. Lead has high affinity for SH groups and this impairs the activity of zinc-ward chemicals like δ-aminolevulinic acid dehydratase (ALAD) which is involved in haem synthesis. Apart from haemoglobin, cytochrome synthesis, steroid metabolism, membrane integrity, synthesis of active metabolite of vitamin D in renal tubular cells (conversion of 1-hydroxyvitamin D to 1,25-hydroxyvitamin D) are also influenced. Tables 3 and 4 give the general signs and symptoms of lead toxicity, and relationship with dose [23]. Lead is a confirmed carcinogen in animals as well as human.

Mercury: It occurs in three forms

Elemental: It exists as a liquid at room temperature, but volatizes readily. It is rapidly distributed in the body through vapour, but poorly absorbed through GI track.

Inorganic: Poorly consumed through GI tract, yet can be caustic. Dermal introduction exposure brings about toxicity.

Organic: It is present in the liquid form. Easily consumed by GI tract, lungs and skin. It can also cumacross placenta and into breast milk.

Anaerobic organisms bio-transform the inorganic structure to methyl mercury which gets bio-aggregated in food chain. It's the most toxic form of mercury. Unfriendly health effects depend upon its chemical form, their span of exposure. Enzymes, receptors, membranes and structural proteins are influenced. There is multiple organ failure. Some of the symptoms of mercury poisoning are summarized here.

At high concentrations, vapour inhalation produces acute necrotizing bronchitis, pneumonitis, and death.

Long term exposure influence central nervous system (CNS).

- Early: insomnia, forgetfulness, anorexia, mild tremor
- Late: progressive tremor and erythrism (red palms), emotional lability, and memory impairment

- Salivation, unreasonable sweating, renal toxicity (proteinuria, or nephrotic syndrome)
- Gastrointestinal ulceration or perforation and haemorrhage are quickly delivered, followed by circulatory collapse.
- Breakdown of mucosal barriers leads to increased absorption and distribution to kidneys (proximal tubular necrosis and anuria).
- Acrodynia (Pink disease) usually from dermal exposure
- Maculopapular rash, swollen and painful extremities, peripheral neuropathy, hypertension, and renal tubular dysfunction.
- Signs progress from paresthesias to ataxia, followed by generalized shortcomings visual and listening, tremor and muscle spasticity, and then coma and death.

Teratogen with large chronic exposure

- Asymptomatic mothers give birth to severely affected infants
- Infants appear normal at birth, but psychomotor retardation, blindness, deafness, and seizures develop over time.

Thus, exposure to inorganic and organic mercury is associated with genotoxicity, Teratogenicity, and embryo toxicity. In the GI track, acute poisoning results in sloughing away of the mucosa with pieces of intestinal mucosa appearing in the stool. In chronic intoxication there is mercury line at the gingival border similar to the 'lead line'.

An established sample of mercury poisoning because of utilization of fish contaminated with methyl mercury from industrial waste is that of Minamata disease, reported from Minamata in Japan in 1956. Symptoms include ataxia, numbness in the hands and feet, general muscle weakness, narrowing of the field of vision and damage to hearing and speech. In some cases madness, loss of motion, coma and death follow within weeks of onset of symptoms. A congenital form of the disease can also influence to the womb.

A second outbreak of Minamata disease occurred in Niigata Prefecture in 1965 [24].

A study conducted at AIIMS, New Delhi showed poor understanding among health professionals, particularly the technical and nursing staff, regarding the hazards of mercury in hospital devices and equipment. These workers continued to prefer mercury containing equipment like thermometers, manometers, etc. despite availability of alternative digital devices.

Cadmium (Cd)

The smoke of tobacco is an important source of cadmium exposure. Smoking one pack a day can absorb 5-10 times the amount of cadmium obtainable through a regular diet. Food is a poor source of Cadmium and it is transported in blood, bound to metallothionin. Highest concentration of Cd is found in kidney and liver. The problem of cadmium toxicity in India is not known till date. Cadmium affects lungs, kidneys, liver and skeletal system. It binds to sulfhydryl groups, displacing other metals from metalloenzymes, disrupting those enzymes.

Cadmium competes with calcium for binding sites on regulatory proteins. Lipid peroxidation has been demonstrated. Cadmium has been classified as a suspected human carcinogen.

Arsenic (As)

Arsenic tends to accumulate in keratin- rich tissues like nails, hair and skin. Inorganic arsenic is converted to organic arsenic (biomethylation to monomethyl arsenic- MMA or DMA) in the liver. This may represent a process of detoxification. 30-50% of inorganic arsenic is excreted in about 3 days through urine. Trivalent forms of As bind to sulfhydryl groups leading to inhibition of enzymatic systems. Arsenic affects energy transduction reactions like Krebs cycle and oxidative phosphorylation, and ATP production is inhibited. Endothelial damage, loss of capillary integrity, capillary leakage, volume loss, finally result in shock. Some of the symptoms of arsenic toxicity are shown in Table 5.

Table 5. Symptoms of arsenic toxicity

Bodily system affected	Symptoms or signs	Time of onset
Systemic	Thirst Hypovolemia, Hypotension	Minutes Minutes to hours
Gastrointestinal	Garlic or metallic taste Burning mucosa Nausea and vomiting Diarrhoea Abdominal pain Hematemesis Hematochezia, melena Rice-water stools	Immediate Immediate Minutes Minutes to hours Minutes to hours Minutes to hours Hours Hours
Hematopoietic system	Haemolysis Hematuria Lymphopenia Pancytopenia	Minutes to hours Minutes to hours Several weeks Several weeks
Pulmonary (primarily in inhalational exposures)	Cough Dyspnea Chest Pain Pulmonary edema	Immediate Minutes to hours Minutes to hours Minutes to hours
Liver	Jaundice Fatty degeneration Central necrosis	Days Days Days
Kidneys	Proteinuria Hematuria Acute renal failure	Hours to days Hours to days Hours to days

Acute exposure can even result in death. Chronic exposure can result in poisoning of the nervous system, liver damage and peripheral vascular disease. This condition is commonly known as ‘black foot disease’. Chronic exposure is associated with leukaemia, kidney and bladder cancers, dermatitis, hyper pigmentation and arsenical keratosis. Arsenic acts as a non-genotoxic carcinogen. However, it affects DNA methylation and repair.

Studies of Giri and colleagues depict that there is a genetic predisposition to arsenic toxicity. Although huge numbers are exposed to arsenic through drinking water, only 15-20% shows arsenic-induced skin lesions. Though both skin lesions and non skin lesions group had higher degree of damage, those with the skin lesions showed greater degree of health effects, and immunological, haematological and genetic damage than those not showing skin lesions [25-26].

Fluoride (F)

Toxicological studies regarding fluoride show that it is absorbed or distributed into two compartments. Fluoride in blood and soft tissues has short half life of few hours, but that in hard tissues like bone and teeth has long half life of eight years. Accumulation in these two tissues is dose and age dependent. Unlimited accumulation of fluoride in bones is the main cause of the disease, skeletal fluorosis. Fluoride toxicity can be acute due to exposure to a single massive dose, as happens with industrial workers (industrial fluorosis) or chronic (endemic fluorosis) due to continuous ingestion of water and food containing high amounts of fluoride. In both the types, teeth and bone are the primary targets. However, fluoride does not spare soft tissues and causes non-skeletal fluorosis. As mentioned above, endemic fluorosis is a dangerous problem in many parts of India.

The characteristic feature of dental fluorosis is dental mottling. The clinical features of skeletal flourosis like muscular skeletal dysfunction, arthralgia, arthritis, fixed flexion deformities, restricted movement of joints, stiffness of the spine, and sometimes paraplegia [27]. In recent years, a variant of skeletal fluorosis – genu valgum or knock knee has been reported from some parts of the world including India, in younger people but its aetiology is not fully understood.

Consumption of high amounts of fluoride through water and food is the main factor in the cause of endemic fluorosis, other factors- probably dietary, also plays a role. In the US, ingestion of even 8 ppm fluoride containing water, over 15 years did not lead to fluorosis but in India 1 ppm levels is considered unsafe.

Management of Pollution from Metals and Minerals

The maxim "prevention is better than cure" applies to ecological contamination as it does to diseases. Dirty environment in any case prompts to disease and ill health. In this way, options should not just be confined to remediation techniques as well as focus on modern methods through substitutions of heavy metal/minerals by refining the current advancement for decreasing the requirement.

Remediation Technologies

Release of heavy metal ions within safe limits occur through three complementary functions: 1) Technological 2) Management (usage) and 3) Regulatory. When the metal is out of the earth's crust, it is hard to return it back into earth despite efforts being made. Innovations for reduction ought to be cost- effective and moderate. Most industries use engineering advancement for remediation based on physico-chemical strategies. Bioremediation routines, utilizing plant and microbial systems have also been developed for detecting pollution as well as for remediation.

Technological functions involve advancement of the treatment scheme whereby the amount and concentration in the waste/effluent is brought within the stipulated safe limits. The structure in which the metal is available will also influence its entry into the food chain and hence the effort should be to convert it to less harmful substances. Different methods are accessible for the treatment of gaseous emissions and liquid effluents to reduce the burden of metals to safe levels. Method for the treatment of gaseous emissions and liquid effluents to reduce the burden of metals to safe levels is accessible. For particulate matter in vapours emissions the advancements like use of cyclones, electrostatic precipitators bag filters, scrubbers independently or in mix. For fluid effluents, physico-chemical procedures like settling, neutralisation, precipitation, flocculation, filtration, etc. are adopted. General strategy is to reduce the metallic burden in gaseous or liquid media by changing over it into solid form and after that either reuse it or set back in to earth after synthetic treatment and altering it in a matrix from where it cannot drain out.

The responsibility of Management is to guarantee that the right advancements are received and to monitor the best possible results. Regulations are made to direct the release of toxic metals in the environment to ensure safety and health of the workers as well as the public. However, the question is how much is safe? What are the safe limits? The current regulatory limits are based on some gross evidence to judge health hazard. More work is expected to determine the safe limits using more refined cellular, molecular and functional parameters. Microbial, Plant, Animal and Human systems have been attempted to detect harmful substances. At present, the rules for administration of tailings from beneficiation plants and slags from smelters which are not ordered as dangerous waste but rather created in large expansive are yet to be arranged.

Proposals for further action:

- Recycling and reprocessing of wastes containing lethal metals should be given greater emphasis not only from environmental and health point of view but also asset preservation measures.
- Monitoring of air, water and soil in the region of the toxic metal processing units should be carried out more rigorously for the particular metal.
- Tailings dumps and process wastes lying in locations close to the processing units need to be remediated on priority.
- Guidelines for proper administration of tailings and slags containing toxic metals ought to be arranged into consideration techno- economic feasibility.
- Health monitoring of workers engaged in the processing of toxic metals/compounds should be carried out regularly.

In India, small commercial enterprises industries which are located in unorganized areas contribute to a great deal of pollution but regular treatment plants can help. As such in India, 22 Common Treatment, Storage and Disposal Facilities in 10 states have been established-7 in Gujarat, 4 in Maharashtra, 3 in UP, 2 in AP, and 1 each in HP, MP, Punjab, Tamil Nadu and West Bengal. For an immeasurable quickly industrializing nation like India this number seems to be very small. In Hyderabad a typical effluent from treatment plant (CETP) has been set up for treating effluents from few little electroplating commercial industries. Proper working of CETSPs must be ensured. Most plants don't work efficiently particularly little commercial ventures would require regular treatment plants, large industries ought to have their own treatment plants and release waste in to sewers after treatment, rather than carrying the burden of the normal treatment plants.

Nagpur NEERI, has built up a few innovations several technologies for water and area sectors for waste management and also a pack for testing the nature of water. For recovery of substantial accumulation focuses should be set up and some discount should be given. As, it seems to be no such system exists and with expanding utilization of CFL bulbs, haphazard disposal can be dangerous. Replacement of CFL bulbs with LED bulbs should be considered. In medical devices like thermometers and BP mechanical assembly, digital devices should replace the mercury-based sphygmomanometers.

To reduce the burden of heavy metal load, Phytotechnologies should be kept up and checked. Microflora connected with plants like endophytic bacteria, rhizosphere organisms and mycorrhizae have the potential to degrade organic compounds in association with plants and this procedure is termed as rhizoremediation. Bioremediation procedures can also be accessed through a multifaceted approach such as natural attenuation, sensing environmental pollution, metabolic pathway designing, applying phyto- and microbial differences to hazardous destinations, plant-endophyte associations and systems biology; plant physiology, agronomy, microbiology, hydrogeology, and building are joined to choose the best possible plant and conditions for a particular site. The specific application will rely upon the mobility, solubility, degradability, and bioavailability of the contaminant(s) of concern [28, 29]. Metal resistance systems and their phytoremediation potential has been researched in excessive variety in wide range of experimental frameworks for example, *Scenedesmus quadricauda, S. bijugatus, Chlamydomonas reinhardtii, Sorghum bicolor, Zea mays, Brassica juncea; Ceratophyllum demersum, Vallisneria americana* and *Lemna trisulca* and phytomass derived itemsproducts as adsorbents for posinous metals [29].

Genetically Modified Plants for Phytoremediation

Both microorganisms and higher plants have been hereditarly modified for detoxification of soils contaminated with alarming level of metals or minerals. Taking a few illustrations: (i) Ralstonia eutropha (a natural inhabitant of soil) was changed utilizing a mouse gene encoding metallothionein, which was communicated on the surface of the cell surface thus helping in sequestering of cadmium from the soil; (ii) Transgenic plants of *Arabidopsis thaliana* were delivered for detoxification of soils polluted with either aluminium, or arsenic or mercury. For aluminium detoxification, a gene encoding citrate synthase (CSb) was used, while for arsenic, genes encoding arsenic reductase (ArsC) and glutamylcysteine synthase (gamma-ECS) were utilized. Similarly, for mercury detoxification, genes encoding mercury reductase and organomercurial lyase (merA and merB) were used [30].

Aside from phytoremediation, plants can also used to detect metal contamination. A system based on the plant *Hordeum vulgare* has been created at Baharampur University by Panda and associates.

Management of Natural Burden of Arsenic and Fluoride

In areas where water has high traces of minerals like arsenic and fluoride, alternative sources- (canal water, rain water harvesting) must be for cultivating. But now the advancement to remove the fluoride from drinking water has been produced.

Table 6. Potential technologies suggested by the National Environment Engineering Research Institute (NEERI), Nagpur[19]

Group 1	**Parameters Groups**	**Range**	**Suggested Technologies**
	Dissolved solids, mg/L Dissolved solids, mg/L Chlorides, mg Cl/l	above 5000 above 2000 above 1000	Water Transportation Water Transportation Water Transportation
Group 2	Sulphates, mg SO4/1 Nitrates, mg NO3/l Fluorides, mg F^-/l Dissolved solids, mg/L Chlorides, mg Cl/l	above 400 above 100 above 1.5 below 2000 above 1000	Reverse Osmosis Electro dialysis Reverse Osmosis
Group 3	Sulphates, mg SO4/1 Nitrates, mg NO3/l Fluorides, mg F^-/l Dissolved solids, mg/L Chlorides, mg Cl/l	above 400 above 100 above 1.5 below 1000 below 250	Electro dialysis
Group 4	Sulphates, mg SO4/1 Nitrates, mg NO3/l Fluorides, mg F^-/l Dissolved solids, mg/L Chlorides, mg Cl/l	below 200 above 45 above 1.5 below 2000 below 1000 below 400	Ion-Exchange Activated Alumina
Group 5	Sulphates, mg SO4/1 Nitrates, mg NO3/l Alkalinity, mg CaCO3 Fluorides, mg F^-/l Dissolved solids, mg/L Chlorides, mg Cl/l	below 45 below 200 above 1.5 below 2000 below 1000	Technology
Group 6	Sulphates, mg SO4/1 Nitrates, mg NO3/l Alkalinity, mg CaCO3 Fluorides, mg F^-/l	below 400 below 45 above 200 above 1.5	Nalgonda Technology for fluoride removal

There are number of villages in the endemic states of India with excessive fluoride content. The water quality assessment undertaken by NEERI during 1970-1990 revealed that physico-chemical parameters of the samples varied widely within the same village in all the 19 endemic states, then surveyed. In view of the outcomes, the water sources are classified into different groups and potential advances recommended Table 6 [19].

Issues Related with Quality of Water

The developing world is worried for the accessibility of safe water for human utilization. The developing nations and quick urbanization has represented a few genuine difficulties to tackle issues related to water in country and ultra region.

The availability of sufficient quantity of water is the basic need for existence. But in the last decade, sustained efforts at international level have enabled access to safe drinking water to about 1 billion additional people in the world. However, quality of water is also of paramount importance. There are constant efforts through worldwide associations and organizations like World Health Organization (WHO) and the International Bureau of Weights and Measures (BIPM) through its Consultative Committee on Quantity of Matter (CCQM) have conceived measures for ensuring nature of safe drinking water and procedure of accomplishing the same through a chain of Testing and Calibration Laboratories in every nation. The estimates must be made as per the globally acceptable strategies and methods with balanced supplies. The labs should mandatorily be licenced by an association like National Accreditation Board for Testing and Calibration Laboratories (NABL) in India. This guarantee that all the estimations made in a nation are traceable to national standards kapt at the National Metrology Institute (NMI) and National Physical laboratory (NCL) of the country. It is the obligation of the NMI to ensure traceability of national principles to the global levels.

WHO and expert group members have identified nearly 100 contaminants and have specified their safe limits for safe drinking water. In India the purity of water is the responsibility of the Bureau of Indian Standards (BIS). The standard (IS) 10500 defines the permissible level of contaminants in the water for human consumption.

To ensure that the determination of contaminants in labs is according to worldwide practices Certified Reference Materials (CRMs), which are utilized to adjust the equipment and to approve the experimental techniques. Certified Reference Materials are arranged with indicated amount of a contaminant in water, similar to a dangerous component for example lead, mercury or arsenic. These are commonly prepared and affirmed by NMIs like NPL in India in collaboration with laboratories which have expertise in this field. In the Indian CRM programme standards of toxic elements and other contaminants like fluoride in water were taken up in the first phase. Effectively more than twenty CRMs have been arranged through the joint endavours of around 30 top labs of the nation. To ensure that the CRMs are accrding to worldwide standard, this group gets involved from time to time in international inter-comparison programme.

It might be specified that International associations like International Council for Science (ICSU), Global Network of Academies (IAP) and others have been effectively included in this field.

The G8 Summit of industrially advanced nations looks for inputs from their science institutions to support projects of valuable societal significance. Science Academies of G8 +5 countries including India after conferences, recommend areas which require urgent action at the highest political level. A meeting of the academies occurred at Paris in March 2011. President N. Sarkozy of France met the Academy Presidents and delegates and mentioned that their suggestions will be considered with all seriousness during the Summit. The Academies have prescribed support in the following two regions:

Water and Health; Education for a Science Based Global Development

The Joint Statements properly marked by Presidents of all the participating Academies have been released simultaneously on 19 May 2011 in their respective countries. INSA had additionally released it and besides media, sent copies to Hon'ble Ministers of External Affairs and Science & Technology. Copies were also sent to Secretaries of Departments of Science & Technology, Biotechnology, Health Research and Scientific and Industrial Research (DG CSIR).

Conclusion

While anthropogenic activities are the significant source of heavy metal pollution, natural sources contribute all together to the weight of arsenic and fluoride. Beside from commercial enterprises, road runoff is also an important source. The harmful components enter the body cheifly through water, food and air. Beautifying agents, dental items, few drugs, especially Ayurvaida and Unani drugs also contribute. More research is expected to assess the extent to which these products influence human health. Public awareness ought to make. There should be checking and control over the concentration of overwhelming metals in beautifying products.

The presence of metals in nano structures should be determined. Toxicity of metals bearing nano particles is an area, where extensive research should be carried out to build up or nullify toxic factors. Susceptibility to danger is influenced by age, physiological status, nutrition status and genetic factors. Further, research is expected to study these interactions. The areas where particular chemical/metallic interactions have known e.g., lead and calcium, fluoride and calcium, populations subjected to these poisonous substances (factory workers, communities living near the factories) should receive periodic health check-up and nutritional support.

Health monitoring of labourers engaged in industries for handling toxic metals/minerals should be carried out frequently and nutritional support where important should be given. Functional consequences which may not be excessively self evident, similar to consequences on reproductive, neurological – cognitive and other functions have to be identified, through more extensive research on animals and human being under controlled conditions.

India frequently utilizes guidelines of well being grown in the western nation. Considering the genetic diversity and rampant malnutrition, would these apply to India? What is the safe level? Is tolerance developed over prolonged exposure adaptation or compromise? Since fixing standards is a very costly and laborious process, regulatory agencies like the MoEF and PCB, tends to follow the standards developed by developing countries, with modifications to suit the local conditions like body weight, nutrition status, etc. The best approach should be followed to modify the standards periodically and upgrade treatment procedure to meet the criteria. Another methodlogy is to create new approach to develop Environment specimen banks (ESB) which can be drawn upon occasionally for testing, by using updated approaches.

There ought to be harmonization of heavy metal measures which are typically hazard based and adopted in developed nations. Dichotomy in standards may not fit in this globalized era. Wherever possible, and techno-economically feasible, source remediation should be polished and encouraged. The cleaner innovations, alternatives to avoid the use of harmful heavy metals and minerals ought to be investigated, recorded and imparted to Indian commercial ventures with an aim adopt cleaner technologies.

Administrative safety standards for emission and discharges from process plants should be strictly implemented. Recycling/reprocessing of wastes containing toxic metals should be given more emphasis not only from environmental and health considerations but also as asset for preservation measure. Observation and study of air, water and soil in the region of the toxic metal processing units should be carried out more strictly for the particular target metal. Local certify research laboratories should be set up to help administrative bodies for examining pollutants in various environmental compartments. Rules for proper management of tailings and slags containing lethal metals should be arranged for looking into techno-economic possibility.

Tailings dumps and process wastes lying in near by areas of the processing units need to be remediated on immediate basis. Phytorestoration improves ecological capital and gives biodiversity of choice suitable to the area where such restoration measures are undertaken. In India, little business in unorganized sector means a lot of pollution. While CETP are needed to help small industries, bigger ones ought to set up their own treatment plants and release the treated effluent in sewers. Subsidies can be provided to industries which invest in clean technologies. There should be continuous research to develop cost- effective technologies for diminishment and replacement. As it seems to be lot of pollution in India is due to obsolete production innovations.

For recovery of heavy metals like mercury from medical devises and CFL bulbs, suitable collection centers should be set up, and some discount should be given. As it may be, no such mechanism exists and with increasing use of CFL bulbs, unplanned and unorganized disposal can be risky.

Attempt should be made to replace CFL bulbs with LED (Light emitting diode) bulbs. Mercury-based medical devices and equipment should be completely eliminated or withdrawn, since digital options are available.

In a matter of records there is utilization of private vehicles-two wheelers, cars, autos, etc. This should change with emphasis on cleaner public transport frameworks to decrease the burden of road run off. CNG should replace petrol and diesel. Use of diesel should be bound to public transport and transport of goods. Manufacture of diesel autos should be stopped. Vested interests are taking the benefit of the subsidy on diesel.

Periodic examination of water quality, especially for detection of fluoride and arsenic is vital in newer alluvium and flood plain areas in distinctive parts of India. Water supplied by urban regions and rural panchayats, should be free of biotic and abiotic toxicants including toxic metals and minerals. Modest devises for purification of water at household level must be created.

Creation of public awareness is very essential. Greater interaction between scientists, technologists and media can help to achieve the target of public health awaerness. Schools and educational institutions can play crucial role in creating awareness. Since India is in developmental transition, the wasteful, consumerist way to life which eats up resources and adds to pollution burden should be discouraged.

We have to remember an old Native American saying that "We didn't acquire Earth from our Ancestors, yet acquired it from our Children". Their future has to be considered, before plundering the earth and polluting it.

REFERENCES

[1] Khan, S., Cao, Q., Zheng, Y. M., Huang, Y. Z. & Zhu, Y. G. (2008). Health risks of heavy metals in contaminated soils and food crops irrigated with wastewater in Beijing, China. *Environmental Pollution*, *152*, 686–692.

[2] Zhang, M. K., Liu, Z. Y. & Wang, H. (2010). Use of single extraction methods to predict bioavailability of heavy metals in polluted soils to rice. *Communications in Soil Science and Plant Analysis*, *41*(7) 820–831.

[3] GWRTAC, "Remediation of metals-contaminated soils and groundwater," Tech. Rep. TE-97-01, GWRTAC, Pittsburgh, Pa, USA, 1997, GWRTAC-E Series.

[4] Kirpichtchikova, T. A., Manceau, A., Spadini, L., Panfili, F., Marcus, M. A. & Jacquet, T. (2006). Speciation and solubility of heavy metals in contaminated soil using X-ray microfluorescence, EXAFS spectroscopy, chemical extraction, and *thermodynamic modeling," Geochimica et Cosmochimica Acta*, *70* (9) 2163–2190.

[5] Adriano, D. C. (2003). Trace Elements in Terrestrial Environments: Biogeochemistry, Bioavailability and Risks of Metals, Springer, New York, NY, USA, 2nd edition.

[6] Maslin, P. & Maier, R. M. (2000). Rhamnolipid-enhanced mineralization of phenanthrene in organic-metal co-contaminated soils," *Bioremediation Journal*, *4* (4) 295–308.

[7] McLaughlin, M. J., Zarcinas, B. A., Stevens, D. P. & Cook, N. (2000). Soil testing for heavy metals," *Communications in Soil Science and Plant Analysis*, *31*, (11–14) 1661–1700.

[8] McLaughlin, M. J., Hamon, R. E., McLaren, R. G., Speir, T. W. & Rogers, S. L. (2000). Review: a bioavailability-based rationale for controlling metal and metalloid contamination of agricultural land in Australia and New Zealand. *Australian Journal of Soil Research*, *38*, (6) 1037–1086.

[9] Ling, W., Shen, Q., Gao, Y., Gu, X. & Yang, Z. (2007). Use of bentonite to control the release of copper from contaminated soils," *Australian Journal of Soil Research*, *45* (8), 618–623.

[10] Pendias, A. K. & Pendias, H. (2001). Trace Metals in Soils and Plants, CRC Press, Boca Raton, Fla, USA, 2nd edition.

[11] Zhao, Q. & Kaluarachchi, J. J. (2002). Risk assessment at hazardous waste-contaminated sites with variability of population characteristics," *Environment International*, *28*(1-2) 41–53.

[12] Bolan, N. S., Ko, B. G., Anderson, C. W. N. & Vogeler, I. (2008). Solute interactions in soils in relation to bioavailability and remediation of the environment," in Proceedings of the 5th International Symposium of Interactions of Soil Minerals with Organic Components and Microorganisms, Pucon, Chile.

[13] Krishnaswamy, K. & Kumar, B. D. (1998). Lead toxicity. *Indian Journal of Paediatrics.*, *35*, 209-216.

[14] Zereini, F., Alt, F., Messerschmidt, J. (2005). et al. *Environ. Sci. Technol.*, *39*, 2983–2989.

[15] Voutsaand, D., Samara, C. (2002). Labile and bio accessible fractions of heavy metals in the airborne particulate matter from urban and industrial areas. *Atmospheric Environment*, 3583-3590.

[16] UNICEF, The State of Art Report on Fluoride in drinking water and resulting endemicity in India. 1999.

[17] Susheela, A. K. (2007). Treatise on Fluorosis, 3rd Revised Edition.

[18] Chakraborti, D., Das, B. & Rahman, M. M. (2009). Status of groundwater arsenic contamination in the state of West Bengal, India: a 20-year study report. *Mol. Nutr. Food Res.*, *53*, 542–551.

[19] Department of Health and Human Services, Public Health Service. Agency for Toxic Substances and Disease registry. Toxicology Profile for Lead. Available at: http://www.atsdr.cdc.gov/toxfaq.html. Accessed August 30, 2008.

[20] Flora, S. J. S., Flora, G. & Saxena, G. (2006). Environmental occurrence, health effects and management of lead poisoning. in Lead chemistry, analytical aspects, environmental impacts and health effects. Elsevier Publication, Netherlands, Cascas SB, Sordo J (eds)., 158–228.

[21] National Institute of Nutrition, Hyderabad, Role of Nutrients in Environmental Toxicity. Annual report 2005–2006, 62–64.

[22] Reddy, Y. S., Pullakhandam, R., RadhaKrishna, K. V. et al. Lead and essential trace element levels in school children: A cross-sectional study. *Annals of Human Biology*, *38*, 372-377, 2011.

[23] Phillips, S. & Balge, M. Heavy Metal Toxicity. UTHSCSA Environmental Medicine Education Program and STEER Program (2007). http://www.aoec.org/resources.

[24] wikipedia.org/wiki/Minamata_disease - Cached – Similar.

[25] De Chaudhuri, S., Ghosh, P., Sarma, N. et al. (2008). Genetic Variants Associated with Arsenic Susceptibility: Study of Purine Nucleoside Phosphorylase, Arsenic (+3) Methyl Transferase and Glutathione S-Transferase Omega Genes. *Environmental Health Perspective*, *116*, 501-505.

[26] Banerjee, M., Sarma, N., Biswas, R. et al. 2008. Deficiency in DNA repair leads to arsenic susceptibility: Evidence from Comet assay and challenge assay. *Int J cancer*, *123*(2), 283-287.

[27] Das, A. A. in text book of Human Nutrition, 3rd edition, Bamji, M.S., Krishnaswamy, K., Brahmam, G. N. V. editors, Oxford IBH, New Delhi, 2010.

[28] Prasad, M. N. V., Freitas, H., Fraenzle, S. et al. 2010. Knowledge explosion in phytotechnologies for environmental solutions. *Environmental Pollution*, *158*, 18–23.

[29] Prasad, M. N. V. (2011). A State-of-the-Art report on Bioremediation, its applications to contaminated sites in India. Ministry of Environment & Forests, Government of India. New Delhi.

[30] Gupta, P. K. (2010). Elements of Biotechnology, 3rd Edition. Rastogi Publications, Meerut.

[31] Handbook on Planning and Implementation of Water Supply Programme in Fluorosis Endemic Areas in India: A Multi Author Contribution. Edited by AK Susheela, 2009.

In: Heavy Metals
Editor: Deepak Pathania
ISBN: 978-1-63484-740-7

Chapter 8

DUBININ-RADUSHKEVICH (DR) ISOTHERM STUDIES OF EQUILIBRIUM SORPTION OF METAL IONS ONTO PHOSPHORIC ACID MODIFIED RICE-HUSK

***Mu. Naushad*[1,*]*, Gaurav Sharma*[2,†]*, Ala'a H. Al-Muhtaseb*[3] *and Rubia Z. Gaur*[4]**

[1]Advanced Materials Research Chair, Department of Chemistry, College of Science, King Saud University, Riyadh, Kingdom of Saudi Arabia
[2]School of Chemistry, Shoolini University, Himachal Pradesh, India
[3]Department of Petroleum and Chemical Engineering, Faculty of Engineering, Sultan Qaboos University, Muscat, Oman
[4]School of Environment and Natural Resources, Doon University, Dehradun, India

ABSTRACT

Adsorption is known as an effective and commercial method for the removal of toxic metal ions and organic pollutants from wastewater. This chapter is written to discuss about the source of heavy metal ions pollution and their removal using low-cost adsorbents, mainly rice husk. Rice husk is one of the low-cost agricultural by-products which have been used as a sorbent material especially to absorb heavy metals. Several activating agents have been reported for chemical activation process; however the most important and commonly used activating agent is phosphoric acid. The chapter provides an insight into various adsorption models laying stress on utility of Dubinin-Radushkevich (DR) isotherm studies of equilibrium sorption of metal ions onto phosphoric acid modified rice-husk.

Keywords: rice husk, heavy metal ions, Dubinin-radushkevich (DR) isotherm, adsorbents

---

[*] Corresponding author address: E-mail: shad81@rediffmail.com (Mu.Naushad); Ph. No.: +96614674198; +919418807170.
[†] gaurav8777@gmail.com (G. Sharma).

1. INTRODUCTION

When the well is dry, we know the worth of water."
Benjamin Franklin, Poor Richard's Almanac, 1746"

The importance of water is realized, as Benjamin Franklin rightly pointed out, when there is scarcity of water. With the constant rise of world population, the demand for clean water is also increasing simultaneously. To be precise, water usage has increased six-fold, as the world population has increased three-fold during the 20th century. On the other hand, rapid industrial growth without adequate planning of water treatment is polluting water day-by-day. In addition, waste water from industries and sewage carries many heavy metals, pharmaceuticals and cosmetics leading to a great threat to water security [1, 2]. Heavy metal ion pollution occurs in many industrial wastewaters such as those produced by electroplating, nuclear power plants, battery manufacturing process, paints and pigments production and municipal and storm water run-off. The main heavy metals which cause metal ion pollution are Cd, Th, Pb, Cr, As, Hg, Cu and Ni [3-8]. Unlike most organic pollutants, heavy metals are generally refractory and cannot be degraded or readily detoxified biologically [9, 10]. These heavy metals consequently enter in the surrounding soils as well as the ground and surface waters. Excessive ingestion of these metals by humans can cause accumulative poisoning, cancer, nervous system damage and ultimately death. Public concern over heavy metal pollution has demanded treatment of such effluents before disposal to the environment. Several techniques have been developed for the qualitative and quantitative determination of such metal ions. These techniques include chemical precipitation, physical treatment such as ion exchange, solvent extraction, reverse osmosis and adsorption [11-16]. Although, reverse osmosis is very effective but it is a cost-prohibitive process as the membranes get easily spoiled and require frequent replacement. Chemical precipitation however is not very suitable when the pollutants are present in trace amounts and also a large amount of sludge is produced [17]. Solvent extraction or electrolytic processes are also available but they are considered to be cost-effective only for more concentrated solutions [12]. But, adsorption is considered to be relatively superior to other techniques because of low cost, simplicity of design and ability to treat wastes in more concentrated form [18]. The adsorption process offers flexibility in design and operation and in many cases. In addition, because adsorption is sometimes reversible, adsorbents can be regenerated by suitable desorption process.

2. ADSORPTION

Adsorption from a solution is a process in which a soluble chemical substance (the adsorbate) is removed from a liquid in contact with a solid surface (the adsorbent). Depending on the nature of forces involved, there are two main types of adsorption, i.e., physical and chemical. Physical adsorption is very similar to a condensation process, and thus, it is exothermic with a heat of adsorption ranges from 8.00-20.00 kJ/mol. The forces involved in physical adsorption are Vander Waals forces between surface atoms and the adsorbate molecules similar to the forces exist between all adjacent molecules. As physical adsorption involves weak forces, therefore it is a reversible process. Physical adsorptions occur at low temperatures. There are several factors that affect physical adsorption. The major factors, which affect physical adsorption, include the surface area of the adsorbent, pore structure of the adsorbent, surface chemistry of the adsorbent, nature of the adsorbate, pH of the solution, and the presence of competing adsorbate. Due to these factors, the physical adsorption is considered to be complex phenomenon. Surface area of the adsorbent is one of the most important factors on which adsorption greatly depend. Chemical adsorption or chemisorptions involves a reaction i.e., transfer and sharing of electrons between an adsorbed species and an adsorbent resulting in a change in the chemical form of the adsorbate. The resulting chemisorbed bond is usually stronger than that derived from the physical Vander Waal forces, and has all the characteristics of the chemical bond. This type of adsorption is accompanied by high heat of adsorption ranges from 40-800 kJ/mol. The chemisorbed bond always localize at the active centers of the adsorbent.

In recent years, considerable attention has been devoted to the study of removal of heavy metal ions from solution by adsorption using agricultural materials. Natural materials that are available in large quantities or certain waste from agricultural operations may have potential to be used as low cost adsorbents, as they represent unused resources, widely available and are environmentally friendly [19-21]. The plenty and accessibility of agricultural by-products such as rice husk make them good sources of cheap raw material for natural adsorbents. Many low cost materials such as barley straw [22], waste tea leaves [23], orange peel [24], hazel nut shell [25], sago waste [26], saw dust [27], waste banana [28], orange peels [28], cocoa shells [29], coffee residue [30], rice husk [31], olive stone waste [32] and grape stalk [33] etc. have been successfully used as adsorbents for the removal of toxic metal ions.

2.1. Rice Husk as Potential Candidature for Heavy Metal Removal

Rice husk (contains about 20% silica) has been reported as a good adsorbent for the removal of many metal ions from aqueous medium. The worldwide annual rice husk output is about 80 million tones and over 97% of the husk is generated in the developing countries. Rice husks are the hard protecting coverings of grains of rice. The husk is made of hard materials, including opaline silica and lignin in order to protect the seed during the growing season. The husk is mostly indigestible to humans. During the milling process, the husks are removed from the grain to create brown rice; the brown rice is then milled further to remove the bran layer to become white rice. Generally, farmers and rice processor often burn the rice husk as wastes, and this releases carbon dioxide (CO_2) into the atmosphere. CO_2 is a well-known greenhouse gas. The increase volume of CO_2 gas in the atmosphere is believed to cause global warming commonly known as greenhouse effect. The ash produced after the rice husks have been burned, is high in silica. A number of possible uses under investigation include:

- Absorbents for oils and chemicals
- As a source of silicon
- Economical substitute for microsilica/silica fumes
- Soil ameliorants
- Aggregates and fillers for concrete and board production.
- As insulation powder in steel mills

- As a release agent in the ceramics industry
- As an insulation material for homes and refrigerants
- As repellents in the form of "vinegar-tar"
- Used as electrode materials for batteries and capacitors

Rice husk is a very cost-effective raw material to be used and it is very easy to obtain. The carbonaceous materials derived from rice husk has fine nanostructure and functional patterns different from the carbonaceous materials obtained by other sources. The literature studies revels that the porous carbon derived from rice husks, in addition to having wide availability, has fast kinetics and appreciable adsorption capacities. The pretreatment of rice husk has been done by number of physical and chemical process as shown in figure below:

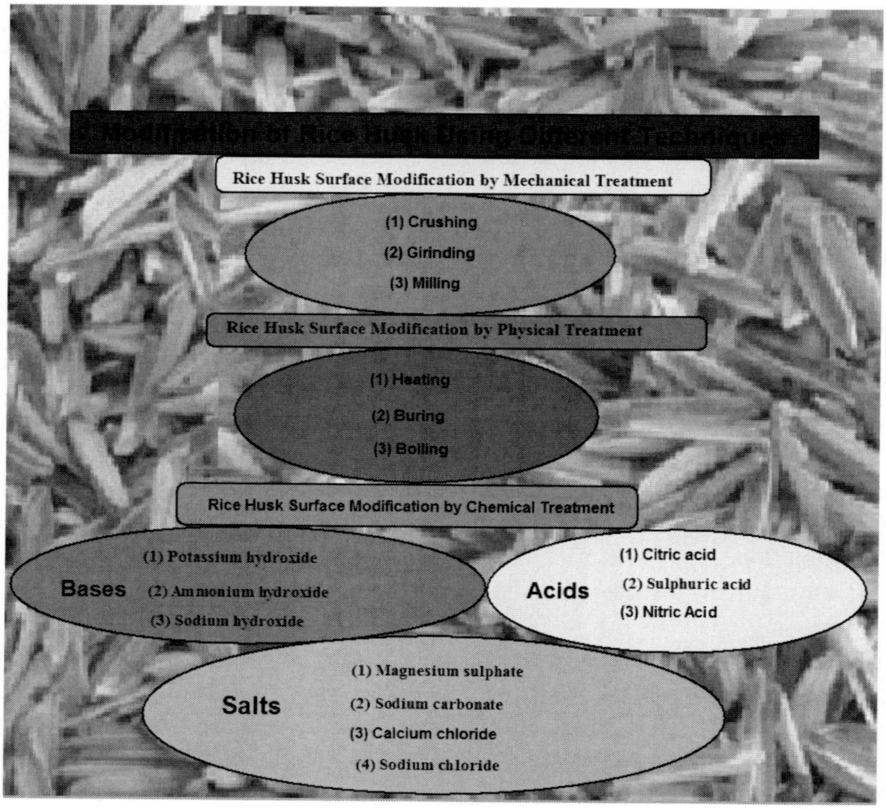

Figure 1. The pretreatment of rice husk has been done by number of physical and chemical process.

Activated carbons are high porosity, high surface area materials manufactured by carbonization and activation of carbonaceous materials, which find extensive use in the industrial sector for adsorption of pollutants from gaseous and liquid streams. Production of activated carbons involves two steps; the carbonization of raw carbonaceous materials in an inert atmosphere and the activation of carbonized carbon. The carbonization consists of a thermal decomposition of the carbonaceous material, eliminating non-carbon species and producing a fixed carbon mass with a rudimentary pore structure. Activation can be carried out by physical (hydrothermal) or chemical means. Activated carbon can be manufactured from carbonaceous material. In physical activation process, generally the precursors are activated with high temperature steam (H_2O) or CO_2 in range of 800-1100°C [34]. Physical activation comprises carbonization of the raw material in an inert atmosphere followed by partial gasification of the resulting char with steam, CO_2 or their mixture [19]. In chemical activation process, the two steps are carried out simultaneously, with the precursor being mixed with chemical activating agents and oxidants such as H_3PO_4 [35-37], K_2HPO_4 [38], Na_2HPO_4, H_2SO_4 [39], K_2CO_3 [34], $ZnCl_2$ [40, 41], $AlCl_3$ [42], KOH [43, 44] and NaOH [45]. As an activation agent, H_3PO_4 is widely used in the preparation of activated carbons from lignocellulosic products and offers some advantages such as non-polluting character and ease of elimination by extraction with water [36]. Moreover, H_3PO_4 inflicted physical and chemical modifications on the botanical structure by penetration, particle swelling, and partial dissolution of biomass, bond cleavage and reformation of new polymeric structures resistant to thermal decomposition [37]. Many studies in the literature have been reported concerning the production of activated carbon from biomass by H_3PO_4 activation [35-37, 46]. Phosphoric acid activation has been applied on a wide variety of cellulosic precursors such as cotton stalks [47], coconut shell [48–50], apple pulp [51, 52], peach stones [53, 54], Arundo donax cane [55], shells of nuts like almond, pecan, English walnut, black walnut and macadamia nut [56-59], acorns and olive seeds [60], peanut hull [61] and sugar cane bagasse [62, 63]. A number of activation procedures have been reported in the literature using phosphoric acid as activating agent. In general, it can be classified either as single-stage or two-stage activation process carried out either in an inert medium or a self-generated atmosphere with efforts made towards developing a high surface area carbon with desired pore size by optimizing the process parameters such as the activation time, activation temperature and impregnation ratio.

3. Adsorption Models

Isotherm adsorption models have been used in waste stream treatment to predict the ability of a certain adsorbent to remove a pollutant down to a specific discharge value. When a mass of adsorbent and a waste stream are in contact for a sufficiently long time, equilibrium between the amount of pollutant adsorbed and the amount remaining in solution will develop. For any system under equilibrium conditions, the amount of material adsorbed onto the media can be calculated using the mass balance Equation:

$$\frac{X}{M} = \frac{(Co - Ce)V}{m} \quad (1)$$

where X/M is the mass of pollutant per mass of media, C_o is the initial pollutant concentration in solution, Ce is the concentration of the pollutant in solution after equilibrium has been reached, V is the volume of the solution to which the media mass is exposed, andMis the mass of the media.

Over the years, a wide variety of equilibrium isotherm models (Langmuir, Freundlich, Brunauer-Emmett-Teller, Redlich-Peterson, Dubinin-Radushkevich, Temkin, Toth, Koble-Corrigan, Sips, Khan, Hill, Flory-Huggins and Radke-Prausnitz isotherm), has been formulated in terms of three fundamental approaches [64, 31, 32]. Kinetic consideration is the first approach to be referred. Hereby, adsorption equilibrium is defined being a state of dynamic equilibrium, where both adsorption and desorption rates are equal. Whereas thermodynamics, being a base of the second approach, can provide a framework for deriving numerous forms of adsorption isotherm models, potential theory, as the third approach, usually conveys the main idea in the generation of characteristic curves. However, an interesting trend in the isotherm modeling is the derivation in more than one approach, thus directing to the difference in the physical interpretation of the model parameters [20, 33, 34]. In general, an adsorption isotherm is an invaluable curve describing the phenomenon governing the retention (or release) or mobility of a substance from the aqueous porous media or aquatic environments to a solid-phase at a constant temperature and pH [35, 36]. Adsorption equilibrium is established when an adsorbate containing phase has been contacted with the adsorbent for sufficient time; beyond this the adsorbate concentration in the bulk solution is in a dynamic balance with the interface concentration. Typically, the mathematical correlation, which constitutes an important role towards the modeling analysis, operational design and applicable practice of the adsorption systems, is usually depicted by graphically expressing the solid-phase against its residual concentration.

Its physicochemical parameters together with the underlying thermodynamic assumptions provide an insight into the adsorption mechanism, surface properties as well as the degree of affinity of the adsorbents.

One among the most widely used isotherms for the description of adsorption equilibrium is the Freundlich isotherm model.

Freundlich isotherm gives the relationship between equilibrium liquid and solid phase capacity based on the multilayer adsorption (heterogeneous surface). This isotherm is derived from the assumption that the adsorption sites are distributed exponentially with respect to the heat of adsorption and was furnished initially by Freundlich [37]. This isotherm is an empirical equation but it is capable of describing the adsorption of organic and inorganic compounds on a wide variety of adsorbents. This equation has the following non linear form:

$$q_e = k_f.C_e^{1/n} \tag{2}$$

Eq. (2) can also be expressed in the linearized logarithmic form:

$$\ln q_e = \ln(k_f) + 1/n(C_e) \tag{3}$$

where k_f (mg g^{-1}) (L mg-1)$^{1/n}$ and n are Freundlich constant and affinity parameter, respectively.

The plot of log(q_e) versus log(C_e) has a slope with the value of 1/n and an intercept magnitude of ln(k_f). ln(k_f) is equivalent to ln(q_e) when C_e equals unity. However, in other case when $1/n \neq 1$, the k_f value depends on the units upon which q_e and C_e are expressed. On average, a favorable adsorption tends to have Freundlich constant n between 1 and 10. Larger value of n (smaller value of 1/n) implies stronger interaction between adsorbent and adsorbate while 1/n equal to 1 indicates linear adsorption leading to identical adsorption energies for all sites. When $n < 0$ implies the solvent has more affinity to surface adsorbent than adsorbate.

Freundlich isotherm has the ability to fit nearly all experimental adsorption–desorption data, and is especially excellent for fitting data from highly heterogeneous sorbent systems.

It is important to mention that Freundlich equation has limitation because it is unable to predict adsorption equilibria data at high concentration. Furthermore, this equation is not reduced to linear adsorption expression at very low concentration and it does not have limit expression at very high concentration. However, researchers rarely face this last problem, because moderate concentration is frequently used in most adsorption studies.

Langmuir adsorption isotherm assumes monolayer coverage of adsorbate over a homogenous adsorbent surface where sorption is assumed to take place at specific homogeneous sites within the adsorbent. Langmuir equation relates the coverage of molecules on a solid surface to concentration of a medium above the solid surface at a fixed temperature. This isotherm is based on three assumptions, namely (i) adsorption is limited to monolayer coverage;(ii) all surface sites are alike and can only accommodate one adsorbed atom and (iii) the ability of a molecule to be adsorbed on a given site is independent of its neighboring sites occupancy. By applying these assumptions, and a kinetic principle (rate of adsorption and desorption from the surface is equal), the Langmuir equation can be written in the following form:

$$q_e = q_{\max} \frac{k_L C_e}{1 + k_L C_e} \tag{4}$$

This equation is often written as:

$$\frac{C_e}{q_e} = \frac{1}{q_e} C_e + \frac{1}{k_L q_{\max}} \tag{5}$$

where, q_e (mg.g^{-1}) and C_e (mg.L^{-1}) are the equilibrium concentrations of solute in solid and liquid phases, respectively. Also, K_L is the Langmuir constant (Lmg^{-1}) and q_{max} is the amount of adsorption corresponding to maximum adsorption capacity (mg.g^{-1}). The value of q_{max} is recognized as the adsorption capacity, which is commonly a measure of adsorption ability of an adsorbent.

If C_{ref} is a reference concentration that is assigned as the highest C_e during an adsorption process, then in this case, the corresponding q_e is set to be q_{ref}. from Eq. (5):

$$\frac{C_{ref}}{q_{ref}} = \frac{1}{q_e} C_{ref} + \frac{1}{k_L q_{\max}} \tag{6}$$

Dividing Eq. (6) by Eq. (5):

$$\frac{q_e}{q_{ref}} = \frac{C_e}{C_{ref}}\left[\frac{1+k_L C_{ref}}{1+k_L C_{ref}(C_e/C_{ref})}\right] \tag{7}$$

Eq. (7) is called the dimensionless form of Langmuir equation, in which $k_L C_{ref}$ is the unique parameter when (q_e/q_{ref}) is plotted against (C_e/C_{ref}).

To determine, if the adsorption process is favorable or unfavorable for the Langmuir type adsorption process, the isotherm can be classified by a term 'R_L,' a dimensionless constant, the separation factor, which is defined by the equation 18 [44]:

$$R_L = \frac{1}{1+k_L C_o} \tag{8}$$

where C_0 is the highest, initial solute concentration in the liquid phase. The R_L value implies that adsorption is either unfavorable ($R_L > 1$), linear ($R_L = 1$), favorable ($0 < R_L < 1$), or irreversible ($R_L = 0$). Based on similar concept, another dimensionless separation factor R_L (also named as the approaching equilibrium factor) is defined as follows:

$$R_L = \frac{1}{1+k_L C_{ref}} \tag{9}$$

Combination of Eqs. (7) and (9) yields:

$$\frac{q_e}{q_{ref}} = \left(\frac{C_e}{C_{ref}}\right)\left[\frac{1}{R_L+(1-R_L)(C_e/C_{ref})}\right] \tag{10}$$

Eq. (10) is the dimensionless form of Langmuir equation in terms of R_L. The characteristic equilibrium curves based on Eq. (20) are shown in Figure7. In the region of favorable adsorption ($0 < R_L < 1$), the equilibrium curve is subdivided into 2 types, slightly curved ($0.4355 < R_L < 0.3164$), or largely curved ($0.2783 < R_L < 0.2045$). The pseudo-rectangular curve was not achieved in the R_L studied.

Temkin and Pyzhev considered the effects of some indirect sorbate/adsorbate interactions on adsorption isotherms and suggested that because of these interactions the heat of adsorption of all the molecules in the layer would decrease linearly with coverage. The Temkin isotherm has been generally applied in the following form:

$$q_e = \left(\frac{RT}{b}\right)\ln(AC_e) \tag{11}$$

and can be linearized as:

$$q_e = B\ln(A) + B\ln(C_e) \tag{12}$$

Where B = RT/b, b is the Temkin constant related to heat of sorption ($J.mol^{-1}$); A is the Temkin isotherm constant ($L.mg^{-1}$), R is gas constant (8.314 $J.mol^{-1}.K^{-1}$) and T is the absolute temperature (K).

Langmuir and Freundlich isotherms do not give any idea about sorption mechanism but the former describes sorption on a single type of sorption sites. In this respect the D-R isotherm is an analogue of Langmuir type but it is more general because it does not assume a homogeneous surface or constant sorption potential.

The equilibrium data can be also applied to the D-R model to determine the type of sorption (physical or chemical). The Dubinin–Radushkevich model is often used to estimate the characteristic porosity and the apparent free energy of adsorption [65]. The linear form of D-R isotherm is presented as the following equation:

$$\ln(q_e) = \ln(q_m) - \beta\varepsilon^2 \tag{13}$$

where q_e is the amount adsorbed per unit dosage, q_m is the theoretical monolayer sorption capacity ($mol.g^{-1}$), β is the constant of the sorption energy (mol^2J^{-2}), which is related to the average energy of sorption per mole of the sorbate as it is transferred to the surface of the solid from infinite distance in the solution and, ε is the Polanyi potential, which is described by equation (24) where T is the solution temperature (K) and R is the gas constant. while ε is Polanyi sorption potential which is equal to:

$$\varepsilon = RT\ln(1+1/Ce) \tag{14}$$

R is a gas constant in kJ mol^{-1} K^{-1}, T is the temperature in Kelvin, and Ce is the equilibrium concentration of sorbate in solution (mg/l) The Polanyi sorption theory [66] postulates fixed volume of sorption sites close to the sorbent surface and existence of sorption potential over these sites. Polanyi sorption potential, ε is the work required to remove a molecule to infinity from its location in the sorption space, independent of temperature. This model assumes the heterogeneity of sorption energies within this space.

The value of mean sorption energy, E (kJ/mol), can be calculated from D-R parameter β as follows:

$$E = \frac{1}{\sqrt{-2\beta}} \tag{15}$$

Table 1. D-R adsorption parameters of different pollutants onto various adsorbent Materials

D-R Adsorption parameters atrazine onto Sheanut shells, activated with H_3PO_4						
Qm(mg $^{g-1}$)		**E(kJ mol^{-1})**		**r^2**		
4.37x10^{-3}		0.8111		0.979		[67]
D-R Adsorption parameters of Cu(II) ions on heat regenerated spent bleaching earth.						
Xm (mol/g)		**β(mol2/kJ2) E**	**(kJ/mol)**	**r^2**		
0.302		-0.0285	4.19	0.9855		[67]
D-R Adsorption parameters of Arsenic onto laterite soil						
As (III)	**Temperature**	**Qm(mg g^{-1})**	**E(kJ mol^{-1})**	**r^2**	**R_L**	
	298	0.220	3.430	0.985	0.843	
	308	0.201	3.172	0.989	0.700	
	318	0.081	3.475	0.961	0.574	
As (V)	298	0.043	8.058	0.995	0.039	[68]
	308	0.037	6.375	0.977	0.092	
	318	0.032	6.455	0.988	0.048	
D-R Adsorption parameters of ammonium onto clinoptilolite						
Temperature		**Qm(mg $^{g-1}$)**	**E(kJ mol^{-1})**	**r^2**		
10°C		19.108	7.024	0.988		
25°C		19.330	7.268	0.989		[69]
40°C		19.406	7.379	0.993		
D-R Adsorption parameters of chromium tri-oxide (VI) at different temperatures on kaolinite surface						
Temperature	**pH**	**X_m (mg.gm^{-1})**	**K_m (mole^2 kJ^{-2})**	**E (kJ mole^{-1})**	**r^2**	
290	4	1.992	1.66065	0.5487	0.9896	
300	4	2.001	1.99557	0.5005	0.9996	
						[70]
310	4	1.967	2.2144	0.4751	0.9983	
290	7	1.928	1.97416	0.50328	0.9931	
290	10	1.866	3.25347	0.39202	0.9994	

D-R Adsorption parameters for copper(II) onto green algae and carbon green algae				
Copper(II) onto GA(dosage g/L)	**QD**	**E**	**r^2**	
2	8.396	0.903	0.7207	
5	3.850	1.162	0.7709	[71]
10	2.231	1.294	0.8181	
Copper(II) onto CGA(dosage g/L)	**QD**	**E**	**r^2**	
2	13.01	2.071	0.8117	
5	5.646	2.261	0.8577	
10	2.635	3.963	0.754	

D-R Adsorption parameters for copper(II) onto NSU CLAY				
qD	**BD (mol2/KJ2)**	**E**	**r^2**	[72]
0.4	0.002	15.8	0.7131	

Table 2. D-R adsorption parameters of different pollutants onto rice husk

D-R Adsorption parameters of Cu onto rice husk				
Temperature(°C)	**Qm(mg $^{g-1}$)**	**E(kJ mol^{-1})**	**r^2**	[73]
25	36.47	14.7	0.917	
30	38.6	13.3	0.903	
40	36.77	14.3	0.936	
45	40.22	14.6	0.903	
D-R Adsorption parameters of different metal ions onto rice husk				
Different Metal Ions	**Qm(m mol $^{g-1}$)**			[74]
Pb	0.15 ± 0.3			
Cd	0.14 ± 0.4			
Zn	0.10 ± 0.2			
Cu	0.10 ± 0.1			
D-R Adsorption parameters of Zn onto rice husk				
qs(mg/g)	**Kad (mol2/kJ2)**	**E(KJ/mol)**	**r^2**	[75]
0.598	1×10^{-6}	0.707	0.792	
D-R Adsorption parameters of Cu onto activated carbon of rice husk				
***Q* mg/g**	**Kad (mol2/kJ2)**	**r^2**		[76]
2016.462	4×10^{-7}	0.8882		

The value of mean sorption energy gives information about chemical and physical sorption. The E value ranges from 1 $kJ.mol^{-1}$ to 8 $kJ.mol^{-1}$ for physical sorption and from 8 kJ mol^{-1} to 16 $kJ.mol^{-1}$ for chemical sorption.

D-R isotherms has been applied to adsorption of different heavy metal ions and organic pollutants. The table 1 shows the adsorption parameters for D-R isotherms.

About 95% silica content of rice husk ash make as good alternative source for the production of zeolite. The adsorption of Lead, Cadmium, Copper and Zinc on rice husks based adsorbents i.e., carbonized rice husk and activated rice husk has been investigated and it was found that activated rice husk has higher potential of metal removal compared to carbonized rice husk. The thermo chemically treated Rice-husk (Partially pyrolysed Rice-husk) can be used as low-cost adsorbent in the removal of phenol.

The table 2 decapitates the adsorption parameters of D-R isotherms of different metal ions onto rice husk.

Pyrolysed rice-husk sorption capacity for phenol can be brought near to that of Commercial Grade Carbon properties by activating it thermally and chemically. It was found that, techniques used by researchers were diverged based on types of rice husk used either, raw and unmodified rice husk or modified rice husk.

Amorphous nanosilica prepared from rice-husk by hydrothermal technique has been effectiviely used as adsorbent for the methylene blue dye. About 65% of methylene blue dye was adsorbed within the first few minute onto nanosilica prepared from rice-husk. The thermodynamic parameters reveled that the adsorption of methylene blue is spontaneous and endothermic process [77]. The zeolite synthesized from agriculture waste by-product in Thailand such as; rice husk ash (RHA) has been used for removal cadmium metal ions from aqueous solution. The result indicated that synthesized zeolites from RHA have high removal capacity close to commercial one of 753.58 mg/g. Thus, the synthesized zeolites from RHA have high efficency to adsorbeb cadmium metal ions from water [78]. White rice husk ash (WRHA) prepared by thermal degradation of rice husks in air, has been used to remove Mn(II) ions from aqueous solutions. The results revealed that the adsorption of Mn(II) ions by WRHA is a chemisorption in nature, where OH^-groups of SiO_2 plays an active role in adsorption by releasing H^+ ions in the solution [79]. The chemically modified Rice husk with hexadecylpyridinium bromide (HDPBr) has been successfully used for adsorption of Cr(VI) [80]. Esterification reactions with hydroxylethylidenediphosphonic acid and nitrilotrimethylene-triphosphonic acid were carried out for chemical modification of rice husk, the results indicated that the functionalized rice husks acted as potential adsorbent of for Au(III) ions form water system. Not only fly ash or nanosilica synthesized form rice husk but also the bio-char fabricated from liquefaction of rice husk with water, could be served as probable low-cost adsorbent for removal of Malachite green dye from aqueous medium [81].

Thus the modification of rice husk has been carried to produce superior materials including elemental Silicon, SiO_2, SiC, Si_3N_4, Mg_2Si and active carbon, etc. [82]. The production of these sophisticated materials depends on the physical and chemical process involved as pyrolysis, and reduction etc. Hence potential of rice husk and its derivatives for heavy metal remediation can be much explored in coming times to develop rice husk based potential multifunctional remedial adsorbents for environmental remediation of organic and inorganic pollutants.

CONCLUSION

The potential of agricultural bio-waste has immense importance for environmental remediation particularly for removal of metal ions and organic pollutant from waste water. The agricultural bio-wastes are economic, easy to use, huge availability and biodegradable in nature. The rice husk is also agricultural bio-waste which has shown to be an efficient adsorbent in its natural and dertivized form. It also has been revealed that, the treatment (i.e., activation) of raw rice husk significantly affect the surface area development, pore structure evolution, and changes in surface functional groups. Apart from that, some new treatment methods such as by physicochemical as well as biological systems are also valuable and interesting to investigate in near future for rice husk applications in wastewater treatment.

ACKNOWLEDGMENTS

The authors extend their sincere appreciation to the Deanship of Scientific Research, College of Science Research Center, King Saud University for the support. One of the authors (G. Sharma) would like to acknowledge the School of Chemistry, Shoolini University, for providing necessary facilities.

REFERENCES

[1] Bogardi, J. J., Dudgeon, D., Lawford, R., Flinkerbusch, E., Meyn, A., Pahl-Wostl, C., Vielhauer, K., Vorosmarty, C., 2012. *Curr. Opin. Environ. Susta inability*, *4,* 35-43.

[2] Kannan T., Verma H., 2012. Applications of polymer based ion exchangers for heavy metal and radioactive waste removal, water purification, protein separation and as a catalyst. *Nova Sci. Pub.,* USA 93-110.

[3] Alshehri S.M., Naushad Mu., Ahamad T., Alothman Z.A., Aldalbahi A. 2014. Synthesis, characterization of curcumin based ecofriendly antimicrobial bio-adsorbent for the removal of phenol from aqueous medium, *Chem. Eng. J.,* 254,181-189.

[4] Naushad Mu., ALOthman Z.A. 2015. Separation of toxic Pb^{2+} metal from aqueous solution using strongly acidic cation-exchange resin: analytical applications for the removal of metal ions from pharmaceutical formulation, *Des. & Water Treat.* 53, 2158-2166.

[5] Shahat A., Awual M.R., Naushad M. 2015. Functional ligand anchored nanomaterial based facial adsorbent for cobalt(II) detection and removal from water samples, *Chem. Eng. J.* 271,155-163.

[6] Naushad M., ALOthman Z.A., Sharma G., Inamuddin 2015. Kinetics, isotherm and thermodynamic investigations for the adsorption of Co(II) ion onto crystal violet modified amberlite IR-120 resin, *Ionics* 21,1453-1459.

[7] Kumar A., Sharma G., Naushad Mu., Thakur S. 2015. SPION/β-cyclodextrin core–shell nanostructures for oil spill remediation and organic pollutant removal from waste water, *Chem. Eng. J.* 280,175-187.

[8] Shahat A., Awual M.R., Khaleque M.A., Alam M.Z., Naushad Mu., Chowdhury A.M.S. 2015. Large-pore diameter nano-adsorbent and its application for rapid lead(II) detection and removal from aqueous media, *Chem. Eng. J.,* 273, 286-295.

[9] Naushad, Mu., ALOthman, Z.A., Khan, M.R., 2013. Removal of bromate from water using De-Acidite FF-IP resin and determination by ultra-performance liquid chromatography-tandem mass spectrometry. *Clean Soil Air and water,* 41 528-533.

[10] ALOthman, Z.A., Naushad, Mu., Inamuddin, 2011. Organic–inorganic type composite cation exchanger poly-o-toluidine Zr(IV) tungstate: Preparation, physicochemical characterization and its analytical application in separation of heavy metals. *Chem. Eng. J.,* 172 369-375.

[11] Naushad, Mu., Al-Othman, Z.A., Islam, M., 2013. Adsorption of Cd^{2+} ion using a new composite cation-exchanger polyaniline Sn(IV) silicate: Adsorption kinetics and thermodynamic studies. *Inter. J. of Env. Scienc. & Tech.* 10, 567-578.

[12] Wan Ngah W.S., Hanafiah, M.A.K.M., 2008. Removal of heavy metal ions from wastewater by chemically modified plant wastes as adsorbents: a review. *Bioresour. Technol,.* 99, 3935-3948.

[13] Bhattacharyya, K.G., Gupta, S.S., 2008. Adsorption of a few heavy metals on natural and modified kaolinite and montmorillonite: a review. *Adv. Colloid Interface Sci.,* 140, 114-131.

[14] Mahitti, P., Fuangfa, U. 2008. Preparation and use of chemically modified MCM-41 and silica gel as selective adsorbents for Hg(II) ions. *J. Hazard. Mater.,* 154, 578-587.

[15] Awual M.R., Hasan M.M., Naushad Mu., Shiwaku H., Yaita T., Peparation of new class composite adsorbent for enhanced palladium(II) detection and recovery, *Sensors and Actuators B: Chem.,* 209 (2015) 790-797.

[16] Awual M.R., Eldesoky G.E., Yaita T., Naushad Mu., Shiwaku H., AlOthman Z.A., Suzuki S. 2015. Schiff based ligand containing nano-composite adsorbent for optical copper(II) ions removal from aqueous solutions, *Chem. Eng. J.,* 279, 639-647.

[17] Jiang, M.Q., Jin, X.Y., Lu, X.Q., Chen, Z.L. 2010. Adsorption of Pb(II), Cd(II), Ni(II) and Cu(II) onto natural kaolinite clay. *Desalination.,* 252, 33-39.

[18] Naushad, Mu., ALOthman, Z.A., Khan, M.R. 2013, Removal of malathion from water using De-Acidite FF-IP resin and determination by UPLC-MS/MS: Equilibrium, kinetics and thermodynamic studies. Talanta, 115, 15-23.

[19] Deans, J.R., Dixon, B.G. 1992. Uptake of Pb^{2+} and Cu^{2+} by novel biopolymers. *Water Res.,* 26, 469-472.

[20] Jha, P. 2010. Potential of rice-husk as an adsorbent for phenol removal from aqueous solutions, *J. Enviro. Res. Develop.,*5, 421-425.

[21] Nhapi, I., Banadda, N., Murenzi, R., Sekomo C.B., Wali, U.G. 2011. Removal of Heavy Metals from Industrial Wastewater Using Rice Husks, *The Open Enviro. Eng. J.,* 4, 170-180.

[22] Pehlivan, E., Altun, T., Parlayıcı, S. 2009. Utilization of barley straws as biosorbents for $Cu^{2+}$and Pb^{2+}ions. *J Hazard Mater.* 164, 982-986.

[23] Mondal, M.K. 2009. Removal of Pb (II) ions from aqueous solution using activated tea waste: Adsorption on a fixed-bed column. *J. Env. Manag.* 90, 3266-3271.

[24] Xuan, Z., Tang, Y., Li, X., Liu, Y., Luo, F. 2006. Study on the Equilibrium, Kinetics and Isotherm of Biosorption of Lead Ions onto Pretreated Chemically Modified Orange Peel. *Bio. Eng. J.* 31, 160–164.

[25] Cimino, G., Passerini, A., Toscano, G. Removal of toxic cations and Cr(VI) from aqueous solution by hazelnut shell. *Water Res.,* 34, 2955-2962.

[26] Quek, S.Y., Wase, D.A.J., Forster, C.F. 1998. The use of sago waste for the sorption of lead and copper. *Water Res.,* 24, 251–256.

[27] Sukla, A., Zhang, Y.H., Dubey, P., Margrave, J.L., Sukla, S.S., 2002. The role of sawdust in the removal of unwanted materials from water. *J. Hazard. Mater.*, 95,137-152.

[28] Annadurai, G., Juang, R.S., Lee, D.J., 2003. Adsorption of heavy metals from water using banana and orange peels. *Water Sci. Technol.* 47, 185-190.

[29] Meunier, N. 2003. Lead Removal from Acidic Solutions by Sorption on Cocoa Shells: Effect of Some Parameters. *J. Environ. Eng. ASCE.,* 693-698.

[30] Boudrahem, F., Aissani-Benissad, F., Aït-Amar, H. 2009. Batch sorption dynamics and equilibrium for the removal of lead ions from aqueous phase using activated carbon developed from coffee residue activated with zinc chloride. *J. Env. Manag.,*90, 3031-3039.

[31] Feng, Q., Lin, Q., Gong, F., Sugita, S., Shoya, M. 2004. Adsorption of lead and mercury by rice husk ash. *J. Coll. Inter. Sci.,* 278, 1-8.

[32] Fiol, N., Villaesscusa, I., Martinez, M., Miralles, N., Poch, J. and Serarols, J. 2006. Sorption of Pb(II), Ni(II), Cu(II) and Cd(II) from aqueous solution by olive stone waste, Sep. Purif. Technol., 50, 132-140.

[33] Martinez, M., Miralles, N., Hidalgo, S., Fiol, N., Villaeseasa, I., Poch, J. 2006. Removal of lead(II) and cadmium(II) from aqueous solutions using grape stalk waste. *J. Hazard. Mater.* 133, 203-211.

[34] El-Sheikh A.H., Newman A.P., 2004. Characterization of activated carbon prepared from a single cultivar of Jordanian olive stones by chemical and physiochemical techniques, *J. Anal. Appl. Pyrol.,* 71, 151-164.

[35] Corcho-Corral, B., Olivares-Marin, M., Fernandez-Gonzalez, C., Gomez-Serrano, V., Macias-Garcia, A. 2006. Preparation and textural characterisation of activated carbon from vine shoots (Vitis vinifera) by H_3PO_4-Chemical activation. *Appl. Surface Sci., 252*, 5961-5966.

[36] Sun, Y., Zhang, J.P., Yang, G., Li, Z.H. 2007. Production of activated carbon by H_3PO_4 activation treatment of corncob and its performance in removing nitrobenzene from water. *Environ. Progress* 26, 78-85.

[37] Alhamed, Y.A. 2008. Phenol removal using granular activated carbon from dates stones by H_3PO_4 activation. *J. Environ. Protect. Ecol.* 9, 417-430.

[38] Aber, S., Khataee, A., Sheydaei, M. 2009. Optimization of activated carbon fiber preparation from Kenaf using K_2HPO_4 as chemical activator for adsorption of phenolic compounds. *Bioresour. Technol.,* 100, 6586-6591.

[39] Gercel, O., Ozcan, A., Ozcan, A.S., Gercel, H.F. 2007. Preparation of activated carbon from a renewable bio-plant of Euphorbia rigida by H_2SO_4 activation and its adsorption behavior in aqueous solutions. *Appl. Surface Sci.,* 253, 4843-4852.

[40] Wang, T.H., Tan, S.X., Liang, C.H. 2009. Preparation and characterization of activated carbon from wood via microwave-induced ZnCl2 activation. *Carbon.,* 47, 1880-1883

[41] Alhamed, Y.A., Bamufleh, H.S. 2009. Sulfur removal from model diesel fuel using granular activated carbon from dates' stones activated by $ZnCl_2$. *Fuel.,* 88, 87-94.

[42] Ioannidou, O., Zabaniotou, A., 2007. Agricultural residues as precursors for activated carbon production - A review. *Renewable & Sustainable Energy Reviews.*, 11, 1966-2005.

[43] Ramos-Fernandez, J.M., Martinez-Escandell, M., Rodriguez-Reinoso, F., 2008. Production of binderless activated carbon monoliths by KOH activation of carbon mesophase materials. *Carbon,* 46, 384-386.

[44] El-Hendawy, A. 2009. An insight into the KOH activation mechanism through the production of microporous activated carbon for the removal of Pb^{2+} cations. *Appl. Surface Sci.,* 255, 3723-3730.

[45] Kobayashi, J., Imamura, T., Ichikawa, M., Kubota, M., Watanabe, F., Kobayashi, N., Hasatani, M., 2006. Production of activated carbon for water vapor adsorption by KOH and NaOH activation. *Kagaku Kogaku Ronbunshu.*, 32, 186-189.

[46] Nakagawa, Y., Molina-Sabio, M., Rodriguez-Reinoso, F., 2007. Modification of the porous structure along the preparation of activated carbon monoliths with H_3PO_4 and $ZnCl_2$. *Microporous and Mesoporous Materials.*, 103, 29-34.

[47] Laine, J., Calafa,t A., Labady, M. 1989. Preparation and characterization of activated carbons from coconut shell impregnated with phosphoric acid. *Carbon,* 27, 191–5.

[48] Laine, J., Yunes, S. 1992. Effect of the preparation method on pore size distribution of activated carbon from coconut shell. *Carbon,* 30, 601–604.

[49] Kirubakaran, J.C., Krishnaiah, K., Sheshadri, S.K. 1991. Experimental study on the production of activated carbon from coconut shells in Kuidized bed reactor. *Indus. and Eng. Chem. Rese.,* 30,2411-2416.

[50] Molina-Sabio, M., Rodriguez-Reinoso, F., Caturla, F., Selles, M.J. 1995. Porosity in granular carbons activated with phosphoric acid. *Carbon,* 33, 1105–1113.

[51] Molina-Sabio, M., Rodriguez-Reinoso, F., Caturla, F., Selles, M.J. 1996. Porosity in combined phosphoric acid–carbon dioxide activation. *Carbon*, 34, 457-462.

[52] Girgis, B.S., Ishak, M.K. 1999. Activated carbon from cotton stalk by impregnation with phosphoric acid. *Material Letters,* 39,107-114.

[53] Ruiz Bevia, F., Prats Rico, D., Marcilla Gomis, A.F. 1984. Activated carbon from almond shells. Chemical activation 1. Activation reagent selection and variables in Kuence. *Indus. and Eng. Chem. Prod. Res. Devel.,* 23, 266-269.

[54] Toles, C.A., Marshall, W.E., Johns, M.M. 1998. Phosphoric acid activation of nutshells from metal and organic remediation: process optimization. *J. Chem. Tech. Bio.*, 72, 255-263.

[55] Toles, C.A., Marshall, W.E., Johns, M.M. 1999. Surface functional groups on acid-activated nutshells. *Carbon,* 37, 1207-1214.

[56] Dastgheib, S.A., Rockstraw, D.A. 2001. Pecan shell activated carbon: synthesis, characterization and application for the removal of copper from aqueous solution. *Carbon,* 39,1849-1855.

[57] La, W.K. 2001. Production of activated carbon from acorns and olive seeds. *Biomass and Bioenergy,* 20,57-62.

[58] Vernersson, T., Bonelli, P.R., Carrella, E.G., Cukierman, A.L. 2002. Arundo donax cane as a precursor for activated carbon preparation by phosphoric acid activation. *Bioresour. Technol.,* 83,95-104.

[59] Garcia, F.S., Alonso, A.M., Tascon, J.M.D. 2001. Porous texture of activated carbons prepared by phosphoric acid activation of apple pulp. *Carbon,* 39:1103-1116.

[60] Garcia, F.S., Alonso, A.M., Tascon, J.M.D. 2002. Pyrolysis of apple pulp: chemical activation with phosphoric acid. *J. Anal. Appl. Pyroly.,* 63,283-301.

[61] Girgis, B.S., Yunis, S.S., Soliman, A.M. 2002. Characterization of activated carbon from peanut hulls in relation to condition of preparation. *Materials Letters,* 57,164-172.

[62] Ahmedna, M., Marshall, W.E., Rao, R.M. 2000. Surface properties of granular activated carbons from agricultural by-products and their effects on raw sugar decolorization. *Bioresour. Technol.,* 71,103-112.

[63] Ahmedna, M., Marshall, W.E., Rao, R.M. 2000. Production of granular activated carbons from select agricultural by-products and evaluation of their physical, chemical and adsorption properties. *Bioresour. Technol.,*71, 113-123.

[64] Gupta, V.K., Pathania, D., Kothiyal, NC., Sharma, G., (2014) Polyaniline zirconium (IV) silicophosphate nanocomposite for remediation of methylene blue dye from waste water. *J. Mol. Liq.,* 190, 139-145.

[65] Nguyen, C., Do, D.D. 2001. The Dubinin–Radushkevich equation and the underlying microscopic adsorption description. *Carbon.,* 39, 1327–1336.

[66] Dą̧browski, 2001. Adsorption from theory to practice. *Advances in Colloid and Interface Science.* 93, 135-224.

[67] Itodo, A.U., Itodo, H.U. 2010. Sorption Energies Estimation Using Dubinin-Radushkevich and Temkin Adsorption Isotherms, *Life Science Journal,* 7, 31-39.

[68] Maji, S.K., Pal, A., Pal, T., Adak, A. 2007. Adsorption Thermodynamics of Arsenic on laterite soil. *J. Surface Sci. Technol.,* 22, 161-173.

[69] Tosun, İ. 2012. Ammonium Removal from Aqueous Solutions by Clinoptilolite: Determination of Isotherm and Thermodynamic Parameters and Comparison of Kinetics by the Double Exponential Model and Conventional Kinetic Models *Int. J. Environ. Res. Public Health, 9*, 970-984.

[70] Al-Kaim, A.F. 2008. Kinetic Study for Adsorption of Chromium tri-Oxide CrO_3 on Kaolinite Surface. *National J. Chem.,* 31, 415-427.

[71] Abdelwahab, O. 2007. Kinetic And Isotherm Studies of Copper (II) Removal From Wastewater Using Various Adsorbents. *Egyptian J. Aquatic Research*, 33, 125-143.

[72] Dawodu, F. A, Akpomie, G. K, Abuh, M. A. 2012. Equilibrium isotherm studies on the batch sorption of copper (ii) ions from aqueous solution unto "nsu clay." *Internat. J. Scient. & Eng. Rese*., 3, 1-7.

[73] Zhang J., Fu, H., Lv, X., Tang, J., Xu, X. 2011. Removal of Cu(II) from aqueous solution using the rice husk carbons prepared by the physical activation process, B*iomass and Bioenergy.,* 35, 464-472.

[74] Akhtar, M., Iqbal S., Kausar A., Bhanger M.I., Shaheen M.A. 2010. An economically viable method for the removal of selected divalent metal ions from aqueous solutions using activated rice husk. *Colloid. & Surf. B: Biointerfaces* 75, 149-155.

[75] Dada A.O., Olalekan A.P., Olatunya A. M., Dada O. 2012. Langmuir, Freundlich, Temkin and Dubinin–Radushkevich Isotherms Studies of Equilibrium Sorption of Zn^{2+} Unto Phosphoric Acid Modified Rice Husk. *IOSR Journal of Applied Chemistry*, 2278-5736, 3, 38-45.

[76] Thajeel, A.S., 2013. Isotherm, Kinetic and Thermodynamic of Adsorption of Heavy Metal Ions onto Local Activated Carbon, *Aquatic Sci. & Tech.,* 1, 1-2.

[77] Tolba, G.M.K., Barakata, N. A.M., Bastaweesy, A.M., Ashour, E.A., Abdelmoez, W., El-Newehy, M.H., Al-Deyab, S.S., Kim H. Y. 2015. Effective and highly recyclable nanosilica produced from the rice husk for effective removal of organic dyes. *J. Indus. Eng. Chem.,* 29 134-145.

[78] Santasnachok, c., Kurniawan, W., Hinode, H. 2015. The use of synthesized zeolites from power plant rice husk ash obtained from Thailand as adsorbent for cadmium contamination removal from zinc mining. *J. Enviro. Chem. Eng.,* 3, 2115–2126.

[79] Tavliev, M. P., Genieva, S. D., Georgieva, V. G., Vlaeva, L.T. 2015. Thermodynamics and kinetics of the removal of manganese(II) ions from aqueous solutions by white rice husk ash. *J. Mol. Liq.,* 211, 938-947.

[80] Ali, I. O., Salama, T.M., Lateef, M.A., Gumaa, H. A., Hegazy, M. A., Bakr M. F. 2015. Chemical modification of rice husk by quaternized hexadecylpyridinium for removal of chromate oxyanions from aqueous solution. *Enviro. Technol. & Innovat.,*4, 110-122.

[81] Leng, L., Yuan, X., Zeng, G., Shao, J., Chen, X., Wu, Z., Wang, H., Peng X. 2015. Surface characterization of rice husk bio-char produced by liquefaction and application for cationic dye (Malachite green) adsorption. *Fuel,* 155, 77-85.

[82] Soltani, N., Bahrami, A., Pech-Canul, M.I., González L.A. 2015. Review on the physicochemical treatments of rice husk for production of advanced materials. *Chem. Eng. J.,* 264, 899-935.

In: Heavy Metals
Editor: Deepak Pathania

ISBN: 978-1-63484-740-7

Chapter 9

URANIUM IN WATER: SOURCES, TOXICITIES AND REMEDIATION

Prabodha Kumar Meher[1,3], Prerna Sharma[1], Akash Khare[2] and Kaushala Prasad Mishra[1,*]

[1]Division of Life Sciences, Research Centre,
Nehru Gram Bharati University, Allahabad, India
[2]Department of Chemistry, Research Centre,
Nehru Gram Bharati University, Allahabad, India
[3]MATS School of Biological and Chemical Sciences,
MATS University, Raipur,
Chhattisgarh, India

ABSTRACT

Uranium is a naturally occurring radioactive heavy metal generally found in trace quantities in various matrices such as soil, rock and water resources. Recent reports on higher level of uranium in water resources at various locations have created interest on determining quantities of uranium in water from different locations and possible implications to health effects. Recent results from our laboratory on determination of uranium in drinking and river water are described. Variations are reported at different locations in Districts of Eastern Uttar Pradesh but they are within permissible limits. Level of uranium in human body beyond permissible limit poses various radiological and chemical hazards. Extensive studies on effects of uranium on microbes, plants and animal models have yielded invaluable results. It was found that chemical effects of uranium were more predominant than its radiological toxicities. The toxicity mechanism of uranium continues enigmatic but involves a wide range of physicochemical and cytological factors. This chapter describes various natural and man-made sources, gives results of biological toxicity studies and outlines current methods for remediation. An indication for future direction research has been outlined.

Keywords: uranium, drinking water, biological toxicities, health hazard, remediation

[*] E-mail: mishra_kaushala@rediffmail.com, pkm046@gmail.com.

INTRODUCTION

Uranium is a radioactive heavy metal of actinide series having atomic number 92 and atomic mass of 238.03 g/mol. It is found in nature in five oxidation states of 2, 3, 4, 5, and 6 [1]. Most prevalent form constitutes hexavalent state in combination with oxygen as $UO2^{2+}$ (Uranyl ion). Uranium occurs in nature in soils, rocks and water. It comprises of three main radionuclide isotopes such as U-234 (0.0054%), U-235(0.72%) and U-238 (99.275%) [2, 3].

These isotopes undergo decay process giving several unstable daughter elements by emitting alpha particle and gamma radiation [4-7]. The final stable element of this process is lead (Pb) (Figure 1). The isotopes of uranium have similar chemical properties but differ in radioactive properties. The least radioactive isotope is U-238 with a half-life of 4.5 billion years. U-234 and U-235 have half life of 2.45 X 10^5 and 7.04 X 10^8 years respectively. Both radiological and chemical hazards pose risk to health when ingested into the body.

Uranium may enter human body through multiple routes depending on the location of source. Regulatory bodies have fixed limits of acceptable level where negligible or no toxic effect occurs.

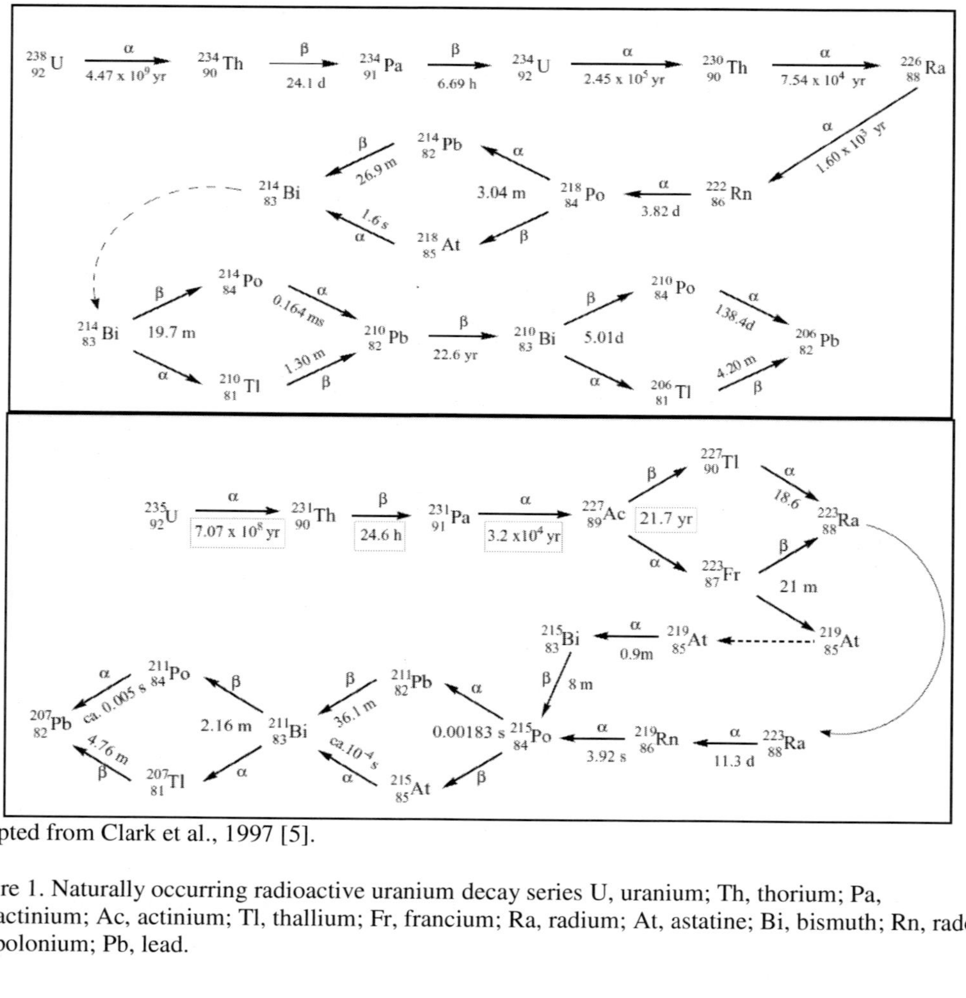

Adapted from Clark et al., 1997 [5].

Figure 1. Naturally occurring radioactive uranium decay series U, uranium; Th, thorium; Pa, protactinium; Ac, actinium; Tl, thallium; Fr, francium; Ra, radium; At, astatine; Bi, bismuth; Rn, radon; Po, polonium; Pb, lead.

However, the toxicological effects become cause of concern when the level of uranium exceeds the permissible limit. This Chapter describes different sources, toxicities from uranium, possible health risks and developing methods for remediation.

SOURCES: NATURAL AND MAN MADE

Uranium is everywhere into the environment and is one of the naturally occurring radioactive and trace heavy metal. It is mostly found in earth crust, soil, plants, rocks and granites. However, uranium concentration mainly depends on the geological features of the region and widely varies from place to place. Uranium found in these sources has a great potential to affect directly and indirectly the living system of the environment. Uranium makes up approximately 2–4 mg/kg of the earth's crust [8]. Although in nature there are more than 100 different uranium ores, uranium typically occurs as the mixed oxide U_3O_8, in amorphous (pitchblende) or crystalline forms (uraninite). Possible sources of uranium include natural distribution and anthropogenic release [9]. From the natural sources, uranium can be deposited on land, mix into soil, and leach into surface water. Mixed uranium in soil and water can be transported to the large distances and even into the underground water stream. It is found in drinking water in well, bore well, hand pipe water. Sometimes higher level of uranium is found in wells drilled in uranium-rich rocks like granites and some mineral deposits [7, 10, 11].

The natural concentrations of uranium found in freshwater systems are highly variable and occurs normally within a range from 0.01μg/l to over 2 mg/l, depending on the geological conditions [21]. Uranium also settles down and adheres to sediments and other particles in water. Uranium may also be added in environment due to release in mining, milling and refining plants, from nuclear industry, spent uranium from war zones, combustion of coal and other fuels and also from the use of phosphate fertilizers in agricultural lands [12]. It exists in air as dust in form of very small particles. The fallout of these particles also contributes to spread of uranium in water, plants, and soil.

DETECTION METHODS

Humans do not have any sense organs to detect any type of radionuclides and associated radiation exposure. As a consequence, it relies primarily on various techniques and instrumental methods for its detection and quantification. Many types of procedures and technologies are available for detection and measurement of various radionuclides and emanating radiation. These methods depend on the two different properties exhibited by radionuclides i.e., radiological properties and chemical properties. Radiological properties depend on the type of electromagnetic radiation emitted by the radioactive atom. However, chemical properties are due to its atomic architecture and chemical nature. To detect various radionuclides the available instrumental methods are like Geiger-Muller counter and scintillation counter; film badges, pocket dosimetry, thermo luminescent dosimetry, gross alpha counter, gross beta counter, NaI-Tl detector and ionization-chamber etc. In brief the Geiger-Muller counter and scintillation counter measures particles emitted by the radioactive element; film badges, pocket dosimeters and thermo luminescent dosimeters measures accumulated doses due to the radiation; and ionization-chamber type instruments measures dose and dose rate [13]. Thus, depending upon individual radiological property, chemical property and the experimental need, various instruments are used to detect the radionuclides in various samples. For example, generally, for detection of alpha emitter radionuclides Alpha counter is used and for detection of beta emitter, Beta counter is used. Also, a capillary electrophoresis (CE) post column radionuclide detector has been developed using commercial phosphor-imaging detector for detection of low-energy β emitters [14]. An extraction method followed by inductively coupled plasma-mass spectrometry (ICP-MS) has also been developed for the detection of various radionuclides in aqueous samples [15]. It is known fairly well that various radionuclides decay at variable rates and via various nuclear transformations (i.e., emission of α, β, and/or γ radiation). Thus, detection of radionuclides depends not only on decay systematic but also on detector technology. Basing on this fact a comparative study on the traditional radionuclide detection by direct atom measurement was not effective compared to inductively coupled plasma-mass spectrometry (ICP-MS). Thus it was reported that inductively coupled plasma mass spectrometry (ICP/MS) was superior to conventional radiation detection techniques for many radionuclides [16]. These methods and techniques work fine for detection of radionuclides in various samples including water. However, a major challenge arises when we deal with the environmental sample, where the possibility of occurrence of these radionuclides may fall to very trace amount. In these cases we need highly sensitive and accurate methods of detection. Highly sensitive scintillation counters, laser and LED fluorimetric, NaI-Tl Detector and Plastic scintillation based counters are used to detect trace amount of radionuclides in various environmental samples including in water. Among these, the mostly used technique for uranium measurement in water sample is Light Emitting Diode (LED) fluorimeter. The fluorimeter works on the principle of measurement of fluorescence in uranium complexes in the aqueous samples. In brief, on excitation of uranium complexes with UV light, it emits green fluorescence, which can be measured by a sensitive Photo Multiplier Tube (PMT). As the fluorescence yield is proportional to the intensity of excitation source and concentration of uranium in the sample, measurement of fluorescence gives the information about concentration of uranium in the sample. Several studies have been reported in literature where the fluorimetric method has been used to detect and quantify uranium in water samples [17].

ENVIRONMENTAL DISTRIBUTION STUDIES

Uranium is ubiquitous in nature and trace amount of uranium is mostly found in the natural water sources. However, the concentration of uranium varies from place to place and it depends on the geological and geographical features. Once the uranium leaches to the water resources, its solubility, transportation and distribution depends on several factors including pH, temperature, ionic strength of water, oxidation reduction potential (ORP) and total dissolved solids (TDS) of water [18]. It is clear that uranium poses both radiological and chemical properties. Chemically, the average uranium concentration in the ocean water is about 3.0 μg/l. And drinking water presents a major part of uranium exposure which works out to. About 85% of the total uranium ingested by people [19]. According to guideline of

United States Environmental Protection Agency (USEPA) and World Health Organization (WHO) minimum contamination level of uranium in drinking water source should be lower than 30 μg/l [20, 21]. However, the distribution pattern of uranium in various water sources around the world is not uniform. If we look at distribution of uranium in India several studies are available in the literature viz. a study on uranium from subsurface of water of Punjab was found to be in the range of $\leq$ 2 to 644 μg/l [22]; about 5 μg/l of uranium was reported in drinking water from few district of Kerala [23]; another recent study from our research group in ground water of Varanasi, Allahabad, Kaushambi and Fatehpur districts of Uttar Pradesh has shown an a range of *11 ± 0.76 to 63.33±2.28 μg/l* of uranium in the water sample. The study further reported that most of the values observed were found below the permissible limits of WHO and USEPA [20, 21]. Furthermore, study on uranium in two different major Indian river water sources have reported to be about 4 μg/l in Yamuna river and about 2 μg/l of uranium was found in Chambal river [24]. In addition, another recent investigation on uranium content in river water by our research group investigated uranium concentration in a long stretch of Alaknanda and Ganges river and it was found to be 2.75 μg/l and 1.86 μg/l respectively. The study finds a decreasing trend of uranium from the origin at hilly area to plain area. The most probable reason indicated was the dilution factor because of additional water streams joining the river system [18]. Whereas another uranium distribution study from our group at Ganges river within the region of Allahabad revelled a comparative seasonal distribution of uranium in river resources. Within the context of regional climate and factors a significant seasonal variation of uranium in the Ganges river water at Allahabad was found [11]. In most of the cases, it has been found that the concentration of uranium distribution of surface water resources is comparatively lower than uranium in the ground water. Among others the important reason for this variation is known to be due to the geology and type of associated rocks in the specific area. Similarly, a number of studies have been carried out to estimate uranium contained and other radionuclides in water system in different parts of the world. Some of the important investigations that include river systems are Columbia river water [25] Ebro river, Northeast Spain [26] and Van Lake Basin, Turkey [27]. These studies were essential and important to evaluate public health, environmental contaminant and consequent ecological effects [28-31].

Humans may be exposed to uranium from drinking water, ingested food, inhaled air and dermal contacts [13]. Unwashed root crops such as potatoes, turnips, and sweet potatoes etc. are the primary source of uranium to the diet. Estimated normal human intake of uranium is 0.9 to 1.5 μg/day. About 0.1–6% of the uranium a person ingests, 0.76–5% of the uranium a person breathes will get into the bloodstream through the gastrointestinal tract and respiratory tract respectively.

A small amount of water soluble uranium compound can be absorbed through the skin [14]. Naturally occurring radionuclides in drinking water sources are the main causes of radiation exposure. Among the radioactive elements, alpha (α) radiation emitting uranium, radium, and their progeny including radon plays a major role [15]. The World Health Organization (WHO), recommended permissible limit of 0.5 Bq/l for gross α or 1 Bq/l for gross β activities.

These activity level generally contributed by natural alpha emitters uranium-238 (U-238), thorium 232 (Th-232), radium-226 (Ra-226), and polonium 210 (Po-210) and beta emitters Ra and Pb. It is suggestive to monitor and control the concentration of radionuclides within

WHO guidance levels i.e., 0.1 Bq/L for Ra; 1 Bq/lfor Ra; 10 Bq/l for U; 100Bq/l for Rn, and 15µg/l for total uranium [16].

Updates on Toxicity Studies

Uranium is ubiquitous in the environment, but has no known metabolic function inside the cells and hence regarded as non-essential element [32]. In fact, natural uranium causes both chemical and radiological hazard to ecosystems, because of its toxicity as a heavy metal and as an alpha-emitting radioisotopes.

It is known that chemotoxicity of uranium is more harmful than radiotoxicity. Among radioisotopes of uranium series, radium and radon are considered major concern for radiotoxicity [33]. In following sections, the major uranium toxicity modes and processes have been described.

Absorption and Excretion Studies

Subsequent to ingestion, rate of absorption vary from the gastrointestinal tract influenced by the various factors such as nature and solubility of uranium compound [33], previously consumed food, nutritional status [34, 35] and the associated intake of oxidizing agents particularly iron in diet (iron(III) ion and quinhydrone [35]. The estimated average human gastrointestinal absorption of uranium is 1–2% [34]. After absorption, uranium rapidly comes into the bloodstream [35-37] and gets associated with the red blood cells (RBCs) [38]. In the blood plasma, uranium forms non- diffusible complex with the albumin protein and diffusible ionic complex uranyl hydrogen carbonate is formed [39]. Uranyl compound has high affinity to carboxyl, hydroxyl and phosphate groups; therefore it combines with cellular proteins and nucleotides to form stable complexes [39]. A recent study has shown that 66% of total absorbed uranium in blood is quickly excreted via urine and remaining uranium subsequently remains deposited in the kidney (12–25%), skeleton (10–15%), and soft tissue [33]. Another study has estimated that approximately 40% of absorbed uranium is found in the muscles, 20% in the skeleton and 10%, 4%, 1% and 0.3% in the blood, lungs, liver and kidneys, respectively [40] and remaining excreted in the urine and faeces once equilibrium reached in the skeleton (Figure 2). However, the daughter product of uranium i.e., radium deposits mainly in the bone [33]. Ingested radon disseminates into the stomach wall, making stomach wall tissues most irradiated due to its short half life (3.8 days) [41]. Urinary excretion in humans has been found approx. 1% of total excretion [42] and the rate of excretion depends on the pH of tubular urine [32].

Some studies have also reported that, in the bones hydroxyapatite complex gets replaced by uranyl ions [33, 39] while a significant amount of uranium accumulates in the liver [35]. It has been observed that the amount of internally accumulated soluble uranium is proportional to the intake of uranium from ingestion or inhalation. Elimination of radioactive uranium was determined in several studies by calculating its half-life inside the body. In a study with animal model, the half-life of uranium in the rat kidney and skeleton has been estimated and found to be ≈15 days, 300 and 5000 days respectively.

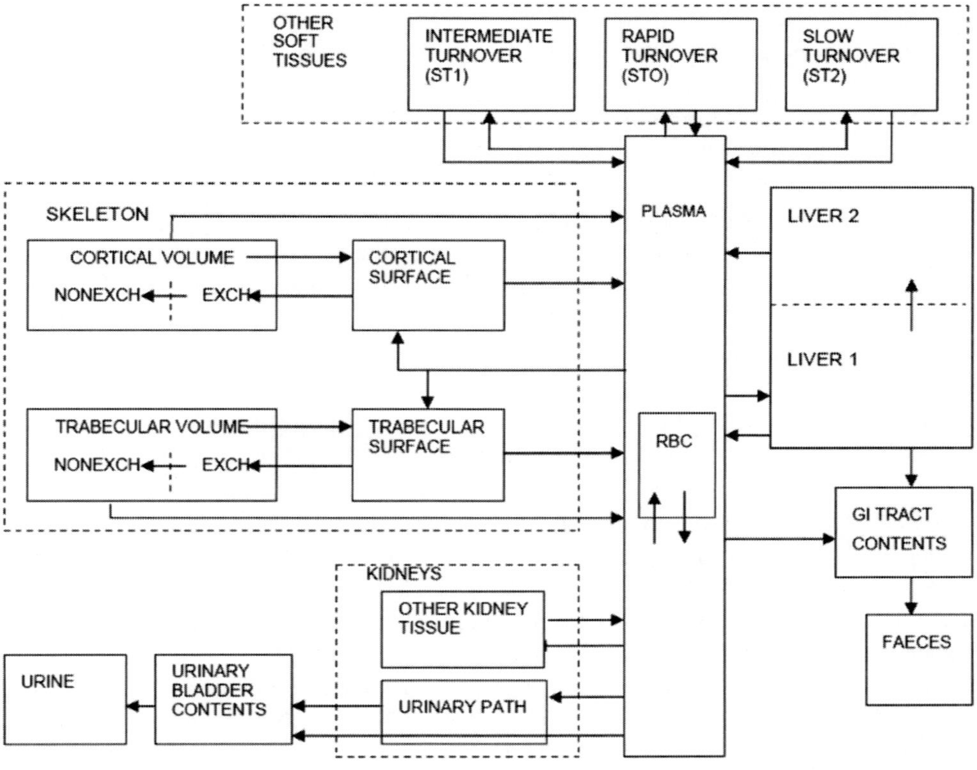

Figure 2. Adapted from the ICRP Publication 69 systemic model for uranium [44] and adapted from Biokinetics model [45, 46].

Another study using a 10-compartment model, half-lives for the clearance of uranium from the rat kidney and skeleton were calculated to be 5–11 and 93–165days, respectively [43]. It was also estimated that overall elimination half-life of uranium under conditions of normal daily intake is between 180 and 360 days in humans [32].

Nephrological Toxicity

Transported uranium in the blood is either deposited in organs of the body or is excreted by the kidney. At the glomerulus, uranium carbonate complexes are filtered. These complexes become more acidic when dissociate as the glomerulate filtrate and then pass along the proximal tubule. Evidently, the uranyl ion released from the complex can react with other components of the filtrate or with components of the tubular membrane. Carbonate complexes may remain dissolved in the filtrate, however, uranyl ions stay bound at the specific ionic sites of the brush border membrane of the proximal tubule. Consequently, soluble forms of uranium can pass into the bladder and are eliminated in urine but the bound forms may adversely alter or destroy renal tubular cells [45]. Uranium binding may affect the sodium ion cellular permeability and therefore increase the transport of glucose, amino acids, and phosphates into the urine [47]. Results based on the *in vitro* animal studies suggested that common renal injury was the damage of proximal convoluted tubules, predominantly in the distal two-thirds [32, 48, 49]. However, damage rate varies with the level of dose [50]. Furthermore, it was also observed that uranyl ion in the blood inhibits both Na+ transport-dependent and Na+ transport-independent ATP utilization with mitochondrial oxidative phosphorylation in the renal proximal tubule [49, 50]. At doses not high enough to destroy a critical mass of kidney cells, some of the cells are replaced with the new epithelial lining which differ morphologically and possibly functionally, from normal epithelium [32, 33]. Some studies provide evidence for tolerance due to frequent uranium exposures [51-53]. This tolerance caused advanced repair process and biochemical changes slowly returned to normal. It was observed that chronic damage to the kidney was the consequence [50]. Later studies found risk of permanent kidney damage on uranium accumulation due to thickening of the glomerular basement membrane [54]. In addition, it was also reported that persistent structural changes in the proximal tubules of rabbits are associated with the kidney's ability to store uranium [55]. A population study with chronic exposure to uranium in drinking water with moderate to high doses concluded that uranium affected kidney function at the proximal tubule [56].

Another study in which population was exposed to well water containing moderate uranium concentration (within permissible limit suggested by WHO: 30 μg/l) were observed with adverse renal effects consistent with modest alterations in proximal tubular function. Uranium in urine was found significantly related to increased fractional excretion of calcium, phosphate and glucose [57]. Other reported biological effects of uranium in kidneys are modifications in renal metabolism of xenobiotics [58], vitamin D homeostasis [59], and iron homeostasis [60]. Accumulation of iron, apoptosis in the tubule interstitial region and oxidative stress were also reported [61-62]. Histopathological studies found lesions in renal tissue generally in the cortical part of the kidneys [63-65].

Reproductive Toxicity

Foetotoxicity has examined due to very high doses exposure for long time. It includes reduced fetal body weight and length both external and internal malformations and increased incidence of developmental variations, reduced ossification and unossified skeletal variations. Embryolethality was not observed at any dose [66]. Uranium exposure also causes teratogenic hazard [67]. Epidemiological case control study of UK Atomic Test Veterans yielded a 9-fold excess of congenital conditions in the children and an 8-fold excess in the grandchildren [68].

Uranium is a development related toxicant. Natural and depleted uranium when given orally or subcutaneously to mice, decreases fertility, develops embryo toxicity, teratogenicity and reduces growth [69, 70]. A study by Raymond-Whish et al. [71] found that uranium in drinking water below the USEPA standard caused estrogen receptor dependent responses in female mice including selective decrease in primary follicles, increased uterine weight, greater uterine luminal epithelial cell height.

Cytotoxic and Genotoxic Effects

Uranium exposure causes decreased cell viability dependence on dose. Decrease in cell cycle kinetics and increase in formation of micronuclei, sister chromatid exchanges and chromosomal aberrations/damages were also reported [72-74]. *In vitro* studies confirmed that uranium exposure modified the gene expression for the cytochromes P450 and glutamate metabolic pathways, cell signaling and trafficking [75-76]. It also caused deregulation of the process of the apoptosis [75]. *In vitro* studies showed that uranium was inducer of the genomic instability and transformed neoplast in osteoblasts [77, 78] consequently altered oxidative metabolism and reduced bone formation, decreased bone volume and delayed healing [75, 79, 80].

Neurotoxicity

Uranium toxicity to the brain or neuro cells has only recently been studied. Inhalation of uranium nanoparticles from the weaponized aerosols provided a direct route to the lower brain through the physiological connections with the nasal passages and olfactory bulb. The French [81] studies were perhaps the first to show the accumulation of uranium in nervous tissue, to which it seems to have an affinity, probably because of the similarity of the uranyl ion to Ca^{++} [82] of the IRSN laboratory in France showed that uranium concentrations in the brains of rats exposed by inhalation were followed an order of olfactory bulb > hippocampus> frontal cortex > cerebellum.

Carcinogenicity

Recent studies have determined the possible association between the uranium in public water and cancer risks [83, 84]. In the study, a slightly increased risk of leukemia in men and of kidney and lung cancer in women was found on exposure of uranium below 20 μg/l [83]. Many epidemiological studies have reported occurrence of cancer due to increased uranium level in drinking water. Twenty eight sites in Iowa towns were studied and increased rates of bladder cancer in males, breast cancer in females, and lung cancer in both sexes were found due to increased radium concentration in water supplies [85]. Another study contradicted this finding and concluded no significant correlation between leukemia cases and radium concentration in water in the study of Iowa towns [86]. Positive significant and non signification association was found between the high Ra^{226} concentration and gross alpha activity respectively in water and bone cancer [87, 88].

Quite the opposite, [89] reported no significant association with radium levels in drinking water supplies. Case control studies showed incidence of digestive cancer and non Hodgkin lymphoma [90, 91]. The cohort study for the incidence of stomach cancer, urinary cancer and leukemia found no significant relation with uranium, radium, radon in well water and also with radiation doses [92-94, 14] did not observed any relation to oral exposure of uranium and human cancer.

However, studies are suggestive for the natural, depleted, or enriched uranium and uranium compounds for the cancer. It appears that uranium can potentially be involved in carcinogenesis, since it emits primarily alpha radiation. Researchers have also shown risk of circulatory system diseases on exposure to soluble uranium. However, these results need confirmation by detailed and deeper investigations [95].

Remediation Methods and Technologies

Natural uranium is ubiquitous in trace amounts. However, due to various natural and anthropogenic activities higher levels of uranium has been found in water resources of different regions in the world. The elevated level of uranium, if ingested may cause a wide variety of disorders to the living biota including human. The need of decontamination or removal of the uranium is required especially in water source having high level of uranium that is being used directly for human consumption and indirectly through agriculture, industries or other uses. Several methods exist for the removal of uranium from water sources but many of them are still under research and development. One of the general procedures of removing uranium from various water sources is coagulation technique. This method involves the coagulation of uranium found in water sources using ferric sulphate or aluminium sulphate. This process of coagulation requires an optimal pH. The procedure promises to achieve 80–95% removal of uranium from the water. However, the methods of lime softening, anion exchange resin or reverse osmosis study have shown about 99% removal of uranium from contaminated water [96]. It is known well that uranium possesses more chemical toxicity than radiological toxicity. Although uranium occurs in the nature in various states the most abundant form of uranium in natural water source is oxidized hexavalent U (VI). Chemically, U (VI) is more hydrophilic and easily soluble in water where as tetravalent uranium U (IV) is less soluble in water. Thus, another procedure to reduce the chemical toxicity of uranium consists in conversion of its hexavalent U (VI) to tetravalent U(IV) form [97]. Thus prevention of uncontrolled dispersion and transport of uranium contaminated water from different water resources to groundwater can be achieved.

Many physiochemical methods are being adopted for removal of uranium from uranium contaminated water. Interestingly, bioremediation of uranium using different microbes offers a good substitute to these physiochemical procedures because of its better efficiency rate and applicability in the *in situ* conditions [98]. Different approaches are being implemented for the bioremediation of uranium. Among these the most general process adopted for bioremediation of uranium is bioreduction. Bioreduction process works by adding an electron donor by stimulation that promotes enzymatic reduction of soluble aqueous U(VI) to sparingly soluble U(IV). Several studies have shown that bioreduction of uranium where the reduction of U (VI) to U (IV) has been promising [99-102]. Apart from bioreduction of U (VI) to U(IV) another available process for remediation of uranium contaminated water is biosorption. Uranium biosorption involves a passive physiochemical process that allows accumulation of uranium radionuclides in the cellular surface of microbial organisms by formation of biopolymers or as dead biomass [103- 105]. Furthermore, microbes possess high surface-volume ratio and they provide a larger surface area for contact with radionuclides. Biosorption in microbes depends on different environmental conditions such as pH, anions, temperature, ionic strength etc. [106, 107]. It has been shown that increase in ionic strength inhibits biosorption capacity [105]. Another potential process of bioremediation of uranium is bioprecipitation. It involves precipitation of uranium involving various types of ligands such as inorganic phosphates. These bioprecipitation reactions involve the cellular machineries like enzymes of the interacting organisms mostly enzymes. Recent studies have shown enzymatic bioprecipitation of different heavy metals as their respective metal phosphates using microbial enzyme phosphatises which is exciting and promising for bioremediation of uranium. A study conducted by over expressing alkaline phosphatase (PhoK) gene in a radioresistant bacterium has shown rapid and efficient precipitation of uranium [98].

CONCLUSION

Many researchers have extensively studied different properties and behavior of uranium. Enormous knowledge has been obtained from various sources, transportation in water, behavior, possible mode of intake and toxicity of uranium. It is known that ingestion of more than permissible limit of uranium has the potential to cause acute to chronic toxicities. The major disorders reported in the literature include kidney dysfunction, neurotoxicity, reproductive disorders, cytotoxicity, genotoxicity and carcinogenesis in living systems. With the advancement of technologies, it has become possible to remove uranium from a wide variety of matrices by using various tools and methods. It is, however, noteworthy that biological methods of uranium remediation offer a great promise and needs to be investigated. It is satisfying to note significant successes in bioremediation of uranium using these microorganisms and technologies. Further studies are highly warranted for successful implementation of technologies and practices to minimize the hazardous health effects of uranium.

Contributions and Conflicts

All authors contributed equally for the book Chapter. There is no conflict of interest.

REFERENCES

[1] Lide D. R., (1994) Uranium, 74th ed., pp. 31–32. Boca Raton, FL: CRC Press.

[2] Greenwood N. N., Earnshaw A. (1984) Chemistry of the elements. Oxford, Pergamon Press.

[3] Weigel F., (1986) Uranium. In: Katz J. J., Seaborg G. T., Morss L. R. (Eds.), The Chemistry of the Actinide Elements, 2nd edition. Chapman and Hall, London, New York (Chapter 5).

[4] Soloveichik G. L. (1993) Actinides: Inorganic and coordination chemistry, vol. 1, pp 2–19. New York: Chapman and Hall, Scientific Data Division.

[5] Clark D. L., Keogh D. W., Neu M. P., and Runde W., (1997) Uranium and uranium compounds, 4th ed., vol. 24, eds. J. I. Kroschwitz and M. Howe-Grant, pp. 639–694. New York: John Wiley and Sons.

[6] Cothern C. R., Lappenbusch W. L. (1983) Occurrence of uranium in drinking water in the U.S. *Health Phys.,* 45:89-99.

[7] Lide D. R., ed. (1992–1993) Handbook of chemistry and physics. Boca Raton, FL, CRC Press.

[8] Agency for Toxic Substances and Disease Registry, (1999), Toxicological Profile for Uranium (Update), GA: Public Health Service.

[9] Craft E. S., Abu-Qare A. W., Flaherty M. M., Garofolo M. C., Rincavage H. L., Abou-Donia M. B., (2004) Depleted and Natural Uranium: Chemistry and Toxicological Effects, *Journal of Toxicology and Environmental Health, Part B,* 7:297–317.

[10] Roessler C. E. et al. (1979) Uranium and radium-226 in Florida phosphate materials. *Health Physics*, 37:267–269.

[11] Khare A., Meher P. K., Sharma P. and Mishra K. P., (2015) Measurement of Uranium in Different Seasons in Ganges River Water in Allahabad Region, *International Journal of Engineering Research Science and Technology,* 4,(1): 179-185.

[12] Sharma P., Meher P. K., Mishra K. P., (2014). Terrestrial gamma radiation dose measurement and health hazard along river Alaknanda and Ganges in India. *Journal of Radiation Research and Applied Sciences,* 7(4): 595–600, doi:10.1016/j.jrras.2014.09.011.

[13] Cember H., Johnson T. E., Introduction to Health Physics, 4th Edition, The McGraw-Hill Companies, New York.

[14] Tracht S. E., Cruz L. A., Stobba-Wiley C. M., Sweedler J. V., (1996), Detection of Radionuclides in Capillary Electrophoresis Using a Phosphor-Imaging Detector, *Anal. Chem.*, 68:3922–3927.

[15] Pepper S. E., Peterman T. R., Tranter T. J.,White B. M., (2009), Detection of radionuclides in aqueous samples using cloud point extraction, *Journal of Radioanalytical and Nuclear Chemistry*, 282:909-912.

[16] Smith M. R., Wyse E. J., Koppenaal D. W., (1992), Radionuclide detection by inductively coupled plasma mass spectrometry: A comparison of atomic and radiation detection methods, *Journal of Radioanalytical and Nuclear Chemistry*, 160: 341-354.

[17] Meher P. K. (2014), Studies on physicochemical and radionuclear properties of different matrices of Ganges river and its effect on microbes: an environmental and biotechnological perspective, Ph.D. Thesis submitted to Nehru Gram Bharati University, Allahabad.

[18] Meher P. K., Sharma P., Kumar A., Gautam Y. P., Mishra K. P., (2015), Post monsoon spatial distribution of uranium in water of Alaknanda and Ganges river, *International Journal of Radiation Research*, 13(1):95-99.

[19] Singh M., Garg V. K., Gautam Y. P., Kumar A. (2013) Spatial mapping of uranium in groundwater and risk assessment around an atomic power station in India. *Environ Engg. and Manag. J.*

[20] USEPA (United States Environmental Protection Agency (2000) office of water, radionuclide's in drinking water.

[21] WHO (World Health Organization) (2011) Guidelines for drinking-water quality - 4th edition, World Health Organization, Geneva.

[22] Ajay K., Usha N., Sawant P. D., Tripathi R. M., Raj S. S., et al., (2011) Risk Assessment for natural Uranium in subsurface water of Punjab state, India. *Human and Ecological Risk Assessment*, 17: 381-393.

[23] Byju S. B., Sunil A., Reeba M. J., Christa E. P., vaidyan V. K., Prasad R., Jojo P. J. (2012), Uranium in drinking water from the south coast districts of Kerala, India. *Iran J. Radiat. Res.*, 10 (1):31- 36.

[24] Rengarajan R., Sarin M. M., Krishnaswami S. (2006) Dissolved uranium and 234U/238U in the Yamuna and the Chambal rivers, India. *Aquatic Geochemistry,* 12:73-101.

[25] Gross M. G., Barnes C., Riel G. K., (1965) Radioactivity of the Columbia river effluent, *Science*, 149(3688):1088-1090.

[26] Pujol L., Sanchez-Cabeza J. A., (2002), Natural radioactivity in surface waters in the right-side tributaries of the Ebro river (Northeast Spain), Advances in Liquid Scintillation Spectrometry Edited by Siegurd Mobius, John Noakes, Franz Schonhofer. 327-334.

[27] Selçuk Z. O., Ceylan H., Doğru M., (2009),Gross alpha and beta radioactivity concentration in water, soil and sediment of the Bendimahi River and Van Lake (Turkey), *Environ. Monit. Assess.*, 148:39-46. doi: 10.1007/s10661-007-0137-x.

[28] Meher P. K., Sharma P., Khare A. and Mishra K. P., (2015) Uranium in Ground water of Eastern Uttar Pradesh, India: A preliminary study, *International Research Journal of Environment Sciences*, Vol. 4(6):70-74.

[29] ATSDR (2013) Toxicological Profile for Uranium. U.S. Department of Health and Human Services, Agency for Toxic Substances and Disease Registry, Division of Toxicology and Human Health Sciences, Environmental Toxicology Branch: Atlanta, GA, USA Available online: http://www.atsdr.cdc.gov/toxprofiles/TP.asp?id=440 andtid=77 (accessed on 15 July 2014.

[30] Canu I. G., Laurent O., Pires N., Laurier D., and Dublineau I., (2011) Health Effects of Naturally Radioactive Water Ingestion: The Need for Enhanced Studies, *Environ. Health Perspect*, 119:16761680. http://dx.doi.org/10.1289/ehp.1003224.

[31] WHO (World Health Organization) (2004) Guidelines For Drinking Water Quality. 3rd edition. Geneva:World Health Organization.

[32] Berlin M., Rudell B. (1986) Uranium. In: Friberg L., Nordberg G. F., Vouk V. B., eds. Handbook on the toxicology of metals, 2nd ed. Amsterdam, Elsevier Science Publishers, pp. 623–637.

[33] Wrenn M. E., Durbin P. W., Howard B. (1985) Metabolism of ingested uranium and radium. *Health Phys.*, 48:601-633.

[34] Sullivan M. F., Ruemmler P. S., Ryan J. L., et al. (1986) Influence of oxidizing or reducing agents on gastrointestinal absorption of uranium, plutonium, americium, curium, and promethium by rats. *Health Phys.,* 50:233-232.

[35] La Touche Y. D., Willis D. L., Dawydiak O. I. (1987) Absorption and biokinetics of U in rats following an oral administration of uranyl nitrate solution. *Health Phys.*, 53(2):147-162.

[36] Konietzka R. (2015) Gastrointestinal absorption of uranium compounds– A review, Regulatory Toxicology and Pharmacology Volume 71, Issue 1, Pages 125–133.

[37] International Commission on Radiological Protection (ICRP) (2007) Publication 100: Human Alimentary Tract Model for Radiological Protection. Oxford:Elsevier Sciences.

[38] Fisenne I. M., Perry P. M. (1985) Isotopic U concentration in human blood from New York City donors. *Health Phys.*, 49:1272-1275.

[39] Moss M. A. (1985). Chronic low level uranium exposure via drinking water-clinical investigations in Nova Scotia. Thesis, Dalhousie University, Halifax, Nova Scotia.

[40] Igarashi Y., Yamakawa A., Ikeda N. (1987) Plutonium and uranium in Japanese human tissues. *Radioisotopes*, 36:433–439.

[41] Hopke P. K., Borak T. B., Doull J., Cleaver J. E., Eckerman K. F., Gundersen C. S., et al. (2000). Health risks due to radon in drinking water. *Environ. Sci. Technol.*, 34(6):921–926.

[42] Singh N. P. et al. (1990) Daily U intake in Utah residents from food and drinking water. *Health Physics*, 59(3):333–337.

[43] Sontag W. (1986) Multicompartment kinetic models for the metabolism of americium, plutonium and uranium in rats. *Hum. Toxicol.,* 5:163-173.

[44] International Commission on Radiological Protection (1995) Age-dependent doses to members of the public from intake of radionuclides: Part 4 Inhalation dose coefficients. ICRP Publication 71. Ann. ICRP 25(3/4). Pergamon Press.

[45] Leggett R. W. (1994) Basis for the ICRP's age-specific biokinetic model for uranium. *Health Phys*., 67, 589–610.

[46] Katz S. A. (2014) The Chemistry and Toxicology of Depleted Uranium, Toxics, 2, 50-78; doi:10.3390/toxics2010050].

[47] Roszell, L. E., Hahn F. F., Lee R. B., Parkhurst M. A. (2009) Assessing the renal toxicity of capstone depleted uranium oxides and other uranium compounds. *Health Phys*., 96, 343–351.

[48] Anthony M. L., Gartland K. P. R., Beddell C. R., et al. (1994) Studies of the biochemical toxicology of uranyl nitrate in the rat. *Arch. Toxicol.,* 68(1):43-53.

[49] Domingo J. L. (1995) Prevention by chelating agents of metal-induced development toxicity. *Reprod. Toxicol*., 9:105–13.

[50] Leggett R. W. (1989) The behavior and chemical toxicity of U in the kidney: A reassessment. *Health Phys.,* 57, 365–385.

[51] Yuile C. L. (1973) Animal experiments. In: Hodge H. C. et al., eds. Handbook of experimental pharmacology. Vol. 36. Uranium, plutonium, transplutonic elements. Berlin, Springer-Verlag, pp. 165–195.

[52] Durbin P. W., Wrenn M. E. (1976) Metabolism and effects of uranium in animals. In: Conference on occupational health experience with uranium. Washington, DC, US Energy Research and Development Administration, pp. 68–129 (available from US National Technical Information Service).

[53] Campbell D. C. C. (1985) The development of an animal model with which to study the nephrotoxic effects of uranium-contaminated drinking water. Halifax, Nova Scotia, Dalhousie University (M.Sc. thesis).

[54] McDonald-Taylor C. K., Bhatnagar M. K., Gilman A., et al. (1992) Uranyl nitrate-induced glomerular basement membrane alterations in rabbits: A quantitative analysis. *Bull. Environ. Contam. Toxicol.,* 48(3):367-373.

[55] McDonald-Taylor C., Singh A., Gilman A., et al. (1997) Uranyl nitrate-induced proximal tubule alterations in rabbits: A quantitive analysis. *Toxicol. Pathol*., 25(4):381-389.

[56] Limson Zamora M., Tracy B. L., Zielinski J. M., et al. (1998) Chronic ingestion of uranium in drinking water: A study of kidney bioeffects in humans. *Toxicol. Sci*., 43(1):68-77.

[57] Kurttio P., Auvinen A., Salonen L., et al. (2002) Renal effects of uranium in drinking water. *Environ. Health Perspect*., 110(4):337-342.

[58] Souidi M., Gueguen Y., Linard C., et al. (2005) In vivo effects of chronic contamination with depleted uranium on CYP3A and associated nuclear receptors PXR and CAR in the rat. *Toxicology,* 214(1-2):113-122.

[59] Tissandie E., Gueguen Y., Lobaccaro J. M., et al. (2007) In vivo effects of chronic contamination with depleted uranium on vitamin D3 metabolism in rat. *Biochim. Biophys. Acta*, 1770(2):266-272.

[60] Berradi H., Bertho J. M., Dudoignon N., et al. (2008) Renal anemia induced by chronic ingestion of depleted uranium in rats. *Toxicol. Sci.,* 103(2):397-408.

[61] Linares V., Belles M., Albina M. L., Sirvent J. J., Sanchez D. J., Domingo J. L. (2006) Assessment of the prooxidant activity of uranium in kidney and testis of rats. *Toxicol. Lett.*, 167(2):152–161.

[62] Taulan M., Paquet F., Maubert C., Delissen O., Demaille J., Romey M. C. (2004) Renal toxicogenomic response to chronic uranyl nitrate insult in mice. *Environ. Health Perspect.*, 112:1628–1635.

[63] Donnadieu Claraz M., Bonnehorgne M., Dhieux B., Maubert C., Cheynet M., Paquet F., et al. (2007) Chronic exposure to uranium leads to iron accumulation in rat kidney cells. *Radiat. Res.*, 167(4):454–464.

[64] Gilman A. P., Villeneuve D. C., Secours V. E., et al. (1998) Uranyl nitrate: 28-day and 91-day toxicity studies in the Sprague-Dawley rat. *Toxicol. Sci.*, 41:117–28.

[65] Ortega A., Domingo J. L., Lloket J. M., et al. (1989) Evaluation of the oral toxicity of uranium in a 4-week drinking water study in rats. *Bull. Environ. Contam. Tox.*, 42:935-941.

[66] Domingo J. L. et al. (1989) The developmental toxicity of uranium in mice. *Toxicology*, 55(1–2):143–152.

[67] Hindin R., Brugge D. and Panekaa B. (2005) Teratogenicity of Depleted Uranium Aerosols; review from an epidemiological perspective. Env. Health. A global Access science source. Aug 26th 4-17.

[68] Busby C., de Messieres M., (2007) British Nuclear test Veterans Association/Green Audit Child Health Study 2007 Aberystwyth: Green Audit.

[69] Domingo J. L. (2001). Reproductive and developmental toxicity of natural and depleted uranium: A review. *Reprod. Toxicol.*, 15:603–609.

[70] Paternain J. L., Domingo J. L., Ortega A., Llobert J. M. (1989) The effects of uranium on reproduction, gestation and postnatal survival in mice. *Ecotoxicol. Env. Safety*, 17: 291-296.

[71] Raymond-Whish S., Mayer L. P., ONeal T. et al. (2007) Drinking water with uranium below the US EPA standard causes estrogen receptor dependent responses in female mice. *Env. Health Perspect.*, 115 (12) 1711-6.

[72] Lin R. H., Wu L. J., Lee C. H., and Lin-Shiau, S. Y. (1993) Cytogenetic toxicity of uranyl nitrate in Chinese hamster ovary cells. *Mut. Res.,* 319:197–203.

[73] Miller A. C., Xu J., Stewart M., Brooks K., Hodge S., et al., (2002) Observation of radiation-induced damage in human cells exposed to depleted uranium: Dicentric frequency and neoplastic transformation as endpoints. *Radiat. Prot. Dosimetry*, 99, 275–278.

[74] Darolles C., Broggio D., Feugier A., et al. (2010) Different genotoxic profiles between depleted and enriched uranium. *Toxicol. Lett.,* 192(3):337-348.

[75] Prat O., Berenguer F., Steinmetz G., Ruat S., Sage N., Quemeneur E. (2010) Alterations in gene expression in cultured human cells after acute exposure to uranium salt: Involvement of a mineralization regulator. *Toxicol. in Vitro,* 24(1):160–168.

[76] Vietti K. R., Lasley S. M. (2007) Stimulusevoked glutamate release is diminished by acute exposure to uranium in vitro. *Neurotoxicol. Teratol.*, 29:607–612.

[77] Miller A. C., Blakly W. F., Livengood D., Whittaker T., Xu J., et al., (1998) Transformation of human osteoblast cells to the tumorigenic phenotype by depleted uranium-uranyl chloride. *Environ. Health Perspect.*, 6:465–471.

[78] Miller A. C., Brooks K., Stewart M., Anderson B., Shi L., McClain D., Page N. (2003) Genomic instability in human osteoblast cells after exposure to depleted uranium: Delayed lethality and micronuclei formation. *J. Environ. Radioact.*, 64:247–259.

[79] Tasat D. R., Orona N. S., Mandalunis P. M., Cabrini R. L., Ubios A. M. (2007) Ultrastructural and metabolic changes in osteoblasts exposed to uranyl nitrate. *Arch. Toxicol.*, 81:319–326.

[80] Guglielmotti M. B., Ubios A. M., Cabrini R. L. (1987) Morphometric study of the effect of a low dose uranium on bone healing. *Acta Stereol.,* 6:357–366.

[81] IRSN (2006). Health consequences of internal contaminations by radionuclides. Comments on the ECRR Report, The Health Effects of Ionizing Radiation Exposure for Radiation Protection purposes. Report DPRH 2005-20. Fontenay aux Roses, IRSN.

[82] Monleau M., Bussy C., Lestaevel P., et al. (2005) Bioaccumulation and behavioural effects of depleted uranium in rats exposed to repeated inhalations. *Neurosci. Lett.*, 390(1):31-36.

[83] Radespiel-Tröger M., Meyer M., (2013) Association between drinking water uranium content and cancer risk in Bavaria, Germany. *Int. Arch. Occup. Environ. Health,* 86, 767–776 (Oct.).

[84] Wagner S. E., Burch J. B., Bottai M., Puett R., Porter D., Bolick-Aldrich S., Temples T., Wilkerson R. C., Vena J. E., Hébert J. R., (2011) Groundwater uranium and cancer incidence in South Carolina. *Cancer Causes Control,* 22, 41–50.

[85] Bean J. A., Isacson P., Hausler W. J. Jr., Kohler J. (1982) Drinking water and cancer incidence in Iowa. I. Trends and incidence by source of drinking water and size of municipality. *Am. J. Epidemiol.*, 116(6):912–923.

[86] Fuortes L., McNutt L. A., Lynch C. (1990) Leukemia incidence and radioactivity in drinking water in 59 Iowa towns. *Am. J. Public Health,* 80(10):1261–1262.

[87] Moss M. E., Kanarek M. S., Anderson H. A., Hanrahan L. P., Remington P. L. (1995). Osteosarcoma, seasonality, and environmental factors in Wisconsin, 1979–1989. *Arch. Environ. Health*, 50(3):235– 241.

[88] Finkelstein M. M. (1994). Radium in drinking water and the risk of death from bone cancer among Ontario youths. *CMAJ*, 151(5):565–571.

[89] Guse C. E., Marbella A. M., George V., Layde P. M. (2002) Radium in Wisconsin drinking water: an analysis of osteosarcoma risk. *Arch. Environ. Health,* 57(4):294–303.

[90] Hirunwatthanakul P., Sriplung H., Geater A. (2006) Radiumcontaminated water: a risk factor for cancer of the upper digestive tract. *Asian Pac. J. Cancer Prev.*, 7(2):295–298.

[91] Witmans M. R., McDuffie H. H., Karunanayake C., Kerrich R., Pahwa P. (2008) An exploratory study of chemical elements in drinking water and nonHodgkin's lymphoma. *Toxicol. Environ. Chem.*, 90(6):1227–1247.

[92] Auvinen A., Salonen L., Pekkanen J., Pukkala E., Ilus T., Kurttio P. (2005) Radon and other natural radionuclides in drinking water and risk of stomach cancer: a case–cohort study in Finland. *Int. J. Cancer,* 114(1):109–113.

[93] Kurttio P., Salonen L., Ilus T., Pekkanen J., Pukkala E., Auvinen A. (2006) Well water radioactivity and risk of cancers of the urinary organs. *Environ. Res.,* 102(3):333–338.

[94] Auvinen A., Kurttio P., Pekkanen J., Pukkala E., Ilus T., Salonen L. (2002) Uranium and other natural radionuclides in drinking water and risk of leukemia: a case–cohort study in Finland. *Cancer Causes Control,* 13(9):825–829.

[95] Canu I. G., Garsi J. P., Caër-Lorho S., Jacob S., Collomb P., Alain A., Laurier D., (2012) Does uranium induce circulatory diseases? First results from a French cohort of uranium workers, *Occup. Environ. Med.,* 2012;69:404-409 doi:10.1136/oemed-2011-100495.

[96] Uranium in Drinking-water: Background document for development of WHO Guidelines for Drinking-water Quality (2004), WHO/SDE/WSH/03.04/118.

[97] Newsome L., Morris K., Lloyd J. R. (2013). The biogeochemistry and bioremediation of uranium and other priority radionuclides, *Chemical Geology*, doi: 10.1016/j.chemgeo.2013.10.034.

[98] Kulkarni S., Ballal A., Apte S. K., (2013). Bioprecipitation of uranium from alkaline waste solutions usingrecombinant Deinococcus radiodurans, *Journal of Hazardous Materials*, 262: 853–861.

[99] Lovley D. R., Phillips E. J. P.,(1992). Reduction of uranium by Desulfovibrio desulfuricans, *Appl. Environ. Microbiol.*, 58: 850–856.

[100] Williams K. H., Bargar J. R., Lloyd J. R., Lovley D. R., (2013). Bioremediation of uranium-contaminated groundwater: a systems approach to subsurface bio-geochemistry, *Curr. Opin. Biotechnol.,* 24:489–497.

[101] Barlett M., Moon H., Peacock A., Hedrick D., Williams K., Long P., Lovley D., Jaffe P., (2012). Uranium reduction and microbial community development in responseto stimulation with different electron donors, *Biodegradation,* 23:535–546.

[102] Begg J. D. C., Burke I. T., Lloyd J. R., Boothman C., Shaw S., Charnock J. M., Morris K., (2011). Bioreduction behavior of U(VI) sorbed to sediments. *Geomicrobiol. J.,* 28:160–171.

[103] Acharya C., Joseph D., Apte S. K., (2009). Uranium sequestration by a marine cyanobac-terium, Synechococcus elongatus strain BDU/75042, *Bioresour. Technol.,* 100:2176–2181.

[104] Akhtar K., Khalid A. M., Akhtar M. W., Ghauri M. A., (2009). Removal and recovery ofuranium from aqueous solutions by Ca–alginate immobilized Trichodermaharzianum, *Bioresour. Technol.*, 100: 4551–4558.

[105] Lu X., Zhou X., Wang T., (2013). Mechanism of uranium (VI) uptake by Saccharomyces cerevisiae under environmentally relevant conditions: Batch, HRTEM, and FTIR studies, *Journal of Hazardous Materials*, 262: 297–303.

[106] Nakajima A., Horikoshi T., Sakaguchi T., (1979) Ion effects on the uptake of uranium by Chlorella regularis, *Agric. Biol. Chem.*, 43: 625–629.

[107] Mapolelo M., Torto N., (2004) Trace enrichment of metal ions in aquatic environmentsby Saccharomyces cerevisiae, *Talanta*, 64:39–47.

In: Heavy Metals
Editor: Deepak Pathania

ISBN: 978-1-63484-740-7

Chapter 10

SLUDGE ASH FOR THE REMEDIATION OF WATER LADEN WITH CU(II) AND PB(II)

Muaz Zakaria, Muhammad Abbas Ahmad Zaini** and Mohd. Azizi Che Yunus

Centre of Lipids Engineering and Applied Research (CLEAR),
Ibnu Sina Institute for Scientific and Industrial Research (ISI-SIR),
Universiti Teknologi Malaysia, Johor Bahru, Malaysia

ABSTRACT

Heavy metals are considered as the major contributor in water pollution, where 70-80% of the illnesses in the developing countries are the result of water contamination. The environmental considerations and increasing costs associated with the use of commercial adsorbent have led to a significant body of research work aimed at manufacturing the low cost adsorbents based on the renewable resources for the remediation of water laden with heavy metals. A promising candidate under this category is the sludge ash of palm oil mill effluent. A number of ash based adsorbents have been tested for the adsorption of heavy metals and organic pollutants from aqueous solution. The selection of ash as potential adsorbent is due to the presence of some minerals such as iron, silica and aluminium which could contributed to the metal ions extraction. This chapter discusses the characteristics and adsorptive properties of sludge ash adsorbent as a promising candidate for heavy metals removal from water.

Keywords: ash adsorbent, adsorption, Cu(II), Pb(II), sludge ash, wastewater treatment

* Corresponding author: MAA Zaini, PhD CEng, CLEAR, ISI-SIR, UTM Johor Bahru, Malaysia, Email: abbas@cheme.utm.my.

1. Introduction

Palm oil industry accounts for the largest biomass waste generation in Malaysia. In palm oil milling, the byproducts consist of shells, empty fruit bunches, pressed fruit fibres (mesocarp fibres) and palm oil mill effluent. The palm oil industry has shown a large potential towards a sustainable future. Recently, many studies have been carried out in the development of palm oil byproducts such as biodiesel, biogas, bioplastic, compost, activated carbon and animal feedstock. In addition, the effluent sludge can also be converted into adsorbent. The sludge contains nearly 60% ash, and therefore it is of interest to understand its adsorptive potential in aqueous solution.

Adsorption is a cost-effective physico-chemical treatment in removing heavy metals from aqueous solutions, and is widely used in effluent treatment processes. According to Barakat [1], adsorption is a process that transfers solute or ions from water to the solid phase via precipitation reactions. The process implies the presence of solid adsorbent that binds molecules by physical attractive forces, ion exchange, and/or chemical binding. It is recommended that the adsorbent used in the adsorption is cheap, abundantly available and regenerable [2]. In recent years, the study on low cost adsorbents derived from agro-wastes has become a subject of considerable interest. Attention has been focused in finding suitable alternatives for commercial adsorbent/activated carbon, and other physico-chemical treatments for heavy metals such as precipitation, ion exchange, liquid membrane extraction, electrode deposition, etc. A promising adsorbent candidate under this category is the sludge ash of palm oil mill effluent.

Heavy metals can cause several disadvantages to the environment. They are major contributor to 70-80% of all illnesses in the developing countries [3]. Naturally, human body needs small amount of some heavy metals such as zinc, calcium, and others. However, if their concentration is in excess than is required, they can give bad implications to the human health. Heavy metals are the most hazardous effluent released by the chemical-intensive industries. They are easily absorbed by the living organisms because of their high solubility in aquatic environments. A large concentration of heavy metals may accumulate in human body via food chain [1-2]. Heavy metal ions are stable and persistent environmental contaminants because they cannot be degraded and destroyed [2].

At present, activated carbon is widely used in the industrial applications and is regarded as a universal adsorbent for the removal of diverse types of pollutants from water [3]. The carbon based adsorbents are known to possess good thermal stability and adsorptive ability for wide range of organic and inorganic pollutants, but in certain situations they may display some limitations. The activated carbon has low adsorption capacity at low adsorbate concentration, thus modifications are needed to improve its uptake capacity. However, the carbon adsorbent is difficult to be modified because of its hydrophobic characteristics. Moreover, the treated carbon adsorbent is relatively expensive due to high material and operating costs. In addition, the treated activated carbon is rather thermally unstable upon chemical modifications consequently making it difficult to be regenerated.

The increasing costs associated with the use of commercial adsorbent have led to a significant body of research work aimed at manufacturing the low cost adsorbents based on the renewable resources. A number of ash based adsorbents have been tested for adsorption of heavy metals and organic pollutants from aqueous solution. The selection of ash as potential adsorbent is due to the presence of some elements such as iron, silica and aluminium which are believed to aid in metals ions extraction. Using ash as adsorbent is an attractive approach because unlike chemical activation/treatment, chemical agents are neither being used nor released during the preparation of adsorbent. Sewage sludge ash [4], fly ash [4-6], chitosan/oil palm ash [7], and oil palm ash [8] are some examples of ash adsorbents. The adsorptive properties of sludge ash of palm oil mill effluent have yet to be understood, and therefore the material is worth to be explored as a promising adsorbent candidate for heavy metals removal.

2. The Origin and Toxicity of Heavy Metals in Water: An Overview

The wastewater streams containing heavy metals such as cadmium, zinc, lead, chromium, nickel and copper are released from the processes to manufacture printed circuit board, scientific instruments and batteries in the electroplating and metal surface treatment industries. Refineries, paints, paper and pulp, and fertilizer industries are among the major contributors of heavy metals in wastewater [9-10]. Heavy metals can also be found in wood processing whereby the use of chromated copper-arsenate in the wood treatment produces arsenic-containing wastewater [1].

Heavy metals exist in water as aqua (dissolved) cations, and they are thermodynamically stable under acidic and neutral environment. As a consequence, heavy metals are prone to accumulate in aquatic creatures, mushrooms and nuts [11]. Examples of heavy metals commonly found in water are copper and lead. The maximum acceptable concentration of copper in drinking water according to the World Health Organization is 1.5 mg/L [9, 11]. Typical drinking water standard for copper is 1.0 mg/L [12], and 0.05-1.5 mg/L for inland surface water [13]. The acceptable concentration of lead in potable water is 0.01 mg/L [14]. Lead is harmful to human central and peripheral nervous systems, and accounts for most pediatric cases of the heavy metals poisoning [15].

Table 1. Toxicities of heavy metals to human health [1]

Heavy metal	Toxicities	Maximum contaminant limit (mg/L)
Arsenic	Skin manifestations, visceral cancers, vascular disease	0.05
Cadmium	Kidney damage, renal disorder, human carcinogen	0.01
Chromium	Headache, diarrhea, nausea, vomiting, carcinogenic	0.05
Copper	Liver damage, Wilson disease, insomnia	0.25
Nickel	Dermatitis, nausea, chronic asthma, coughing, human carcinogen	0.20
Zinc	Depression, lethargy, neurological signs and increased thirst	0.80
Lead	Damage the fetal brain, diseases of the kidneys, circulatory system and nervous system	6×10^{-3}
Mercury	Rheumatoid arthritis, and diseases of the kidneys, circulatory system and nervous system	3×10^{-5}

Heavy metals can enter the water bodies through the industrial and domestic activities, and from acidic rain breaking down soils and releasing heavy metals into the lakes, rivers and groundwater. They are toxic, and tend to accumulate in human tissue, fat cells and organs. Heavy metals can harm central nervous function and bring damage to the blood composition, lungs, kidneys and liver. Long-term exposure may result in the Alzheimer- and Parkinson-like diseases and may even cause cancer. Table 1 summarizes the toxicities and health implications due to the heavy metals poisoning.

3. Summary of the Heavy Metals Removal Strategies

Because heavy metals are toxic and non-biodegradable, it is therefore imperative to remove them from the water bodies. The methods to remove metal ions from aqueous solution include ion exchange, reverse osmosis, filtration, solvent extraction, evaporation, adsorption, electrochemical treatment, chemical oxidation/reduction and precipitation by pH adjustment [16-17].

Microfiltration is of not much use for wastewater treatment because of the large pore size of the filter media as compared to that of metals ions. Usually, this method is used to remove metal hydroxide precipitate (insoluble). The process is economically feasible but sometimes becomes expensive due to the cost of chemicals (for precipitation) for satisfactory removal of heavy metals. Microfiltration for heavy metals removal is pH dependent and it generates concentrated sludge in large quantity. In addition, the hydroxide molecules cause frequent clogging to the membrane pores making the separation of limited use. The main drawbacks are high working pressure, high energy consumption, high membrane cost and short membrane life.

Electrochemical method is also used to remove heavy metals apart from its wide application as a tertiary treatment to produce high-quality water. The removal of heavy metals can be achieved either by electro-oxidation with non-soluble anodes or by electro-coagulation using consumable materials. The removal efficiency is dependent on the anode material and the working potential. The disadvantages of this process are high electricity cost, sludge production and secondary pollution due to indirect oxidation.

Ion-exchange has several advantages for the heavy metals removal: high treatment capacity, high removal efficiency and fast kinetics. Ion-exchange resin, either synthetic or natural has the specific ability to exchange cations with contaminants in the wastewater. The strong sulfonic acid groups (— SO_3H) and weak carboxylic acid groups (— COOH) are the major players for exchanging the cations. The hydrogen ions in acid groups of the resin can serve as exchangeable ions with the metal cations. The ionic charge also plays an important role in the ion-exchange process. The uptake of heavy metal ions by the ion-exchange resins is dependent on pH, temperature, concentration of heavy metals and contact time [18].

The aforementioned removal strategies are expensive and only effective for streams containing high concentration of heavy metals. Adsorption is a preferred removal technique for lower concentration of heavy metals. Activated carbon is a widely used adsorbent in adsorption process. Because the price of commercial activated carbon is expensive, attention has now been directed on various natural solid precursors that are able to remove water contaminants at low cost.

The production of alternative sorbents to replace the costly activated carbon has intensified in recent years. The cost and removal performance are the important criteria for the selection of adsorbent materials in adsorption. Adsorbent can be considered low-cost if it is abundant in nature and requires simple processing [6]. The adsorbent precursor can be the by-products or waste materials from agricultural, industrial or domestic activities. These raw materials represent potentially economical alternative adsorbents for the heavy metals removal.

4. Adsorption of Heavy Metals

Adsorption refers to a process where the solute is concentrated on a solid surface from its liquid surroundings, forming a molecular or atomic film. The factors affecting the adsorbate (solute)-adsorbent interaction in adsorption include the particle size, specific surface area and pore volume of adsorbent, surface chemistry of adsorbent, concentration of solute in solution, solution pH, temperature and contact time.

The adsorption can be categorized into physical and chemical adsorption. The physical adsorption occurs when the weak inter-particle bonds, namely the van der Waals, hydrogen or dipole-dipole exist between the adsorbate and adsorbent. The physical adsorption is easily reversible. On the other hand, the chemical adsorption (chemisorption) occurs when the strong intra-particle bonds such as the covalent or ionic are present between the adsorbate and adsorbent due to exchange of electrons. The chemisorption is an irreversible process due to its strong solute binding.

In general, the fundamental of heavy metal removal by adsorption may be assumed to involve four steps: 1) Bulk diffusion- the migration of solute from the bulk of the aqueous solution to the surface of adsorbent; 2) Film diffusion- the diffusion of solute through the boundary layer to the surface of the adsorbent; 3) Pore diffusion/intra-particle diffusion- the solute transport from the surface to within the pores of the particle; 4) Adsorption/chemical reaction- the adsorption of solute at an active site on the surface of material via ion-exchange, complexation or chelation. Generally, the bulk diffusion and adsorption are assumed to be rapid, and therefore are not the rate determining steps [19].

The specific surface area of adsorbent is important because it increases the number of sites available for adsorption. A smaller particle size increases the external surface area of adsorbent. However, to enhance the accessibility of the solute to the internal pores, the adsorbent should possess larger micropore and mesopore volume. The surface functional groups and metals oxides (minerals) within the adsorbent may also interact with the heavy metals in solution [4, 10, 20]. The metals cations are attracted to the adsorbent due to opposite charge or anionic surface. The polar groups such as silicon dioxide may also play some role in removing the heavy metals via cationic exchange [13]. Thus, it is essential to establish the physical and chemical characteristics of the adsorbent to understand the removal mechanisms.

The concentration is a common process parameter to show the ratio of active sites to the total metal ions in aqueous solution [21]. A higher initial concentration provides a greater driving force to overcome the mass transfer resistances in the film outside and in the pore inside the adsorbent. The removal of metal ions increased linearly with increasing the initial concentration to a point where the uptake becomes constant due to the surface saturation, whereas the rate of uptake capacity decreased with increasing the concentration. The solution pH affects the adsorption and the surface charge of adsorbent [9, 17]. At low pH, the cations compete with H^+ ions in the solution for active sites, and at the same time the adsorbent surface would become protonated and repel the cations. As a consequence, the concentration of heavy metal ions to bind the unpaired electrons or lodge on the adsorbent surface decreases with increasing the concentration of H^+ ions. Increasing the solution pH would result in the basification of adsorbent surface. Generally, this would lead to a higher attraction of heavy metals and possible precipitation attributed to the reaction between the metals cations and OH^- ions [9].

Adsorption is now recognized as an effective and economic method for the treatment of wastewater-containing heavy metals. The adsorption process offers flexibility in the design and operation and in many cases will produce a high-quality treated effluent. In addition, because adsorption is sometimes reversible, the adsorbent can be regenerated by suitable desorption techniques [22].

5. Characteristics of Sludge Ash

Palm oil mill effluent (POME) sludge was obtained from the anaerobic pond at Felda Taib Andak palm oil mill, Kulai, Johor state of Malaysia. It was burnt to ash in a furnace at 800°C for 2 h. The combustible organic matter in sludge is 40.4 wt % and the ash is 59.6 wt %. The carbon content is relatively small (19.1 wt %) to be converted into activated carbon [23]. The ash was soaked separately in 5 M of HCl and 5 M of NaOH to remove leachable non-organic elements and impurities, and to modify the surface chemistry of ash. The treatment procedures are expected to improve the adsorption performance of ash [24]. The adsorbents were characterized using ICP-MS for ash composition, FTIR for possible surface functional groups and the pH of the point of zero charge.

The adsorption of heavy metals is usually influenced by the physical texture and surface chemistry of adsorbent. The characterization of adsorbent usually serves as a guide to better understand the nature and interaction between the adsorbate and adsorbent in the adsorption process. The inductively coupled plasma mass spectrometry (ICP-MS) is an analytical tool for the inorganic elemental determination. Table 2 shows the composition of sludge ash and its treated derivatives after the acid digestion.

The composition of minerals in palm oil mill effluent sludge (So), sludge ash (Ao), acid-treated ash (A11) and alkali-treated ash (A12) is dependent on the treatment strategies. When the sludge (So) is burnt to ash (Ao) at 800°C, nearly half of the iron is liberated along with the organic constituents. It is suggested that the iron element is mainly composed of iron (III) compounds. The removal of a considerable fraction of iron element from sludge (So) gives rise to the fraction of calcium, magnesium and lead elements in ash (Ao). It indicates that these elements are thermally stable. In acid treatment (for the preparation of A11), almost all minerals are partially de-mineralized, except for lead, barium and potassium.

This is in agreement with other related study [25]. The result suggests that these minerals are relatively stable, not leachable and resistive to acidic environment. On the other hand, A12 displays a large removal of iron upon alkali treatment. The composition of A12 signifies that magnesium and lead are stable after alkali treatment.

Table 2. Inorganic constituents in sludge ash adsorbents

Inorganic (wt %)	So	Ao	A11	A12
Copper	0.6	0.6	0.1	0.3
Manganese	1.4	1.6	0.1	1.7
Zinc	1.1	0.9	0.7	0.2
Calcium	2.4	5.6	0.6	7.1
Magnesium	3.6	15.1	1.1	15.2
Lead	8.6	30	66.5	64.1
Barium	0.1	0.1	0.3	0.1
Iron	81	44.7	25.8	8.6
Potassium	1.2	1.4	4.8	2.7

For both acid and alkali treatment, lead was found to be the most stable element in ash that is resistant to acid and alkali. For both ash-treated adsorbents, there is a possibility for some minerals/elements to leach out from the materials in adsorption at varying pH. However, the release could be considered trivial as these minerals are usually not categorized as major pollutants to the environment. Although the minerals are structured within the adsorbent, some of them tend to form particles. Calcium, magnesium, lead and iron are present in high amount in adsorbents tested.

Table 3. Peaks assignment

Wave length	Functional groups	So	Ao	A11	A12
3600 – 3200	O — H stretching	✓	✓	✓	✓
3500 – 3100	N — H stretching (amides)	✓	✓	✓	✓
3000 – 2850	C — H stretching	✓	×	×	×
3100 – 3000	C = C stretching	✓	×	×	×
1680 –1630	C = O	✓	×	×	×
1350 – 1000	C — N	✓	×	×	×
~1200	Si — CH_3	✓	×	×	×
~1150	Si — O — Si	✓	✓	×	×
~1050	S = O	×	✓	✓	✓

The Fourier transform infra-red (FTIR) spectroscopy was used to obtain information on the nature of possible interactions between the functional groups of adsorbents and metal ions. The FTIR spectra of sludge (So), sludge ash (Ao) and the ash adsorbents (A11 and A12) are shown in Figures 1 and 2, and the peaks assignment is summarized in Table 3. This characterization also serves as evident that the surface modification has taken place post the treatment procedures.

From Figures 1 and 2, and Table 3, it is obvious that considerable fractions of functional groups are removed when the sludge is burnt into ash at high temperature. In general, the sludge ash and its derived adsorbents possess similar spectra but with different peaks intensity after the treatment procedures, implying the alteration of surface chemistry.

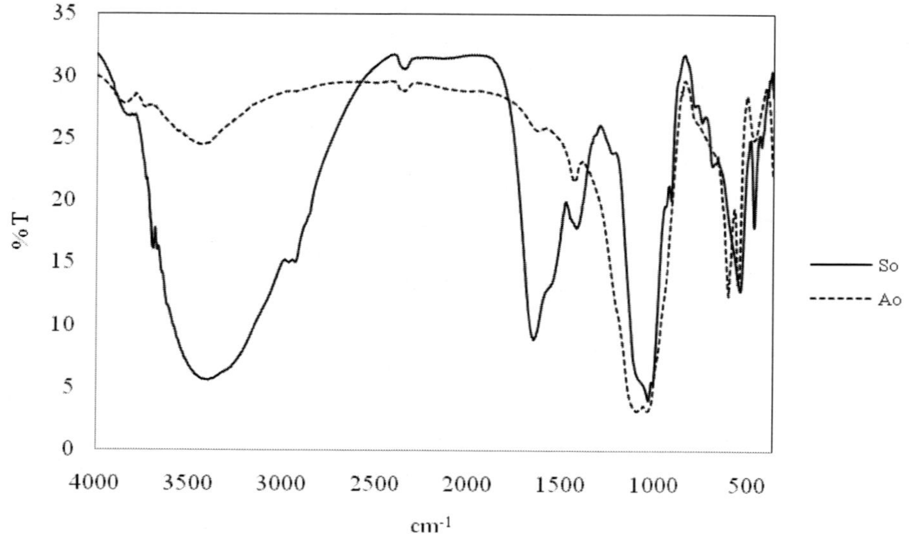

Figure 1. FTIR spectra of palm oil mill effluent sludge and sludge ash.

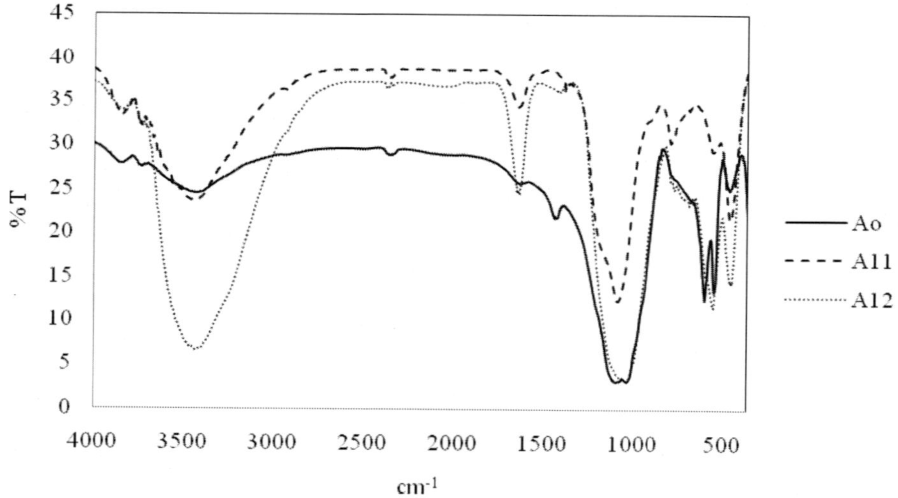

Figure 2. FTIR spectra of sludge ash and its adsorbent derivatives.

The peaks around 3400 cm^{-1} are usually due to O — H stretching [26]. The peaks around 3500 and 3100 cm^{-1} for So and Ao could be due to the overlapping of the N — H bond of the amino group and O—H bond. A broad absorption band between 3600 and 3200 cm^{-1} is assigned to the phenolic or carboxylic groups [27-28], and amino group [28]. The broadening of the O — H stretching vibration can be attributed to the hydrogen bonding between different functionalities. The C — H stretching of C — H, C — H_2 and C — H_3 present in the form of cellulose, lignin and hemicelluloses can be seen in the peak region of 2900 to 2880 cm^{-1} for So spectrum.

The C=O stretching of carboxylic acid that exists with the intermolecular hydrogen bond was observed at 1680 and 1630 cm^{-1}. Rahman and Islam [29] reported that copper(II) adsorbed onto maple wood sawdust either by ion exchange mechanism or hydrogen bonding due to the presence of — COOH and — OH groups. It was reported that metal ions would also bound to mass biomass due to the presence of — OH, — NH_2 and — COOH groups. The surface functional groups could give some advantages to the uptake of metal ions as compared to specific surface area.

The pH of the point of zero charge (pH_{pzc}) is the pH when the surface charge of the adsorbent becomes zero. It was determined via a method called pH drift, by plotting the final pH (equilibrium) of adsorbent against the initial solution pH. The pH_{pzc} is the point when the value of equilibrium pH is equal to the initial pH. The adsorbent that is submerged in an acidic solution where the solution pH is lower than the pH_{pzc} increased the attraction of anions towards the sorbent. This is explained by the additional protons donated from acidic solution which causes the sample surface to become positively charged. On the other hand, whenever the adsorbent is exposed in a solution of higher pH than its pH_{pzc}, the attraction of cations increases due to the negatively charged surface. Figure 3 shows the pH drift for So, A11 and A12.

From Figure 3, the pH_{PZC} for So is 5.22, while for A11 and A12 the values are 4.20 and 7.80, respectively. From the viewpoint of surface charge, it can be explained that the adsorption of cations would be favourable when done in solution with pH higher than 5.22, 4.20 and 7.80 for So, A11 and A12, respectively.

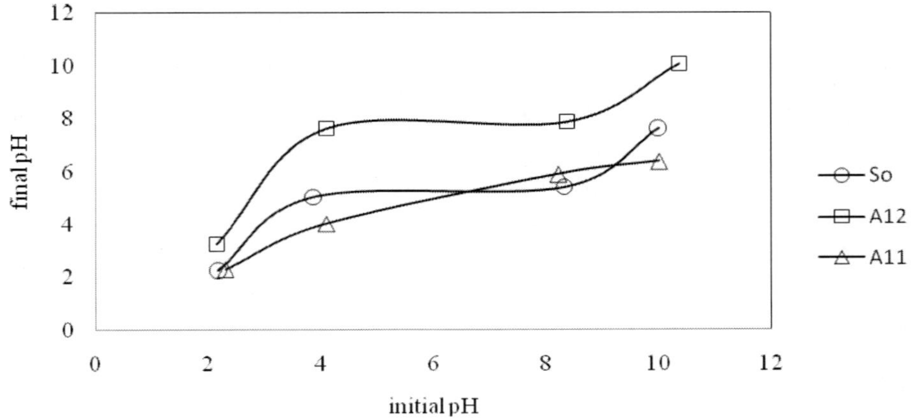

Figure 3. pH_{PZC} for So, A11 and A12 using pH drift method.

6. Cu(II) and Pb(II) Removal by Sludge Ash

Copper (II) and lead (II) solutions of various initial concentrations were prepared from metal salts of $CuCl_2 \cdot 2H_2O$ and $PbCl_2$, respectively. The stock solutions were diluted to the desired concentration with distilled water. The batch adsorption was performed by adding 0.5 g ash adsorbent in 150 mL heavy metal solutions. The flasks were tightly stoppered and agitated on an orbital shaker at room temperature and 150 rpm for 24 h. The residual concentration was measured using atomic absorption spectroscopy (model PerkinElmer Analyst400). The heavy metals removal at equilibrium, q_e (mg/g), was determined as

$q_e = (C_o\text{-}C_e)V/W$, where C_o and C_e (mg/L) are the liquid-phase concentrations of heavy metal at initial and equilibrium, respectively. V (L) is the volume of the solution, and W (g) is the mass of adsorbent used. The pH of heavy metal solution was adjusted using dilute HCl and NaOH to evaluate the effect of solution pH on heavy metals removal by ash adsorbents. The rate of adsorption was also evaluated by measuring the residual concentration of heavy metals at different time intervals.

The batch adsorption was carried out to evaluate the possible application of sludge ash in the adsorption of heavy metals. The adsorption of copper(II) and lead(II) was examined under different parameters, namely solution pH, contact time and initial concentration. Because the porosity of sludge ash is somewhat insignificant as compared to that of commercial adsorbent or activated carbon, it is expected that the metals ions can be removed by chemical interaction [23-24, 30]. Suitable isotherm and kinetics models were established to represent the behavior of adsorption.

6.1. Effect of Solution pH

Adsorption of heavy metals is strongly dependent on initial pH, as it can affect the magnitude of positive or negative charge on the adsorbent surface [9]. The solution pH is commonly identified as one of the most important parameters affecting metal ions adsorption. Figures 4 and 5 show the effect of solution pH on the removal of heavy metals by sludge ash adsorbents. The purpose of having acid treatment for ash is to modify the adsorbent surface to be an oxidizing agent which can promote the reduction of cations and improve the removal of heavy metals [3]. However, A11 demonstrates a lower removal of copper (II) and lead (II) than A12 at lower solution pH of 4.0 and 6.0.

It is likely due to the decreasing of negative sites, and competition between H^+ ions and metal ions for sorption sites at lower pH. This is also partly due to the fact that the hydrogen ions themselves are strongly competing adsorbate. A lower pH leads to the protonation of adsorbent surface rather than metal binding. In other words, the sorbent surface becomes more positively charged thus reducing the attraction between sorbent and metal cations [31].

The effect of chemical treatment reveals that the adsorption of copper(II) and lead(II) is enhanced when the sludge ash was washed with sodium hydroxide. Similar finding was reported by Febrianto et al. [32]. According to Asadi et al. [33], the thermo-chemical reaction with NaOH has resulted in the increase of negative sites and cationic exchange capacity of the adsorbent material, leading to the increase of metal sorption capacity. The equilibrium pH is affected by the acidity and alkalinity of the solution and the surface charge of the sludge ash adsorbents.

Solution pH can influence the speciation of metal ions, the polarization of adsorbent and the ionization of surface sites [34]. The removal of heavy metals by A11 and A12 is comparable when the solution pH was adjusted to 8.4 and 10.1. It is suggested that the removal of heavy metals at these two pH values is driven by precipitation of metal ions to form insoluble solids instead of adsorption. Woolard and co-workers [35] reported that lead(II) adsorbed increased as pH increased. However, the divalent metal ions namely Cu^{2+} and Pb^{2+} begin to precipitate to form $Cu(OH)_2$ and $Pb(OH)_2$ at pH 6.6, thus decreasing the soluble aqua ions [36]. Because the metals ions are susceptible to precipitate at a higher pH due to increasing OH^- concentration, it is difficult to ascertain that the removal is primarily due to the adsorption by ash adsorbents.

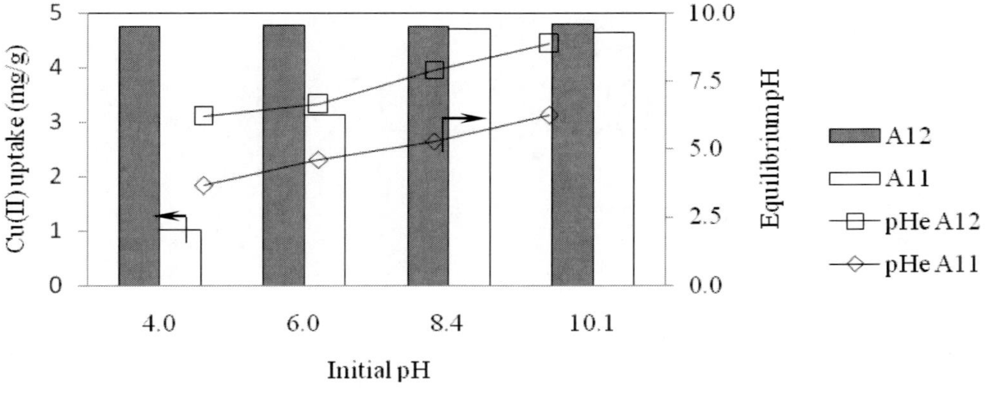

Figure 4. Effect of pH on the removal of copper(II) by sludge ash adsorbents (C_o = 48 mg/L Cu(II), W = 0.5 g, t = 72 h).

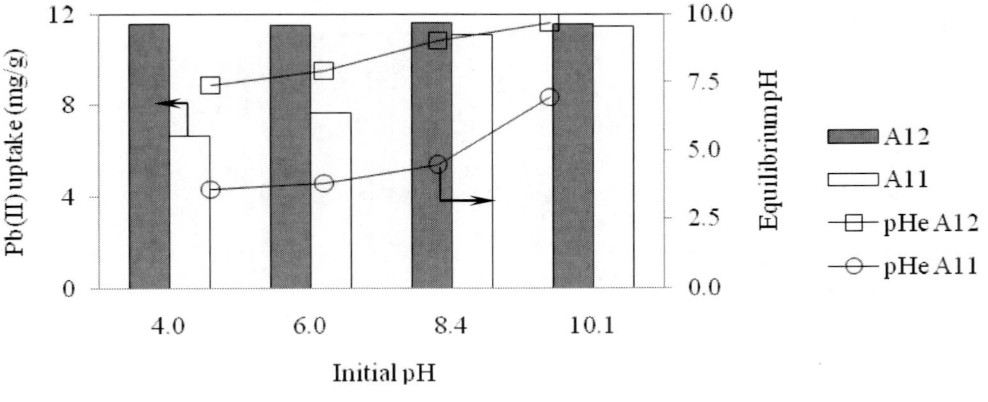

Figure 5. Effect of pH on the removal of lead(II) by sludge ash adsorbents (C_o = 117 mg/L Pb(II), W = 0.5 g, t=72 h).

It was found that the adsorption of heavy metals onto A12 is pH-independent. The equilibrium pH was found to increase from 6.6 to 8.9 and 8.0 to 9.7 with increasing initial pH for Cu(II) and Pb(II) removal, respectively. From the operational point of view, pH-independent adsorption is favourable as there is no need for pH adjustment to improve the removal efficiency. Yet, there is a considerable doubt as whether the removal is mainly triggered by precipitation of metal aqua ions because of higher equilibrium pH. Stable removal for ash treatment with NaOH was also explained by Woorlard et al. [35].

6.2. Effect of Concentration

This part discusses the effect of initial concentration on the adsorption of copper(II) and lead(II) in aqueous solution by the sludge ash adsorbents. Figure 6 displays the equilibrium uptake of Cu(II) by A11 and A12.

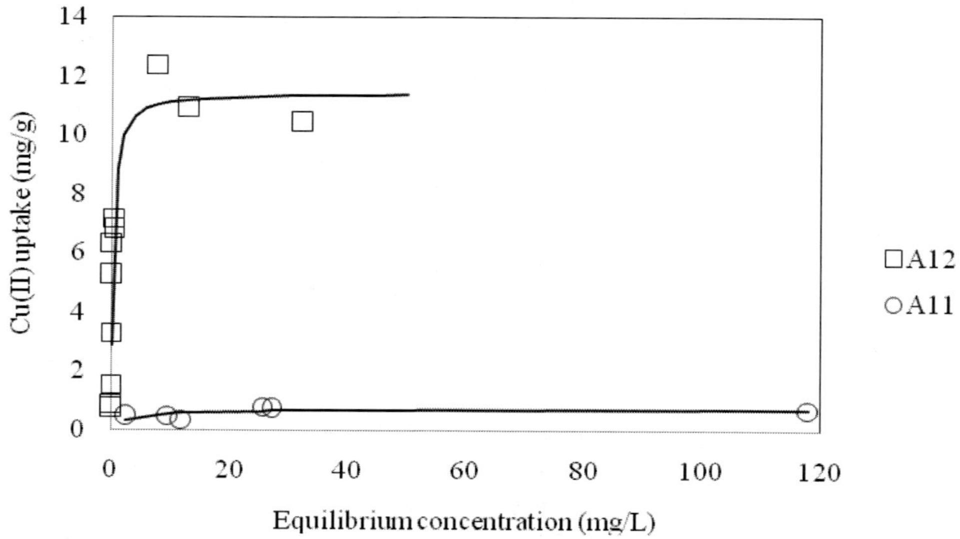

Figure 6. Equilibrium removal of copper(II) by sludge ash adsorbents (unadjusted pH = 5.2, t = 72 h).

The metals uptake is particularly dependent on initial concentration, C_o. The amount of copper(II) ions adsorbed increased with increasing initial concentration, and reached a saturation state of maximum uptake at a higher concentration. The transfer of metal ions from the bulk solution to the adsorbent surface is a dynamic process and increases with increasing the metal ions concentration. With increasing the concentration, all active sites are accommodated, and results in a maximum adsorption capacity.

From Figure 6, A12 exhibits a maximum uptake of 12.4 mg/g (or 0.195 mmol/g), a 17 times greater than that of A11 (0.72 mg/g). The equilibrium pH was recorded as 3.9 to 4.1 and 5.7 to 8.0 for A11 and A12, respectively. For A12, the equilibrium pH decreased with increasing copper(II) concentration. A12 releases considerable amount of OH^- ions, thus increasing the pH of copper(II) solution. According to the Le-Chatelier's principle, as the concentration of copper(II) ions increases in the presence of abundance OH^- ions, the precipitation becomes accelerated and the equilibrium position is shifted to the right. The equilibrium pH decreased because more OH^- ions have been consumed to form solids precipitate at a higher copper(II) concentration.

It is important to establish the most appropriate correlation for equilibrium curve in order to optimize the design of an adsorption system. The Langmuir isotherm describes the homogeneous surface assuming that all adsorption sites have identical affinity and that adsorption at one site would not affect the adjacent site; each adsorbate molecule is located on a single site and hence monolayer formation of adsorbate on the adsorbent surface. The correlations between q_e and C_e based on the Langmuir expression for A11 and A12 are

$q_e = 0.262\ C_e/(1 + 0.358\ C_e)$ and $q_e = 38.8\ C_e/(1 + 3.39\ C_e)$, respectively. The predicted values of maximum uptake by Langmuir model are 0.73 and 11.4 mg/g for A11 and A12, respectively. In addition, A12 displays a greater adsorption intensity of 3.39 L/mg, a nearly ten times higher than that of A11.

Figure 7 shows the equilibrium removal of lead(II) by A11 and A12.

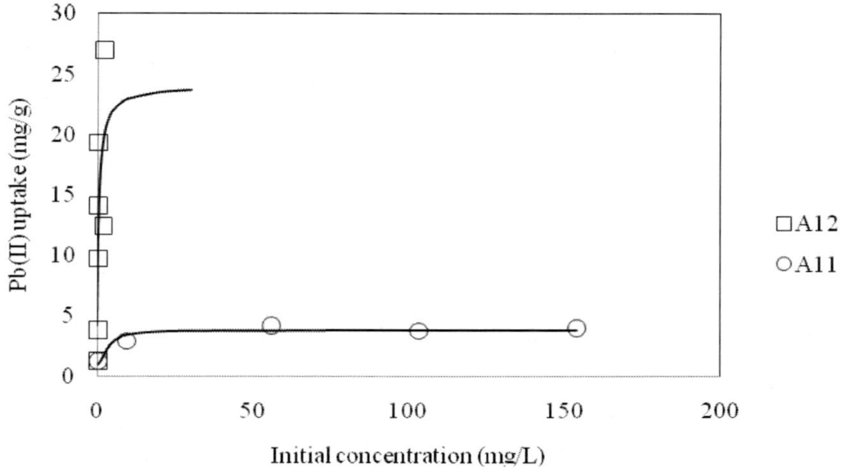

Figure 7. Equilibrium removal of lead(II) by sludge ash adsorbents (unadjusted pH = 5.2, t = 72 h).

The Langmuir expressions for lead(II) removal were determined as $q_e = 3.33\ C_e/(1 + 0.849\ C_e)$ and $q_e = 50.8\ C_e/(1 + 2.11\ C_e)$ for A11 and A12, respectively. The predicted values of maximum uptake capacity and adsorption intensity are 3.92 mg/g (or 0.0189 mmol/g) and 0.849 L/mg, and 24.1 mg/g (or 0.116 mmol/g) and 2.11 L/mg for A11 and A12, respectively. The profile of lead(II) uptake by A12 forms an abscissa of 0.31 mg/L for initial concentration between 10 and 200 mg/L. The vertical trend suggests that the removal is promoted by precipitation. In general, the removal trend by A11 and A12 for lead(II) is identical with that for copper(II) removal. On a molar basis, however A11 exhibits a 1.6 times greater removal of lead(II) as compared to that of copper(II), while copper(II) is more prone to form solid precipitates in the presence of A12.

An attempt has been made to compare the removal of heavy metals by the palm oil mill effluent sludge (So). Figures 8 and 9 illustrate the removal comparison between So and A12. For the selected concentrations studied, A12 displays a greater removal of Cu(II) than So, while the two adsorbents reveal a comparable performance for Pb(II) removal. The equilibrium pH for heavy metals removal by So ranged between 3.8 and 5.1 indicates that the removal is driven by adsorption via the role of surface functional groups [23]. A comparable performance as shown in Figure 9 also suggests that the untreated sludge is also useful for lead(II) removal, as it minimizes the cost implications required for the treatment. The removal performance of methylene blue and heavy metals by untreated ash has been reported elsewhere [24, 30], and it resembles that of A11 suggesting that acid treatment did not contribute much in improving the surface properties of ash.

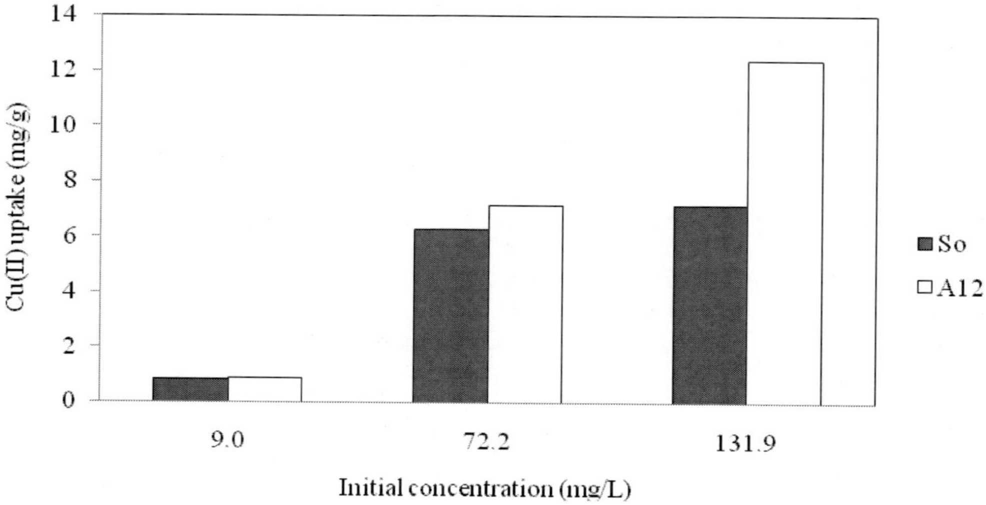

Figure 8. Effect of initial concentration on copper(II) removal by So and A12 (unadjusted pH = 5.2, t = 72 h).

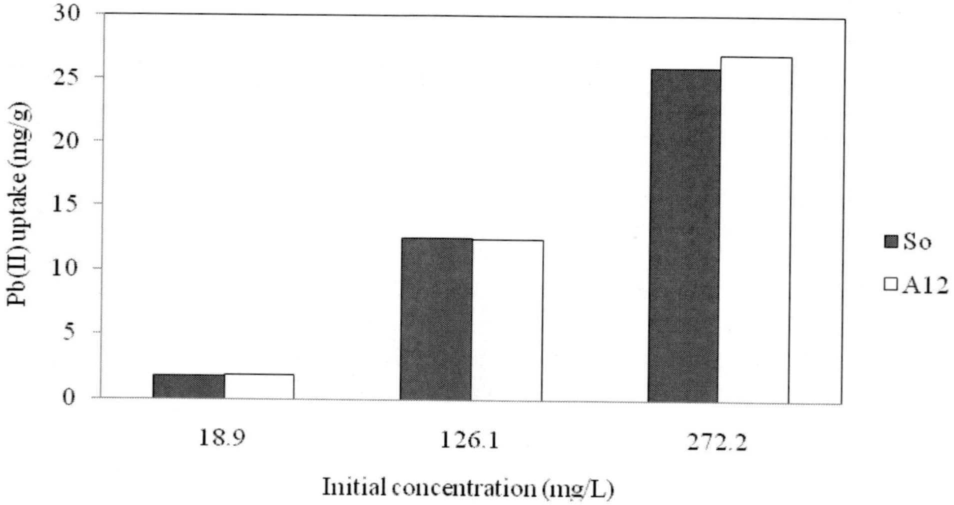

Figure 9. Effect of initial concentration on lead(II) removal by So and A12 (unadjusted pH = 5.2, t = 72 h).

6.3. Effect of Contact Time

Contact time to achieve steady-state equilibrium is an important factor in adsorption as it depends on the nature of adsorbent and adsorbate. The effect of contact time was performed at concentration at which the adsorption reached a maximum removal of heavy metals. The pseudo-second order model that describes the chemical adsorption was used to analyze the rate of adsorption.

Figures 10 and 11 show the adsorption rate by adsorbents for copper(II) and lead(II) removal, respectively. In general, a rapid increasing uptake was observed during the first 120 minutes, thereafter the rate (gradient) of adsorption decreased gradually most probably due to the active sites being occupied before the equilibrium is reached.

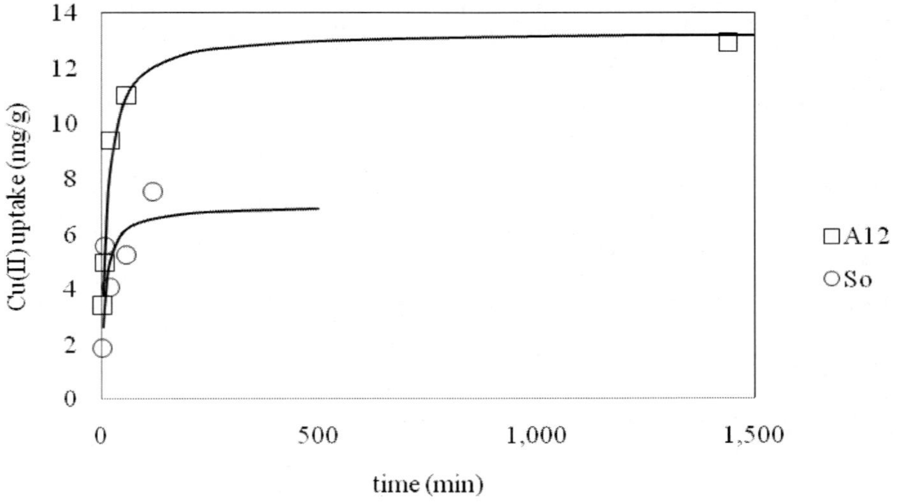

Figure 10. Rate of Cu(II) uptake by So and A12 (C_o = 154 mg/L, unadjusted pH = 5.1, V = 0.15 L, W = 0.5 g).

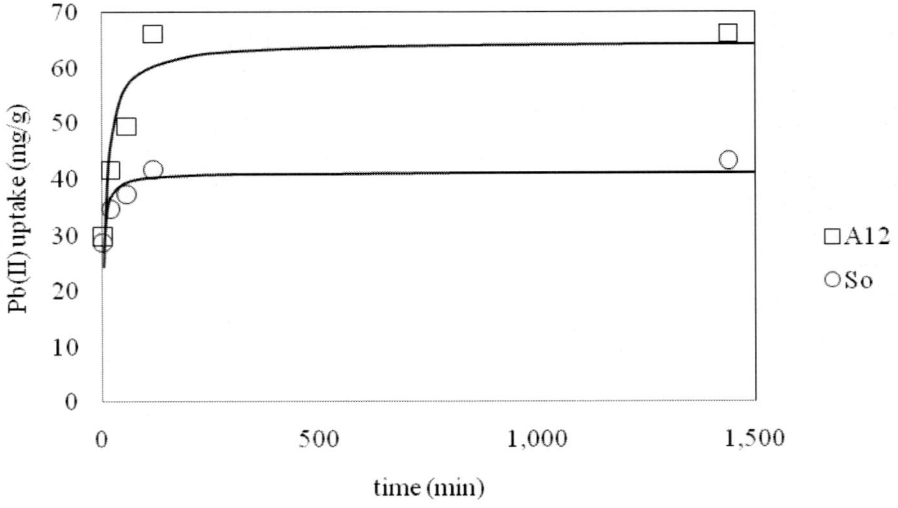

Figure 11. Rate of Pb(II) uptake by So and A12 (C_o = 360 mg/L, unadjusted pH = 5.2, V = 0.1 L, W = 0.5 g).

It can be explained that the amount of heavy metals adsorbed increased with an increased in contact time, and attained the maximum removal after 120 minutes. The pseudo-second order model fitted well with the adsorption data. The rate expressions for copper(II) removal were determined as $q_t = 1.05\ t/(1 + 0.079\ t)$ and $q_t = 0.846\ t/(1 + 0.12\ t)$ for A12 and So, respectively. For lead(II) removal, the rate expressions are $q_t = 7.78t/(1 + 0.12\ t)$ and $q_t = 16.5t/(1 + 0.40\ t)$ for A12 and So, respectively. The values of rate constant are tabulated in Table 4. Although A12 displays a greater removal of copper(II) than So, the rate constant of the former is nearly three times lower than that of the latter. With respect to copper(II) removal, the result suggests that the rate of precipitation (A12) is faster than that of adsorption (So). Similar pattern was observed for lead(II) removal. Febrianto et al. [32] also reported similar finding for distillery sludge and its-treated derivatives.

Table 4 also shows the pseudo-second order rate constant for heavy metals removal by some adsorbents. Increasing initial concentration usually results in decreasing value of rate constant [32]. According to Arrhenius expression, the decreasing value of rate constant could be due to the elevating activation energy for adsorption to take place when the adsorbate concentration increases.

Table 4. Rate constant of the pseudo-second order model for some adsorbents

Solute	Adsorbent	C_o (mg/L)	Pseudo-second order model		Reference
			q_e (mg/g)	k_2 (g/mg.min)	
Cu(II)	Tea waste	200	19	0.013	[32]
		100	14	0.017	
	Waste coconut buttons	100	38.6	0.002	[37]
	Palm oil mill effluent sludge (So)	154	7.05	0.017	This work
	NaOH-treated sludge ash (A12)		13.3	0.006	
Pb(II)	Tea waste	200	35	0.009	[32]
		100	19	0.028	
	Waste coconut buttons	100	45	0.004	[37]
	Distillery sludge	100	38	0.007	[32]
	HCl-treated distillery sludge		50.3	0.006	
	NaOH-treated distillery sludge		52.4	0.003	
	Palm oil mill effluent sludge (So)	360	41.1	0.01	This work
	NaOH-treated sludge ash (A12)		64.7	0.002	

7. Comparison of the Removal Performance with Other Adsorbents

Table 5 summarizes the maximum uptake capacity of heavy metals as predicted by the Langmuir model for some adsorbents. The palm oil mill effluent sludge and NaOH-treated sludge ash demonstrate a comparable removal performance with some adsorbents reported in the literature. Most of the studies, however are silent on the solution pH upon equilibrium for the maximum removal. Thus, it is rather difficult to confirm whether the removal was due to the adsorption, precipitation or a hybrid adsorption-precipitation process. For this purpose, close monitoring of equilibrium pH at various initial concentrations studied is imperative. In addition, the rate constant also plays an important role in the removal of heavy metals. Fast rate of adsorption is essential especially when dealing with large volume of waste to be treated.

Table 5. Lead(II) and copper(II) removal by some adsorbents

Adsorbent	Maximum uptake capacity (mg/g)		References
	Pb(II)	Cu(II)	
Zeolite	1.6	1.6	[38]
Bentonite	22	17.9	[39]
Fly ash	18.8	18.8	
Tea waste	65	48	[40]
Fly ash	2.96	3.52	[41]
Walnut sawdust	4.48	-	[42]
Peanut hull carbon	-	65.6	[43]
Sawdust carbon	-	5.73	[44]
Hazelnut shell AC	-	58	[11]
HCl-treated sludge ash (A11)	3.92	0.73	This work
NaOH-treated sludge ash (A12)	24.1	11.5	

Conclusion

Palm oil mill effluent sludge is a solid residue from the anaerobic treatment facility at palm oil milling. Because of its high ash content (59.6 wt %), it would be interesting to explore the adsorptive characteristics of sludge for heavy metals removal. The sludge ash was treated with acid and alkali.

The acid-treated sludge ash did not show much improvement in the adsorption capacity of heavy metals as compared to the untreated sludge ash. On the other hand, the NaOH-treated exhibits a greater removal of heavy metals, and it is likely due to the precipitation of metal aqua ions in the profusion of OH^- ions. It is also interesting to observe a faster heavy metals removal due to precipitation as opposed to that of adsorption (via the role of surface functional groups).

Acknowledgments

This work was part of M. Zakaria's Masters Dissertation at the Universiti Teknologi Malaysia. Authors gratefully acknowledge the Malaysia Ministry of Higher Education and Universiti Teknologi Malaysia for the research (flagship) grant #03G07.

REFERENCES

[1] Barakat, M. A., 2011. New trends in removing heavy metals from industrial wastewater. *Arab. J. Chem.* 4(4), 361–377.

[2] Demirbas, A., 2008. Heavy metal adsorption onto agro-based waste materials: A review. *J. Hazard. Mater.* 157(2-3), 220-229.

[3] Bhatnagar, A., Sillanpaa, M., 2010. Utilization of agro-industrial and municipal waste materials as potential adsorbents for water treatment-A review. *Chem. Eng. J.* 157, 277–296.

[4] Nowak, B., Aschenbrenner, P., Winter, F., 2013. Heavy metal removal from sewage sludge ash and municipal solid waste fly ash-A comparison. *Fuel Process. Technol.* 105, 195-201.

[5] Gonzalez, A., Moreno, N., Navia, R., Querol, X., 2011. Development of a non-conventional sorbent from fly ash and its potential use in acid wastewater neutralization and heavy metal removal. *Chem. Eng. J.* 166(3), 896-905.

[6] Bailey, S. E., Olin, T. J., Bricka, R. M., Adrian, D. D., 1999. A review of potentially low-cost sorbents for heavy metals. *Water Res. 33*(11), 2469-2479.

[7] Hasan, M., Ahmad, A. L., Hameed, B. H., 2008. Adsorption of reactive dye onto cross-linked chitosan/oil palm ash composite beads. *Chem. Eng. J.* 136(2-3), 164-172.

[8] Zainudin, N. F., Lee, K. T., Kamaruddin, A. H., Bhatia, S., Mohamed, A.R., 2005. Study of adsorbent prepared from oil palm ash (OPA) for flue gas desulfurization. *Sep. Purif. Technol.* 45(1), 50–60.

[9] Ofomaja, A. E., 2010. Equilibrium studies of copper ion adsorption onto palm kernel fibre. *J. Environ. Manage.* 91(7), 1491-1499.

[10] Boujelben, N., Bouzid, J., Elouear, Z. 2009. Adsorption of nickel and copper onto natural iron oxide-coated sand from aqueous solutions: Study in single and binary systems. *J. Hazard. Mater.* 163(1), 376-382.

[11] Demirbas, E., Dizge, N., Sulak, M. T., Kobya, M., 2009. Adsorption kinetics and equilibrium of copper from aqueous solutions using hazelnut shell activated carbon. *Chem. Eng. J.* 148(2-3), 480-487.

[12] Chua, H. 1998. Effects of trace chromium on organic adsorption capacity and organic removal in activated sludge. *Sci. Total Environ.* 214(1-3), 239-245.

[13] Ahmaruzzaman, M., Gupta, V. K. 2011. Rice husk and its ash as low-cost adsorbents in water and wastewater treatment. *Ind. Eng. Chem. Res.* 50(24), 13589-13613.

[14] Zaini, M. A. A., Yohihara, K., Okayama, R., Machida, M., Tatsumoto, H. 2008. Effect of out-gassing of $ZnCl_2$-activated cattle manure compost (CMC) on adsorptive removal of Cu(II) and Pb(II) ions. *Tanso.* 234, 220-226.

[15] Wang, L-H., Lin, C-I. 2008. Adsorption of lead(II) ion from aqueous solution using rice hull ash. *Ind. Eng. Chem. Res.* 47(14), 4891-4897.

[16] Cochrane, E.L., Lu, S., Gibb, S.W., Villaescusa, I. 2006. A comparison of low-cost biosorbents and commercial sorbents for the removal of copper from aqueous media. *J. Hazard. Mater.* 137(1), 198-206.

[17] Shah, J., Jan, M. R., Haq, A. U., Zeeshan, M. 2015. Equilibrium, kinetic and thermodynamic studies for sorption of Ni (II) from aqueous solution using formaldehyde treated waste tea leaves. *J. Saudi Chem. Soc.* 19(3), 301-310.

[18] Gode, F., Pehlivan, E. 2006. Removal of chromium (III) from aqueous solutions using Lewatit S 100: The effect of pH, time, metal concentration and temperature. *J. Hazard. Mater.* 136 (2), 330-337.

[19] Unlu, N., Ersoz, M. 2006. Adsorption characteristics of heavy metal ions onto a low cost biopolymeric sorbent from aqueous solutions. *J. Hazard. Mater.* 136(2), 272-280.

[20] Pan, S-C., Lin, C-C., Tseng, D-H. 2003. Reusing sewage sludge ash as adsorbent for copper removal from wastewater. *Resour. Conserv. Recy.* 39(1), 79-90.

[21] Mousavi, H.Z., Seyedi, S.R. 2011. Nettle ash as a low cost adsorbent for the removal of nickel and cadmium from wastewater. *Int. J. Environ. Sci. Tech.* 8(1), 195–202.

[22] Fu, F., Wang, Q. 2011. Removal of heavy metal ions from wastewaters: A review. *J. Environ. Manage.* 92(3), 407-418.

[23] Zaini, M. A. A., Zakaria, M., Mohd.-Setapar, S.H., Che-Yunus, M.A., 2013. Sludge-adsorbents from palm oil mill effluent for methylene blue removal. *J. Environ. Chem. Eng.* 1, 1091-1098.

[24] Zaini, M.A.A., Yee-Cher, T., Zakaria, M., Kamaruddin, M.J., Mohd.-Setapar, S.H., Che-Yunus, M.A., 2014. Palm oil mill effluent sludge ash as adsorbent for methylene blue dye removal. *Desalin. Water Treat.* 52(19-21), 3654-3662.

[25] Erto, A., Giraldo, L., Lancia, A., Moreno-Pirajan, J.C., 2013. A comparison between a low-cost sorbent and an activated carbon for the adsorption of heavy metals from water. *Water Air Soil Poll.* 224, 1531-1533.

[26] Kuncoro, E. P., Fahmi, M. Z., 2013. Removal of Hg and Pb in aqueous solution using coal fly ash adsorbent. *Procedia Earth and Planetary Science*, 6, 377-382.

[27] Pan, J-H., Liu, R-X., Tang, H-X., 2007. Surface reaction of Bacillus cereus biomass and its biosorption for lead and copper ions. *J. Environ. Sci.* 19(4), 403-408.

[28] Singh, R., Chadetrik, R., Kumar, R., Bishnoi, K., Bhatia, D., Kumar, A., Bishnoi, N.R., Singh, N., 2010. Biosorption optimization of lead(II), cadmium(II) and copper(II) using response surface methodology and applicability in isotherms and thermodynamics modeling. *J. Hazard. Mater.* 174(1-3), 623-634.

[29] Rahman, M.S., Islam, M.R., 2009. Effect of pH on isotherm modelling for Cu(II) ions adsorption using maple wood sawdust. *Chem. Eng. J.* 149 (1-3), 273-280.

[30] Zaini, M. A. A., Zakaria, M., Alias, N, Zakaria, Z. Y., Johari, A., Mohd.-Setapar, S.H., Kamaruddin, M. J., Che-Yunus, M. A., 2014. Removal of heavy metals onto KOH-activated ash-rich sludge adsorbent. *Energy Procedia* 61, 2572-2575.

[31] Rao, R. A. K., Khan, M. A., 2009. Biosorption of bivalent metal ions from aqueous solution by an agricultural waste: kinetics, thermodynamics and environmental effects. *Colloid Surfaces A.* 332 (2-3), 121-128.

[32] Febrianto, J., Kosasih, A. N., Sunarso, J., Ju, Y. H., Indraswati, N., Ismadji, S., 2009. Equilibrium and kinetic studies in adsorption of heavy metals using biosorbent: A summary of recent studies. *J. Hazard. Mater.* 162(2-3), 616-645.

[33] Asadi, F., Shariatmadari, H., Mirghaffari, N., 2008. Modification of rice hull and sawdust sorptive characteristics for remove heavy metals from synthetic solutions and wastewater. *J. Hazard. Mater.* 154(1-3), 451-458.

[34] Eslamzadeh, T., Nasernejad, B., Bonakdar-Pour, B., Zamani, A., Esmaail-Bygi, M., 2004. Removal of heavy metals from aqueos solution by carrot sesidues. *Iranian Journal of Science & Technology, Transaction A 28*(A1) 161-167.

[35] Woolard, C.D., Petrus, K., van-der-Horst, M., 2000. The use of a modified fly ash as an adsorbent for lead. *Water SA* 26(4), 531-536.

[36] Zaini, M. A. A., 2010. *Heavy Metal Removal from Aqueous Solution Using Activated Carbons Rich in Nitrogen Content*. PhD thesis, Chiba University, Japan.

[37] Anirudhan, T.S., Sreekumari, S.S., 2011. Adsorptive removal of heavy metal ions from industrial effluents using activated carbon derived from waste coconut buttons. *J. Environ. Sci.* 23(12), 1989-1998.

[38] Babel, S., Kurniawan, T.A., 2003. Low-cost adsorbents for heavy metals uptake from contaminated water: a review. *J. Hazard. Mater.* 97(1-3), 219-243.

[39] Visa, M., 2012. Tailoring fly ash activated with bentonite as adsorbent for complex wastewater treatment. *Applied Surface Sci.* 263, 753-762.

[40] Amarasinghe, B.M.W.P.K., Williams, R.A., 2007. Tea waste as a low cost adsorbent for the removal of Cu and Pb from wastewater. *Chem. Eng. J.* 132(1-3), 299-309.

[41] Daci, M.N., Daci, N.M., Zeneli, L., Gashi, S.T., Hoxha, D., 2011. Coal ash as adsorbent for heavy metal ions in standard solutions, industrial wastewater and streams. *International Journal of Ecohydrology and Hydrobiology* 11(1-2) 129.

[42] Bulut, Y.,Tez, Z., 2007. Removal of heavy metals from aqueous solution by sawdust adsorption. *J. Environ. Sci.* 19(2), 160-166.

[43] Periasamy, K., Namasivayam, C., 1996. Removal of copper(II) by adsorption onto peanut hull carbon from water and copper plating industry wastewater. *Chemosphere* 32(4), 769-789.

[44] Kalavathy, M. H., Karthikeyan, T., Rajgopal, S., Miranda, L. R., 2005. Kinetic and isotherm studies of Cu(II) adsorption onto H_3PO_4 activated rubber wood sawdust. *J. Colloid Interface Sci.* 292 (2), 354-362.

EDITOR'S CONTACT INFORMATION

Dr. Deepak Pathania
Professor and Head
Department of Chemistry
Shoolini University, Solan,
Himachal Pradesh, India -173212
Email: dpathania74@gmail.com

INDEX

#

A

C

D

E

F

G

H

I

J

K

L

M

N

Q

R

S

V

W

Y

Z